INTRODUCTORY
DIFFERENTIAL GEOMETRY
FOR PHYSICISTS

INTRODUCTORY DIFFERENTIAL GEOMETRY FOR PHYSICISTS

A VISCONTI

Universite de Provence and Centre de Physique Theorique de Marseille

World Scientific
Singapore • New Jersey • London • Hong Kong

Published by

World Scientific Publishing Co. Pte. Ltd.

P O Box 128, Farrer Road, Singapore 9128

USA office: Suite 1B, 1060 Main Street, River Edge, NJ 07661

UK office: 73 Lynton Mead, Totteridge, London N20 8DH

INTRODUCTORY DIFFERENTIAL GEOMETRY FOR PHYSICISTS

ISBN 9971-50-186-4
9971-50-187-2 (pbk)

Printed in Singapore by
Singapore National Printers Ltd

CONTENTS

CHAPTER I — LEVEL 0

CHAPTER II — LEVEL 1

CHAPTER IV

Fibre Spaces   372

CHAPTER I — LEVEL 0
The Intuitive Approach — Theory of Surfaces (19th Century)

Level 0 will mean that we are in the framework of elementary geometry developed by Euclid in his *Elements*, which ruled for about twenty centuries in mathematics and in physics. The ambient physical space $\mathcal{E}_3$ contains points, lines and planes as basic concepts, and these elements have their well-known properties. Euclid's postulate maintaining that there exists a single parallel to a given straight line Δ through a given point is here accepted without any argument. Let us consider two points A and B of $\mathcal{E}_3$: they determine a unique straight line Δ and the segment AB belongs to this line. The vector **AB** which could also be called double-point (A, B) is characterized by its length, which is the length of the segment AB, by its direction which is the direction of Δ, and its orientation, which is the orientation going from the origin A to the endpoint B. Given this simple concept, one sees immediately a generalization: one can substitute for the double-point or vector the notion of **free vector**. Two free vectors **AB** and **A′B′** are identified if they are equipollent, i.e. if they have the same length, the same direction and the same orientation. This is as much as one can do if one does not introduce the notion of equivalence class of representative **AB**, a concept which is rather abstract for the physical intuition underlying Euclidean geometry.

1

Another important concept is the following: given a curve C and the free vector $\mathbf{AB} = \mathbf{u}$ connected with this curve, one can by the postulate of Euclid consider at every point of this curve a vector that is equipollent to $\mathbf{u}$. One says then that the vector $\mathbf{u}$ undergoes parallel displacement along C: then, by Euclid's postulate, one finds again the same vector after coming back to the starting point. The extension of this concept to an arbitrary space (manifold) will be one of the keys to differential geometry.

In this chapter and the following one, we shall use without fear the notions of infinitesimally small objects. We shall consider infinitesimally close points or infinitesimal vectors: dx, dy, dz, df will denote infinitesimal quantities. It is good to notice, however, that the notion of differential will be profoundly changed at Level 2; dx, dy, dz will then be differential 1-forms. The reader can go quickly through this chapter, except for paragraphs 3 and 5 where the main notions of differential geometry are described.

Let us also remark that the names "differential geometry" and "infinitesimal geometry" have been used interchangeably. The first one is now in more general use.

1. *Curves*

According to Euclid's definition, a curve is a line without width. In order to have a more elaborate formulation we have first to find a mathematical model of the physical ambient space in our laboratory. An adequate model is $\mathcal{E}_3$: we choose an arbitrary point O and an orthonormal basis $\mathbf{i}, \mathbf{j}, \mathbf{k}$ (or else $\mathbf{e}_1, \mathbf{e}_2, \mathbf{e}_3$) bound to 0. Every point $M \in \mathcal{E}_3$ is then described in a unique way by the triplet of real numbers (x, y, z) which are the components of the vector $\mathbf{OM}$ in the frame $(0, \mathbf{i}, \mathbf{j}, \mathbf{k})$. In this way one defines a one-to-one correspondence between $\mathcal{E}_3$ and $\mathbb{R}^3$.

Let us assume that the coordinates (x, y, z) depend on a real parameter λ: we then say that they form a **parametric representation** of a curve C.

$$x = f(\lambda); \quad y = g(\lambda); \quad z = h(\lambda). \tag{1.1}$$

If the functions f, g, h are differentiable as many times as necessary we shall say "vaguely" that they are honest in a certain interval $I \subset \mathbb{R}$. In order to economize on symbols, we shall write

$$\mathbf{x} = \mathbf{x}(\lambda): \quad x = x(\lambda), \quad y = y(\lambda), \quad z = z(\lambda); \tag{1.2}$$

this notation is admittedly not very good, but it has been extensively used for a long time. Finally, if we eliminate λ (when possible) between the three preceding equations taken in pairs, we can obtain a different representation of C as intersection of two surfaces:

$$F(\mathbf{x}) = F(x, y, z) = 0 \,; \qquad G(\mathbf{x}) = G(x, y, z) = 0 \,. \qquad (1.3)$$

The **tangent** to C at M is carried by the vector $d\mathbf{x}/d\lambda$; if certain differentiability conditions are satisfied, this tangent is unique; however, the **normal** to C at M is not unique, but there exists a unique **normal plane** to C at M.

Let M and M' be two infinitesimally close points of C, and let the respective tangents be $\mathbf{t}$ and $\mathbf{t}'$. There are two possible cases: $\mathbf{t}$ and $\mathbf{t}'$ belong to the same plane or they do not. In both cases, Euclid's postulate allows us to draw through M a vector equipollent to $\mathbf{t}'$: the plane $(M, \mathbf{t}, \mathbf{t}')$ is the **osculating plane** of C. If this plane does not depend on the points M and M', we say that C is a **plane curve**; otherwise it is a **space curve**. The osculating plane allows us to consider two normal directions that play a privileged role: the first one is orthogonal to the osculating plane and it is the **binormal** of the unit vector $\mathbf{b}$; the second one is in the osculating plane and it is the **principal normal** $\mathbf{n}$ (see Fig. 0.1). We direct $\mathbf{n}$ in the direction where C is concave and construct a frame $(\mathbf{t}, \mathbf{n}, \mathbf{b})$ which is called the **moving frame** associated to any point $M \in C$. The curves we study here are **rectifiable** which means that if A and B are two points of C corresponding to the values λ_0 and λ_1 of the parameter λ (formula (1.2)), then it is possible to inscribe into C a polygonal line of extremities A and B and infinitesimally small sides. The length element ds is by definition

$$ds = \|d\mathbf{x}\| = \sqrt{dx^2 + dy^2 + dz^2} \qquad (1.4)$$

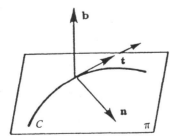

Fig. 0.1. Planar curve: its osculating plane is its own plane.

and the arc length AB is given as the integral of ds between λ_0 and λ_1.

One obtains the **intrinsic equation** of the curve if λ is assumed to be the parameter s and the equation of C can then be written as $\mathbf{x} = \mathbf{x}(s)$. The unit tangent vector is

$$\mathbf{t} = \frac{d\mathbf{x}}{ds}; \quad \mathbf{t}^2 = 1. \tag{1.5}$$

Let us also introduce the unit vector $\mathbf{n}$ by

$$\frac{d\mathbf{t}}{ds} = \frac{d^2\mathbf{x}}{ds^2} = \frac{\mathbf{n}}{R}, \tag{1.6}$$

$R(\mathbf{x})$ is then a function of M, it is the **curvature radius** of C at M and the **curvature** at M is $1/R$.* One sees that $\mathbf{n}$ is orthogonal to $\mathbf{t}$ (which is a unit vector) and that in addition $d\mathbf{t} = \mathbf{t}' - \mathbf{t}$ is in the osculating plane (see Fig. 0.2); consequently $\mathbf{n}$ is the principal normal. In addition

$$\frac{1}{R} = \frac{\|d\mathbf{t}\|}{ds} = \frac{d\sigma}{ds}, \tag{1.7}$$

where $d\sigma$ is the infinitesimal angle between $\mathbf{t}$ and $\mathbf{t}'$, as can be seen if we identify $d\mathbf{t}$ with the arc of circle of center M and going through the extremities of $\mathbf{t}$ and $\mathbf{t}'$ (see Fig. 0.2). We shall not derive here the Serret-Frenet formulas which will not be used in the sequel but we may notice that for a plane curve, $\mathbf{b}$ is constant.

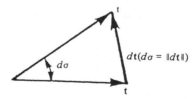

Fig. 0.2. Curvature radius: $R = ds/d\sigma = ds/\| dt\|$ with $\| d\boldsymbol{\varepsilon}\| = d\sigma$ — the curvature is $1/R$.

---

*Later on, following the general use, we shall denote by R the curvature of a surface.

2. *Surfaces — Topological invariants*

The concept of curve as it was introduced in the preceding paragraph is an example of one-dimensional manifold contained in $\mathcal{E}_3$. A surface S will be defined as a two-dimensional manifold contained in $\mathcal{E}_3$. With the interpretation of the preceding paragraph, we shall write the equation of the surface S as

$$S : \mathbf{x} = \mathbf{x}(u,v) : x = x(u,v), \quad y = y(u,v) \quad z = z(u,v), \tag{2.1}$$

where u and v are real parameters, and the three functions x, y, z are supposed to be differentiable *ad libitum*. If the elimination of u, v between these three equations is possible one defines a surface by the relation

$$F(\mathbf{x}) = F(x, y, z) = 0. \tag{2.2}$$

In order to study the local properties of a surface S at a point M, the simplest method is to study properties of curves of S going through M. In particular, if we take v as constant $(v = v_0)$, one finds a curve $\mathbf{x} = \mathbf{x}(u, v_0)$ which goes through M_0. There is another curve going through M_0, namely $\mathbf{x} = \mathbf{x}(u_0, v)$ and, clearly, M_0 is the point of S corresponding to the values u_0, v_0 of u and V. These two curves are called the **coordinate curves** in M_0; they are parametrized as in paragraph 1. The tangent of S at the point M is not unique (there are as many tangents in M as there are curves going through M); however the normal of unit length $\mathbf{N}$ is unique: if S is defined as in (2.2), one sees that $\mathbf{N}$ is colinear to ∇F. Using the parametric representation (2.1), one notices that at the point $M(u, v)$ the tangents of the coordinate curves are respectively $\partial \mathbf{x}/\partial u = \mathbf{x}_u$ and $\partial \mathbf{x}/\partial v = \mathbf{x}_v$: the two notations $\mathbf{x}_u, \mathbf{x}_v$ are explicit and simple. The normal $\mathbf{N}$ is then colinear to the outer product $\mathbf{x}_u \wedge \mathbf{x}_v$.

Given the normal $\mathbf{N}$ at the point M, the **tangent plane** to S at M is well-defined. Elementary considerations allow us to write down its equation. We shall not go further at this point, but we have to notice that this concept of tangent plane — which will be supplemented later by the notion of cotangent plane — is one of the main concepts of differential geometry at Level 2.

Let us consider a simple example: let S be the sphere of radius R with its center at the origin

$$x^2 + y^2 + z^2 = R^2 : \quad R > 0, \tag{2.3}$$

let us parametrize the sphere with the help of polar coordinates

$$x = R \sin\theta \cos\varphi, \quad y = R\sin\theta\sin\varphi, \quad z = R\cos\theta : \begin{cases} 0 < \theta < \pi \\ 0 < \varphi \le 2\pi \end{cases}.$$
$$(2.4)$$

Notice that if we want to establish a bicontinuous one-to-one correspondence between the sphere and $\mathbb{R}^2$ it is necessary to exclude the two poles N and S which correspond to $\theta = 0$ and $\theta = \pi$, because at these points the value of φ can be arbitrary: there is no one-to-one bicontinuous correspondence between the set of the points on the sphere and a rectangle θ, φ with sides π and 2π. We shall come back to this point particularly at Level 2 (Chap. 3) when we study differentiable manifolds. The representation

$$x = \frac{Ru}{\sqrt{1 + u^2 + v^2}}, \quad y = \frac{Rv}{\sqrt{1 + u^2 + v^2}}, \quad z = \pm\frac{R}{\sqrt{1 + u^2 + v^2}} : \quad R > 0$$
$$(2.5)$$

describes the upper or lower half-sphere depending on the sign of z.

Let S be given in parametric form and let λ be real: if we introduce a couple of "honest" functions φ_1, φ_2 and if we write $u = \varphi_1(\lambda), v = \varphi_2(\lambda)$ we obtain a one-parameter manifold which is a curve C on the sphere S. Let us consider the points M and M' which correspond to the values (u, v) and $(u + du, v + dv)$; the curve element $\mathbf{MM'}$ (or $d\mathbf{x}$) is of length

$$ds^2 = d\mathbf{x}^2 = (\mathbf{x}_u du + \mathbf{x}_v dv)^2 = E du^2 + 2F du dv + G dv^2, \qquad (2.6)$$

where

$$E = \mathbf{x}_u^2, \quad F = \mathbf{x}_u \cdot \mathbf{x}_v, \quad G = \mathbf{x}_v^2, \qquad (2.7)$$

here we are using notations introduced by Gauss himself.* The normal vector of unit length at the point M can be written as

$$\mathbf{N} = \frac{\mathbf{x}_u \wedge \mathbf{x}_v}{\|\mathbf{x}_u \wedge \mathbf{x}_v\|} \qquad (2.8)$$

and this expression can be easily expressed, as function of E, F, G since

$$\|\mathbf{x}_u \wedge \mathbf{x}_v\|^2 = (\mathbf{x}_u^2)(\mathbf{x}_v^2) - (\mathbf{x}_u \cdot \mathbf{x}_v)^2 = EG - F^2. \qquad (2.9)$$

---

*Remark that

$$du = \frac{d\varphi_1}{d\lambda}d\lambda, \quad dv = \frac{d\varphi_2}{d\lambda}d\lambda.$$

The quadratic form associated with ds^2 is the **first form of Gauss**; it is of fundamental importance in Riemannian geometry where it will be used with a different notation (see formula (3.11)).

Let us note two other formulas which can be useful: the first one gives the surface area dS of the elementary parallelogram formed by two elements $d\mathbf{x} = \mathbf{x}_u du$ and $\delta\mathbf{x} = \mathbf{x}_v dv$, starting from M and corresponding to two coordinate lines going through M. We have

$$dS = \|d\mathbf{x} \wedge \boldsymbol{\delta}\mathbf{x}\| = \|\mathbf{x}_u \wedge \mathbf{x}_v\| du dv = \sqrt{EG - F^2}\, du dv \,. \tag{2.10}$$

The second formula allows us to calculate the sine of the angle between two infinitesimal elements $\Delta\mathbf{x} = \mathbf{x}_u du + \mathbf{x}_v dv$ and $D\mathbf{x} = \mathbf{x}_u \delta u + \mathbf{x}_v \delta v$, i.e. the angle of two arbitrary curves going through M

$$
\begin{aligned}
\sin\alpha &= \frac{\|(\mathbf{x}_u du + \mathbf{x}_v dv) \wedge (\mathbf{x}_u \delta u + \mathbf{x}_v dv)\|}{\|\Delta\mathbf{x}\|\|D\mathbf{x}\|} \\
&= \frac{\|\mathbf{x}_u \wedge \mathbf{x}_v\| |du\delta v - \delta u dv|}{\|\Delta\mathbf{x}\|\|D\mathbf{x}\|} \\
&= \frac{\sqrt{EG - F^2}|du\delta v - \delta u dv|}{\|\Delta\mathbf{x}\|\|D\mathbf{x}\|} \,,
\end{aligned}
\tag{2.11}
$$

a short calculation and a short moment of reflexion can replace the details omitted here. An interesting application of the preceding formula can be the problem of calculating $\sin\alpha$ for the coordinate curves: $v = v_0$ associated with $du \neq 0, dv = 0$ i.e., $ds = \sqrt{E}\, du$ and $u = u_0$ associated with $\delta u = 0, \delta v \neq 0$, i.e., $ds = \sqrt{G}\, dv$. Because of (2.11) one has

$$\sin\alpha = \frac{\sqrt{EG - F^2}|du\delta v|}{ds\delta s} = \sqrt{\frac{EG - F^2}{EG}} \,. \tag{2.12}$$

This shows that the system of coordinate curves is orthogonal: ($\alpha = \pi/2$), if $F = 0$. (See Fig. 0.5.) As an example, we may consider the sphere S^2 with its parallels and its meridians.

We come now to the concept of total curvature which plays a central role in Riemannian geometry. Consider the vector $\mathbf{N}$, orthogonal to S in M and the normal sections, defined as the intersection of S by planes containing $\mathbf{N}$. In all classical books one can find a discussion of those planar sections (see Fig. 0.3); one proves there that the radius of curvature of these sections

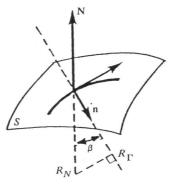

Fig. 0.3. Curvature of a surface Meunier's theorem.

takes minimum value R_m for one of the sections, and a maximum value R_M for another one. (See Problems 0.3 and 0.4.)

The **total curvature** K of S at the point $M(u, v)$ is defined by[*]

$$K(u, v) = \frac{1}{R_m R_M} , \qquad (2.13)$$

the two corresponding directions in the tangent plane to S at M are called **principal directions**. For instance, the curvature K vanishes for the plane; it is constant and positive for a sphere of radius R and its value is then $1/R^2$. This quantity also has a geometrical significance: let dS be a surface element of S. From an arbitrary point $A \in \mathcal{E}_3$ let us draw parallels to the normals to S along its boundary. If $d\Sigma$ is the solid angle obtained in this way, one can show that

$$K(u, v) = \frac{d\Sigma}{dS} \qquad (2.14)$$

in complete analogy with (1.7). We shall not give here the proof of these results, which will be found in another form in the next paragraph and in the Problems 0.3 and 0.4. Let us finally notice that, among all the curves that can be traced on S, there exist some that have remarkable properties.

---

[*]The notations may be somewhat confusing: the curvature of a given manifold $\mathcal{M}_N$ will be noted later on by R (formula (10.9), Level 1, for instance), then $K = R/2$. On the other hand, we denoted in formula (1.7), Level 0 by R the radius of curvature of any curve C ($R = ds/d\sigma$). This is the general use which we follow.

For instance:

a. the **lines of curvature** are tangent at each point to a principal direction.

b. the **geodesics** which are curves on S such that the integral $\int_{AB} ds$ is extremum and where AB is a path on S having A as origin and B as end. We shall discuss this problem in Sec. 4 as a variational problem with subsidiary conditions, in the framework of Euler-Lagrange equations. Those curves have interesting properties: for instance their principal normals are orthogonal to S. Furthermore, in mechanics the trajectories of free particles are geodesics. In non-Euclidean geometry they play the part of straight lines.

With the introduction of the total (or Gaussian) curvature we progress along — often unpredictable — directions which lead us towards the global point of view. We shall now state without proof and without any attempt at rigour three fundamental theorems related to total curvature. First of all, Gauss' "Theorem egregium": it will be formulated precisely at Level 1. We shall be content at present with an intuitive and experimental approach. Let us imagine a part of a surface S made out of a thin metallic film that can be bent without being torn. A deformation of S into S_1 does not change the length of paths traced on S: the deformation is an **isometry** and we also say that S and S_1 can then be applied one on the other. Such a deformation preserves certain quantities, for instance the total curvature. Geodesics on S correspond to geodesics on S_1, and the angles between two curves on S is preserved. Secondly one has the Gauss-Bonnet theorem: let Δ be a connected part of S, bounded by a geodesic polygon made out of n geodesics. Denoting by $\alpha_1 \dots \alpha_N$ the internal angles, one can show that

$$\sum_{}^{n} \alpha_j = \int_{\Delta} K(u,v) dS(u,v) + (n-2)\pi \ . \tag{2.15}$$

In particular, it is interesting to consider surfaces with constant curvature K_0: in that case

$$\sum_{}^{n} \alpha_j = K_0 S_{\Delta} + (n-2)\pi \ , \tag{2.16}$$

where S_{Δ} is the area of the geodesic polygon. In Euclidean space we can give three examples of surface of constant curvature: first the plane with 0 curvature, then the sphere with constant curvature $(1/R^2)$ and finally the pseudo-sphere with negative curvature. This last surface is a surface

of revolution obtained from a curve called tractrix which rotates around
zz': this curve is defined as the curve such that a segment of the tangent
between the tangency point M and the intersection T of this tangent with
$z'z$ should be of constant length. (See Fig. 0.4.) One can show that its
curvature is negative and constant.

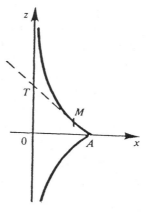

Fig. 0.4. Tractrix. The point A is a retrogression point and $MT = a$.

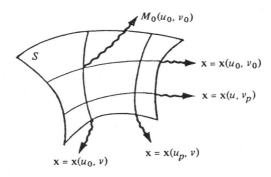

Fig. 0.5. Coordinate curves: $\mathbf{x}=\mathbf{x}(u,v_p); \mathbf{x}=\mathbf{x}(u_p,v)$. The u_p and v_p variables range over
all the values in the intervals $\{u_a,u_b\}$ and $\{v_a,v_b\}$ so to cover all S. The point M_0 is at
the intersection of the curves: $\mathbf{x}=\mathbf{x}(u,v_0)$ and $\mathbf{x}=\mathbf{x}(u_0,v)$.

Let us now consider a geodesic triangle of a surface of constant curvature
and let us use (2.16), which now becomes

$$\sum_{}^{3} \alpha_j = K_0 S_\Delta + \pi \ . \tag{2.17}$$

For a plane $K_0 = 0$ the sum of its internal angles is equal to π; for a sphere one has $K_0 > 0$ and $\Sigma\alpha_j > \pi$. Finally for a surface with $K_0 < 0$ one has $\Sigma\alpha_j < \pi$. One obtains so the results of Euclidean, Riemannian and Lobatchewski geometry respectively.

We should mention another special case: the plane with curvature $K_0 = 0$ is clearly simple. What are the surfaces that can be mapped on a plane? The answer is simple. They are the developable surfaces generated by the motion of a straight line Δ which is called the generating line, and such that their tangent plane along Δ is constant: their total curvature vanishes. This is for instance the case with a cylinder and a cone. One can similarly show that the surfaces of constant (positive or negative) curvature can respectively be mapped on a sphere or on a pseudo-sphere.

The above remarks do not exhaust the usage that can be made of K: the deformations that we have considered preserve length. But what would happen if we replaced the thin metallic film by a rubber film. The deformation would then be called **topological**. The study of those transformations will lead us to invariants associated to surfaces that can be deformed. Those invariants can be given a physical interpretation and the introduction of the concept of triangulation will be of help in this context. A part of a surface is said to be **triangulizable** and **orientable** if it can be covered by a finite network of triangles, where each grid is oriented according to the convention used in electricity when one considers network of lines at which one applies Kirchoff laws. If a surface S is given, it is not automatically clear that it is triangulizable. One can show that this happens for closed surfaces, such as a sphere S^2, an ellipsoid, torus. Triangulizable surfaces give rise to interesting theorems: suppose that we deform continuously a surface without tearing or wrinkling it, then the initial surface and the deformed surface are said to be homeomorphic; this terminology will be carefully studied at Level 2. It is capital to determine the topological invariants, i.e. quantities or numbers associated to a surface which do not change by deformations. Such an invariant is for instance the dimension of a manifold.* Another invariant concept is the triangulizability of a surface; a triangulizable surface stays triangulizable after a continuous deformation. We may also consider the Euler-Poincaré characteristic (or index) which is another invariant: let us triangulate a closed surface S. Let F be the number of faces, A the

---

*A manifold is a generalization of the concept of surface, which is a 2-dimensional manifold embedded into a 3-dimensional space.

number of edges and s the number of vertices of the triangulization mesh covering S. The number*

$$\chi(S) = F - A + s \,, \tag{2.18}$$

not only is independent of the chosen triangulation but is also a topological invariant; two surfaces with the same characteristic are said to be topologically equivalent. We shall not try here to give the proofs. Let us notice that this index was already considered by Descartes and Euler who showed that its value is 2 for any convex polyhedron. Let us calculate χ for a few closed surfaces as drawn in Figs. 0.6 and 0.7.

In Fig. 0.6 the first object is a tetrahedron, the second one consists of two tetrahedrons glued by their basis. One can obtain the third one by deforming the lower tetrahedron until it coincides with the basis of the first. The fourth one is a sphere and the fifth is a half-sphere which can be obtained by deforming the lower half-sphere until it is brought into coincidence with the upper half-sphere. A simple calculation shows that the $\chi(S)$ have the common value 2. Those surfaces are topologically equivalent, i.e. they can be transformed into one another by a continuous deformation.

The situation is different in Fig. 0.6 with a torus T^2: an immediate way of triangulating it consists in starting with a rectangular piece of paper with the two diagonals, and to transform this paper successively by wrapping it first into a cylinder and then into a torus. If we count the number of faces, edges and vertices we obtain $\chi(T^2) = 0$. The torus is not topologically equivalent to the five surfaces just described; in particular it is not equivalent to S^2. It will be useful to extend those considerations a little. Let us define a handle as a torus cut as in Fig. 0.7. Given a sphere S^2, we can graft on it a handle. Let us deform the sphere by a first depression between the sides of the torus. By continuing one can transform this object into a torus. Consequently a torus and a sphere with one handle are topologically equivalent. By gluing two handles, we obtain a 2-holed torus with two holes: we can also obtain it by considering a sphere with two handles and

---

*We may also mention another important invariant. Consider a domain $\mathcal{D}_c \subset \mathcal{E}_3$, a continuous and derivable function $V(\mathbf{x})$ in D (called potential) and a field of forces $\mathbf{X}(\mathbf{x}) = -\nabla V(\mathbf{x})$. Let A and B be two points of D and all possible paths (with a tangent at each of their points) joining A to B. It is well-known that the work of $\mathbf{X}(\mathbf{x})$ along any of these paths is a constant number. In other words, the work of the field $\mathbf{X}(\mathbf{x})$ is an invariant for a continuous deformation of a given path AB. One can state a similar result for the flux, using Stokes theorem (see Sec. 12, Level 2).

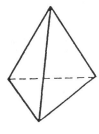 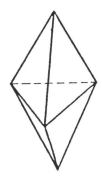 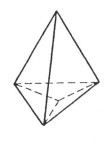

Tetrahedron
$F = 4$, $A = 6$, $s = 4$

Two tetrahedrons with
a common base
$F = 6$, $A = 9$, $s = 5$

The tetrahedron below has
collapsed into the base of
the above tetrahedron
$F = 6$, $A = 9$, $s = 5$

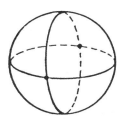 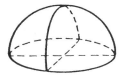

Sphere S_2
$F = 8$, $A = 12$, $s = 6$

The half-sphere below has collapsed
into the equation of the sphere S_2
$F = 8$, $A = 12$, $s = 6$

$$F - A + s = 2$$

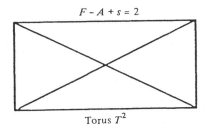

Torus T^2

Fig. 0.6.

applying the deformation procedure defined above. If we define the **genus** g of a closed surface by $\chi(s) = 2(1-g)$, we can show that g is the number of handles to be inserted. One should notice by the way that these remarks are perhaps not just an amusement of the mind; according to some cosmological theories, right after the "big bang" and during a given time interval going from 10^{-30} to 10^{-40} sec, the manifold that represented the universe just born was of a handle type, each handle representing an elementary particle. This is, at least, the theory supported by J. A. Wheeler.

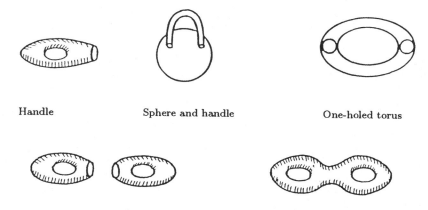

Handle   Sphere and handle   One-holed torus

Two handles make a double-holed torus

Fig. 0.7.

One should point out as well that Gauss' curvature which was introduced through metric considerations will lead us to a simple topological invariant. For instance, let us consider the formula (2.15) that we shall write for a curvilinear triangle Δ and as follows:

$$\int_{\Delta} K(u,v)dS(u,v) = \sum^{3} \alpha_j - \pi = 2\pi - \sum^{3}(\pi - \alpha_j) \ . \qquad (2.19)$$

Now let us apply this formula to every triangle on a triangulated closed surface and let us take the sum. One can then realize that the three terms contribute to the three terms in Eq. (2.18) and that it finally leads to the remarkable relation

$$\oint_{S} K(\mathbf{x})dS(\mathbf{x}) = 2\pi\chi(S) \ . \qquad (2.20)$$

This theorem has important consequences on which we cannot dwell here, let us simply conclude that depending on whether $\oint K\,dS$ is positive or equal to zero, S is topologically equivalent to a sphere S^2 or to a torus T^2 (if $K \neq 0$).

3. *Geometry on a surface or Riemannian geometry*

Suppose for a while the existence of infinitely flat beings on a surface S which are able to think, to build up abstractions and to perform some modelization of their experience: could they build a geometry? For us human beings living in a 3-dimensional world, the use of the third dimension leads us to a model where the universe of these flat beings is embedded in the 3-dimensional space of Euclid which is supposed to adequately describe our environment. This possibility leads to capital simplifications: we can indeed define vectors of the bi-point type with their origin at S and tangent to S, move these vectors along S by using Euclid's postulate and the notion of equipollent bi-point vectors. In contradistinction to our point of view, all these possibilities would be denied to the thinking infinitely flat beings. The only kind of geometry they would have been able to construct will be the object of the study of Level 2 where we will see that after a convenient generalization of our notion of space which will lead to the concept of manifold and to a new definition of a vector on a manifold, we shall be able to consider intrinsically the geometry of S. This was B. Riemann's point of view, the founder of modern differential geometry, a frame into which A. Einstein developed later on his theory of general relativity. In all fairness, one should emphasize also that Riemann contemplated the possibility of experimenting with the physical space to see if Euclid's model is the only one model adapted to the description of the physical space in which our universe is embedded. Unfortunately, at his time, the lack of experimental facts and also the insufficiency of experimental apparatus and methods could not lead to any definite answer. Riemann also tried to deduce from local properties other general and global ones: all that was done after the introduction of the ds^2 of a surface, the cornerstone of the geometry of surfaces.

In the present paragraph, we shall study the simplest form of Riemannian geometry, the one which deals with surfaces S embedded in an affine space $\mathcal{E}_3$, i.e. 2-dimensional manifolds defined in $\mathcal{E}_3$. We will start with two simple surfaces: the plane and the sphere, a curved surface with constant curvature. The method we are going to use is the method of the

moving frame: we will associate to each point $M \in S$ a frame and study its evolution when we go from one point of S to another one. This is a very intuitive idea that E. Cartan adapted adequately for the general study of manifolds. Most of the concepts and definitions we shall come across will be of use later on.

Let us start with a plane embedded in $\mathcal{E}_3$ and parametrized in polar coordinates

$$\mathbf{x} = \begin{pmatrix} x \\ y \end{pmatrix} = \begin{pmatrix} u \cos v \\ u \sin v \end{pmatrix} : u = r, v = \varphi ; \tag{3.1}$$

its coordinate curves at $M \in S$

$$\begin{cases} x = u_0 \cos v \\ y = u_0 \sin v \end{cases} , \qquad \begin{cases} x = u \cos v_0 \\ y = u \sin v_0 \end{cases} , \tag{3.2a}$$

are respectively circles with their center at the origin and straight lines going through the same origin. Their tangent vectors

$$\mathbf{e}_1(M) = \mathbf{x}_u = \begin{pmatrix} \cos u \\ \sin v \end{pmatrix} , \qquad \mathbf{e}_2(M) = \mathbf{x}_v = \begin{pmatrix} -u \sin v \\ u \cos v \end{pmatrix} \tag{3.2b}$$

are orthogonal. We then remark that if M is fixed, then all points of the considered plane can be parametrized with respect to the frame $\{\mathbf{e}_1(M),$ $\mathbf{e}_2(M)\}$: in the vocabulary of Level 2, one says that a single **chart** is enough for the parametrization of the whole plane. The corresponding ds^2 reads

$$ds^2 = du^2 + u^2 dv^2 . \tag{3.3}$$

Let us now consider a sphere S^2 of radius 1 and the origin of $\mathcal{E}_3$ as its center; we will parametrize S^2 with spherical coordinates

$$\mathbf{x} = \begin{pmatrix} x \\ y \\ z \end{pmatrix} = \begin{pmatrix} \sin u \cos v \\ \sin u \sin v \\ \cos u \end{pmatrix} : u = \theta, v = \varphi . \tag{3.4}$$

The coordinates curves are then

$$\begin{pmatrix} x \\ y \\ z \end{pmatrix} = \begin{pmatrix} \sin u_0 \cos v \\ \sin u_0 \sin v \\ \cos u_0 \end{pmatrix} ; \quad \begin{pmatrix} x \\ y \\ z \end{pmatrix} = \begin{pmatrix} \sin u \cos v_0 \\ \sin u \sin v_0 \\ \cos u \end{pmatrix} ; \tag{3.5}$$

the first curve is a parallel of S^2 with z_0 as one of its coordinates and $|\sin u_0|$ as its radius, the second curve is the intersection of the plane $y = x \tan v_0$ with S^2: it is a meridian of the sphere. Their tangents

$$\varepsilon_1 = \begin{pmatrix} \cos u \cos v \\ \cos u \sin v \\ -\sin v \end{pmatrix} , \qquad \varepsilon_2 = \begin{pmatrix} -\sin u \sin v \\ \sin u \cos v \\ 0 \end{pmatrix} \qquad (3.6)$$

are again orthogonal, but all points of S^2 cannot be parametrized with respect to $\{\varepsilon_1(M), \varepsilon_2(M)\}$, i.e. a single chart is not enough for the parametrization of S^2. We shall come back (see below) to this point and draw important consequences. The ds^2 of S^2 is well-known and reads

$$ds^2 = du^2 + \sin^2 u dv^2 . \qquad (3.7)$$

We notice here that the two ds^2 given respectively by (3.3) and (3.7) can be brought to common form by writing

$$ds^2 = du^2 + \chi^2(u, v) dv^2 , \qquad (3.8)$$

where the expressions of $\chi^2(u, v) : \chi^2 = u^2, \chi^2 = \sin^2 u$ correspond respectively to the plane and to the sphere S^2.

At that point, let us be more systematic with our notations. We shall write

$$u = u^1, \quad v = u^2 ; \quad \mathbf{e}_i \cdot \mathbf{e}_j = g_{ij}(u^1, u^2) = g_{ji}(u^1, u^2), \qquad (3.9)$$

then the ds^2 given by (3.8) and valid for the plane and the sphere takes the form*

$$ds^2 = g_{ij} du^i du^j \; : i = 1, 2$$
$$j = 1, 2, \qquad (3.10)$$

and this formula should be read using the known **summation rule** under which a summation is to be taken on every couple of indices (dummy indices) denoted by the same letter and which are disposed the one up and the other down. Comparison between (3.8) and (3.10) shows that

$$g_{11} = 1, \quad g_{22} = \chi^2(u^1, u^2), \quad g_{12} = g_{21} = 0 . \qquad (3.11)$$

---

*These notations will be modified at the Levels 1 and 2: the vector x will be denoted by **OM** with $\{X^1...X^N\}$ as components, the parameters $\{u, v\}$ or $\{u^1, u^2\}$ will be denoted by the N-uplet $\{x^1...x^N\}$.

Before starting some new calculations, let us introduce some rules which will appear here as simple recipes and will be fully justified at Level 1, Sec. 8. We will collect these prescriptions under two headings:

A: We shall consider the N by N matrix field $G(x)$ with $g_{ij}(x)$ as elements, $G(x)$ should be symmetric $g_{ij} = g_{ji}$.

The points x will be chosen in a domain where $G(x)$ is regular, i.e. its determinant $g(x) \neq 0$. The elements $g^{ij}(x)$ of its inverse G^{-1} may be expressed through Cramer's rule

$$g^{ij}(\mathbf{x}) = \frac{G^{ij}(\mathbf{x})}{g(\mathbf{x})} \tag{3.12}$$

and $g^{ij}(x)$ is the minor relative to $g_{ij}(x)$. The G matrix represents the metric tensor (precise definition at Level 1), g^{ij} and g_{ij} are respectively its **contravariant and covariant components**. Furthermore from $GG^{-1} = G^{-1}G = 1$, one also obtains

$$g^{ij}g_{jk} = g_{ij}g^{jk} = \delta^i_k = \delta^k_i \ . \tag{3.13}$$

B: The next important rule of the tensor algebra is the one concerning the **raising and lowering of the indices**: let us consider two sets of functions denoted by $A_{ijk...}(x)$ and $A^{ijk...}(x)$. The procedure of raising the index i goes as follows

$$A_{ijk} \rightarrow A^i_{jk} = g^{il}A_{ljk} \tag{3.14a}$$

and the one for the lowering of i:

$$A^{ijk} \rightarrow A^{jk}_i = g_{il}A^{ljk} \ . \tag{3.14b}$$

The reader has to consider these rules as simple recipes: they will be justified at Level 1 showing, after an accurate definition of the term tensor, that the tensorial character is invariant while raising or lowering an index. We may now come back to our main subject of discussion, i.e. the study of a surface S embedded in $\mathcal{E}_3$. Let $M(x)$ and $M(x + dx)$ be two points of S which are infinitely near to each other and consider two frames $\{e_1, e_2, e_3\}$ at x and $\{e'_1, e'_2, e'_3\}$ at $x + dx$; we want to calculate $e'_i - e_i$. Since the manifold S is supposed to be embedded in $\mathcal{E}_3$, we use Euclid's postulate and draw at M three vectors equipollent at the primed frame: we are then led to calculate at M the differences $\mathbf{e}'_i - \mathbf{e}_i$. It should be pointed out, by the way, that such a procedure could not be applied if the existence of the embedding space $\mathcal{E}_3$ was denied.

Then for any generic basis $\{\mathbf{e}_1, \mathbf{e}_2, \mathbf{e}_3\}$ one has

$$d\mathbf{e}_i = \frac{\partial \mathbf{e}_i}{\partial u j} du^j = \partial_j \mathbf{e}_i du^j \ . \tag{3.15}$$

If we consider the plane embedded in $\mathcal{E}_3$, the vector $\partial_j \mathbf{e}_i$ is a vector of this plane and with respect to the basis (3.2b) it is a linear combination of $\mathbf{e}_1, \mathbf{e}_2$.

$$\partial_j \mathbf{e}_i = \partial_j \frac{\partial \mathbf{x}}{\partial u^i} = \frac{\partial^2 \mathbf{x}}{\partial u^j \partial u^i} = \Gamma^k{}_{ij}(\mathbf{x}) \mathbf{e}_k \ ; \tag{3.16}$$

its components Γ^k_{ij} are called **Christoffel symbols or affine connection coefficients** or Γ-symbols. They can readily be expressed as functions of the g_{ij}: indeed from (3.9), one obtains

$$\partial_k \{\mathbf{e}_i, \mathbf{e}_j\} = (\partial_k \mathbf{e}_i) \cdot \mathbf{e}_j + \mathbf{e}_i \cdot \partial_k \mathbf{e}_j = \partial_k g_{ij} \ . \tag{3.17}$$

We now introduce the Γ-symbols by using (3.16):

$$\Gamma^l{}_{ik} \mathbf{e}_l \cdot \mathbf{e}_j + \mathbf{e}_i \cdot \Gamma^l{}_{kj} \mathbf{e}_l = g_{jl} \Gamma^l{}_{ik} + g_{il} \Gamma^l{}_{kj} = \partial_k g_{ij} \ . \tag{3.18}$$

But the rule about the lowering of an index allows to consider

$$\Gamma_{j,ik} = g_{jl} \Gamma^l{}_{ik} \tag{3.19a}$$

and to bring (3.18) into the form

$$\Gamma_{j,ik} + \Gamma_{i,kj} = \partial_k g_{ij} \ . \tag{3.19b}$$

However the differentiations ∂_i and ∂_j being commutable, the Γ in (3.16) are symmetric with respect to their two lower indices

$$\Gamma^k{}_{ij} = \Gamma^k{}_{ji} \tag{3.20}$$

and also

$$\Gamma_{k,ij} = \Gamma_{k,ji} \ . \tag{3.21}$$

We now permute the indices in (3.19b) to obtain two other relations of the same kind. Taking a linear combination corresponding to the sum of the two first equations and subtracting the last one, we obtain

$$\Gamma_{j,ik} = \frac{1}{2}(\partial_i g_{jk} + \partial_k g_{ij} - \partial_j g_{ki})\,. \tag{3.22}$$

Let us consider again the de_i of formula (3.15) which we will express by introducing the Γ symbols

$$de_i = \Gamma^k{}_{ij} e_k du^j\,. \tag{3.23}$$

This is a total differential: consequently it satisfies the integrability conditions

$$\partial_l\{\Gamma^j{}_{ik} e_j\} = \partial_k\{\Gamma^j{}_{il} e_j\}\,; \tag{3.24}$$

written out explicitly this expression becomes with the help of (3.16)

$$(\partial_l \Gamma^h{}_{ik} + \Gamma^j{}_{ik}\Gamma^h{}_{jl})e_h = (\partial_k \Gamma^h{}_{il} + \Gamma^j{}_{il}\Gamma^h{}_{jk})e_h\,,$$

where the $\{e_i\}$ denote a basis. This expression can be written as

$$R^h{}_{ilk}(u^1, u^2) = 0 \tag{3.25}$$

by defining

$$R^h{}_{ilk} = \partial_l \Gamma^h{}_{ik} - \partial_k \Gamma^h{}_{il} + \Gamma^j{}_{ik}\Gamma^h{}_{jl} - \Gamma^j{}_{il}\Gamma^h{}_{jk}\,. \tag{3.26}$$

The R^h_{ilk} are the components of the famous **Riemann-Christoffel curvature tensor**, a central concept of differential geometry and of general relativity and the name "tensor" will be justified at Level 1. All formulas that we have just obtained keep their validity without any change of notations in a hyperplane defined in $\mathcal{E}_N$. Up to now, we considered the plane only, but what can we say about the sphere? The discussion above cannot now be applied – it breaks down precisely at formula (3.16). Indeed, in the case of the plane, the four vectors $\partial_j e_i(x)$ associated with the point M belong to the considered plane, hence (3.15) and (3.16). In the case of the sphere S^2, the same four vectors are elements of the affine space $\mathcal{E}_3$ and vectors $\{\varepsilon_1(\mathbf{x}), \varepsilon_2(\mathbf{x})\}$ of (3.6) cannot clearly build the base of the vector space E_3 associated to $\mathcal{E}_3$.

However the Γ-symbols and the Riemann tensor can still be defined respectively by (3.21) and (3.26), but they do not vanish and hence can characterize locally the difference between flat and curved surfaces. It is also clear that such statements which are intuitive but rather formal will be justified later on.

We now want to calculate those symbols for a ds^2 given by (3.10) but with a diagonal G matrix, then

$$ds^2 = \Sigma g_{il}(du^i)^2 \ . \tag{3.27}$$

It is then easy to see that the Γ-symbols take a simplified expression given by

$$\Gamma^i{}_{jk} = 0 : i \neq j \neq k \ , \qquad \Gamma^i{}_{kk} = -\frac{1}{2g_{il}}\frac{\partial g_{kk}}{\partial u^i} \ : i \neq k$$

$$\Gamma^k{}_{jk} = \frac{1}{2g_{kk}}\frac{\partial g_{kk}}{\partial u^j} \ : \ j \neq k, j = k \tag{3.28}$$

where we do not sum over the index k for the last formula.

Let us consider a surface S, 2-dimensional manifold: among the $2^3 = 8$ Christoffel symbols, only 6 do not vanish.

$$\Gamma^1{}_{11} = \frac{1}{2g_{11}}\frac{\partial g_{11}}{\partial u^1} \ , \quad \Gamma^1{}_{12} = \Gamma^1{}_{21} = \frac{1}{2g_{11}}\frac{\partial g_{11}}{\partial u^2} \ , \quad \Gamma^1{}_{22} = -\frac{1}{2g_{11}}\frac{\partial g_{22}}{\partial u^1} \ ,$$

$$\Gamma^2{}_{11} = \frac{1}{2g_{22}}\frac{\partial g_{11}}{\partial u^1} \ , \quad \Gamma^2{}_{12} = \Gamma^2{}_{21} = \frac{1}{2g_{22}}\frac{\partial g_{22}}{\partial u^1} \ , \quad \Gamma^2{}_{22} = -\frac{1}{2g_{22}}\frac{\partial g_{22}}{\partial u^2} \ ,$$

$$\tag{3.29a}$$

to which one associates

$$\Gamma_{1,11} = \frac{1}{2}\frac{\partial g_{11}}{\partial u^1} \ , \quad \Gamma_{1,12} = \Gamma_{1,21} = \frac{1}{2}\frac{\partial g_{11}}{\partial u^2} \ , \quad \Gamma_{1,22} = -\frac{1}{2}\frac{\partial g_{22}}{\partial u^1} \ ,$$

$$\Gamma_{2,11} = \frac{1}{2}\frac{\partial g_{11}}{\partial u^1} \ , \quad \Gamma_{2,12} = \Gamma_{2,21} = \frac{1}{2}\frac{\partial g_{22}}{\partial u^1} \ , \quad \Gamma_{2,22} = \frac{1}{2}\frac{\partial g_{22}}{\partial u^2} \ .$$

$$\tag{3.29b}$$

Further simplifications occur for a ds^2 of the form

$$ds^2 = g_{11}(du^1)^2 + g_{22}(du^2)^2 \ , \tag{3.8}$$

with

$$g_{11} = 1 \ , \quad g_{22} = K^2 \ , \quad g^{11} = 1 \ , \quad g^{22} = \frac{1}{g_{22}}\frac{1}{\chi^2} \ ;$$

formulas (3.29) simplify to

$$\Gamma^1{}_{22} = -\frac{1}{2}\frac{\partial g_{22}}{\partial u^1}, \quad \Gamma^2{}_{12} = \Gamma^2{}_{21} = \frac{1}{2g_{22}}\frac{\partial g_{22}}{\partial u^1}, \quad \Gamma^2{}_{22} = \frac{1}{2g_{22}}\frac{\partial g_{22}}{\partial u^2},$$

$$\Gamma_{1,22} = -\frac{1}{2}\frac{\partial g_{22}}{\partial u^1}, \quad \Gamma_{2,12} = \Gamma_{2,21} = \frac{1}{2g_{22}}\frac{\partial g_{22}}{\partial u^2}, \quad \Gamma_{2,22} = \frac{1}{2}\frac{\partial g_{22}}{\partial u^2};$$

$$(3.30)$$

only 3 Christoffel symbols do not vanish.

We now go over to the Riemann tensor for the same surface S: to $R^j{}_{ilk}$ there corresponds, by lowering the index j, the following tableau (see also Problem 0.15)

$$R_{jilk} = g_{jh}R^h{}_{ilk} = \partial_l\Gamma_{j,ik} - \partial_k\Gamma_{j,il} + \Gamma^h{}_{ik}\Gamma_{j,hl} - \Gamma^h{}_{il}\Gamma_{jhk} . \qquad (3.31)$$

The advantage of this last form is its antisymmetry with respect to the couples (j, i) and (l, k); only 4 of its $16 = 2^4$ components are different from 0:

$$R_{12,12} = R_{2121} = -R_{1221} = R_{2112} \qquad (3.32)$$

and

$$R_{1212} = \frac{\partial \Gamma_{1,22}}{\partial u^1} - \Gamma^2{}_{21}\Gamma_{1,22} = -\frac{\partial^2 g_{22}}{(\partial u^1)^2} + \frac{1}{2g_{22}}\left(\frac{\partial g_{22}}{\partial u^1}\right)^2 . \qquad (3.33)$$

In the case of the plane ($\chi = u^1$) the curvature tensor vanishes identically. In the case of the sphere ($\chi = \sin u^1$)

$$R_{1212} = -2\cos 2u^1 + 2\cos^2 u^1 = 2\sin^2 u^1 . \qquad (3.34)$$

There are two other tensor fields that appear in general relativity; the **Ricci tensor** which is obtained by contraction between h and l in R^h_{ilk},

$$R_{ik} = R^l{}_{ilk} = g^{lm}R_{lmik} \qquad (3.35)$$

and the **scalar curvature**

$$R = g^{ik}R_{ik} = g^{ik}g^{lm}R_{limk} . \qquad (3.36)$$

The symbol R introduced by (3.36) is the curvature related to the Gaussian curvature $K(u, v)$ and is not the radius of the sphere under consideration.

Let us now come back to a surface with a ds^2 given by (3.8). The scalar curvature R is given by (3.36):

$$\left(\frac{\partial g_{22}}{\partial u^1}\right)^2 - 2g_{22}\frac{\partial^2 g_{22}}{(\partial u^1)^2} - 2R(g_{22})^2 = 0 . \tag{3.38a}$$

With $\sqrt{g_{22}} = \chi$ and $u^1 = u, u^2 = v$, Eq. (3.38a) becomes

$$\frac{\partial^2 \chi(u,v)}{\partial u^2} + \frac{1}{2}R(u,v)\chi(u,v) = 0 . \tag{3.38b}$$

For a plane, $\chi = u$ and from the preceding equation $R = 0$.

We may now consider the sphere with radius 1, we notice first of all that

$$R(u,v) = -\frac{2}{\chi(u,v)}\frac{\partial^2 \chi(u,v)}{\partial u^2} , \tag{3.39}$$

taking then $\chi = \sin u$, one has $R = 2$.

On the other hand, a direct calculation of Gauss curvature starting from formula (2.13) (see also Problem 0.8) shows that $K = 1$ then $K = R/2$, a result which is general.

Let us repeat, once more, that a flat surface corresponding to $R(u,v) = 0$ is not necessarily a plane. In order to have a plane all the R_{ijkl} should vanish; we stated already that a cone and a cylinder are flat surfaces corresponding to $R = 0$. Another distinction between flat and Euclidean surfaces goes as follows: a flat surface corresponds to $ds^2 = \varepsilon_{ij}dx^i dx^j, \varepsilon_{ij} = \pm 1$ while an Euclidean one to $ds^2 = \Sigma(dx^i)^2$; both definitions are consistent with the remarks above.

4. *Geodesics*

Let S be a surface and $\mathbf{x} = \mathbf{x}(u^1,u^2)$ its parametric equation. A curve C on S is obtained, by supposing u^1 and u^2 to be honest functions of a real variable t, corresponding then to a parametric equation $x = x(u^1(t),u^2(t))$. Let A and B be two points of C corresponding to the values t_I and t_{II} of t. An infinitesimal arc of C has the length ds, given by (3.12).

$$ds^2 = g_{ij}(u^1,u^2)du^i du^j > 0$$

so that the length of the arc AB of C is

$$s_{\widehat{AB}} = \int_{t_I}^{t_{II}} \sqrt{g_{ij}\dot{u}^i\dot{u}^j}\,dt = \int_{t_I}^{t_{II}} \frac{ds}{dt}\,dt . \tag{4.1}$$

A geodesic going through the points A and B will be defined as the curve associated with the functions $u^1(t), u^2(t)$ such that $s_{\widehat{AB}}$ is minimum. This is a typical problem of variational methods. As a matter of fact, we shall deal with a simpler problem: we will be looking for a curve going through A and B, for which $s_{\widehat{AB}}$ is an extremal, i.e. its first variation $\delta s_{\widehat{AB}}$ should vanish.

We first of all remind the reader that such a problem is basic in analytical dynamics: we describe the motion of a system of points as solutions of the **Euler-Lagrange** system of differential equations of second order. One considers indeed a set of N honest real functions of the real parameter t: $\{q^1(t), \ldots, q^N(t)\}$ and another function called the **Lagrangian**

$$L(q^1(t) \ldots q^N(t); \dot{q}^1(t) \ldots \dot{q}^N(t)) = L(q(t); \dot{q}(t)) , \qquad (4.2)$$

where $\dot{q}^k(t) = dq^k(t)/dt$ and one introduces the **action**

$$\mathcal{A} = \int_{t_I}^{t_{II}} L(q(t); \dot{q}(t)) dt . \qquad (4.3)$$

Then one looks for functions $q^k(t)$ for which the action is an extremum, i.e. a minimum, a maximum or is stationary. After some analytical manipulations known to all physicists, one obtains the Euler-Lagrange system of differential equations

$$\frac{\partial L}{\partial q^k} - \frac{d}{dt}\frac{\partial L}{\partial \dot{q}^k} = 0, \quad k = 1 \ldots N ; \qquad (4.4)$$

one should remark that in the calculation of the partial derivatives of L the q^k and $\dot{q}^k$ functions are to be considered as independent and also that the nature of the extremum of $\mathcal{A}$ is an involved problem.

We may now apply all these considerations to the problems of geodesics. We may write

$$s_{\widehat{AB}} = \int_{t_I}^{t_{II}} \frac{ds}{dt} dt = \int_{t_I}^{t_{II}} L(u(t), \dot{u}(t)) dt , \qquad (4.5)$$

where

$$\frac{ds}{dt} = L(u(t), \dot{u}(t)) = \sqrt{g_{ij}(u(t))\dot{u}^i \dot{u}^j} ,$$

the partial derivatives in (4.4) have the following expression

$$\frac{\partial L}{\partial u^k} = \frac{\frac{\partial g_{ij}(u)}{\partial u^k}\dot{u}^i\dot{u}^k}{ds/dt}\ ,$$

$$\frac{\partial L}{\partial \dot{u}^k} = \frac{2g_{ki}(u)\dot{u}^i}{ds/dt}\ ,$$

and after some transformation, one obtains the Euler-Lagrange system of equations

$$0 = \frac{\frac{\partial g_{ij}}{\partial u^k}\dot{u}^i\dot{u}^j}{ds/dt} - \frac{d}{dt}\left\{\frac{2g_{ki}\dot{u}^i}{ds/dt}\right\}$$

$$= 2\left(-\frac{g_{ki}\ddot{u}^i}{ds/dt} + \frac{1}{2}\frac{\partial g_{ij}}{\partial u^k}\cdot\frac{\dot{u}^i\dot{u}^j}{ds/dt} - \frac{\partial g_{ki}}{\partial u^j}\cdot\frac{\dot{u}^i\dot{u}^j}{ds/dt} + g_{kl}\dot{u}^i\frac{d^2s/dt^2}{(ds/dt)^2}\right)\ ,$$

or

$$g_{ki}\ddot{u}^i - \frac{1}{2}\frac{\partial g_{ij}}{\partial u^k}\dot{u}^i\dot{u}^j + \frac{\partial g_{ki}}{\partial u^j}\dot{u}^i\dot{u}^j - g_{ki}\dot{u}^i\frac{d^2s/dt^2}{ds/dt} = 0$$

and also

$$g_{ki}\ddot{u}^i + \frac{1}{2}\left(\frac{\partial g_{ki}}{\partial u^j} + \frac{\partial g_{kj}}{\partial u^i} - \frac{\partial g_{ij}}{\partial u^k}\right)\dot{u}^i\dot{u}^j - g_{ki}\dot{u}^i\frac{d^2s/dt^2}{ds/dt} = 0\ . \qquad (4.6)$$

We may obtain a more compact form of (4.6) by introducing the Γ-symbols as given by (3.22):

$$g_{ki}\ddot{u}^i + \Gamma_{k,ij}\dot{u}^i\dot{u}^j - g_{ki}\dot{u}^i\frac{d^2s/dt^2}{ds/dt} = 0\ , \qquad (4.7)$$

we multiply both sides by g^{nk} and contract over k to obtain

$$\ddot{u}^n + \Gamma^n_{ij}(u)\dot{u}^i\dot{u}^j - \dot{u}^n\frac{d^2s/dt^2}{ds/dt} = 0\ . \qquad (4.8)$$

The last term of the right-hand side depends on the kind of parametrization we choose for the curve $\widehat{AB}$. With the special choice of the intrinsic parametrization where $t = s$, one has a simpler form than (4.8):

$$\frac{d^2u^n}{ds^2} + \Gamma^n_{ij}(u)\frac{du^i}{ds}\frac{du^j}{ds} = 0\ ; \qquad (4.9)$$

we notice that one could choose for the parameter t any linear function of s. As a simple example, consider the ds^2 given by (3.8):

$$ds^2 = (du^1)^2 + \chi^2(u)(du^2)^2 , \tag{4.10}$$

where $\chi(u) = \chi(u^1, u^2)$. The Γ-symbols are given by the formulas (3.28), (3.29), (3.30) and the (4.9) equations read

$$\begin{aligned}
&\ddot{u}^1 - \chi \frac{\partial \chi}{\partial u^1} \dot{u}^2 = 0 \\
&\ddot{u}^2 + \frac{2}{\chi} \frac{\partial \chi}{\partial u^1} \dot{u}^1 \dot{u}^2 + \frac{1}{\chi} \frac{\partial \chi}{\partial u^2} (\dot{u}^2)^2 = 0 ,
\end{aligned} \tag{4.11}$$

where $\dot{u}^k = du^k/ds$. A special solution is: $\dot{u}^2 = 0$ and we deduce

$$\ddot{u}^1 = 0 , \quad u^1 = As + B . \tag{4.12}$$

We also remark that

$$(\dot{u}^1)^2 + \chi^2 \dot{u}^2 = 1$$

is a first integral of the system.

The two following points are in need of some attention: first, we did not take into account (as boundary conditions) the condition on the curve C: it is to go through the points A and B and second, we did not care about the nature of the extremum of $s_{\overset{\frown}{AB}}$. If the first problem is a classical one (with a solution in a limited number of cases), the second one is one of the main problems of variational calculus. Indeed one has very often very simple local statements which become very involved if one tries to formulate them globally. We do not intend studying such problems completely; we shall give two very simple examples to illustrate their solutions.

Let us begin with the simplest of such problems: the one concerning geodesics when the considered manifold is a plane. Then the ds^2 is of the form (Cartesian coordinates) $ds^2 = dx^2 + dy^2$ and the geodesic between A and B is the straight line AB and its length is minimum.

Let us now consider a cylinder generated by a straight line cutting a circle C of radius 1 and orthogonal to its plane. (See Problem 0.11.) Its ds^2 in cylindrical coordinates is

$$ds^2 = d\varphi^2 + dz^2$$

and has the same form as the ds^2 of the plane. Also its total curvature vanishes (see same Problem 0.11) as does the curvature of the plane. But the behaviour of the geodesics of these two surfaces will be completely different: for the cylinder, since $\chi = 1$, one gets from (4.11)

$$\ddot{\varphi} = \ddot{z} = 0 \; ,$$

then a solution will be

$$\varphi = \alpha s + \beta \; , \quad z = \gamma s + \delta \; , \tag{4.13a}$$

with $\alpha, \beta, \gamma, \delta$ real constants. In cylindrical coordinates after elimination of s, one gets

$$z = \frac{\gamma}{\alpha}\varphi + \frac{\alpha\delta - \beta\gamma}{\alpha} \; ; \tag{4.13b}$$

it is a helix drawn on the cylinder, its slope is $2\pi\gamma/\alpha$. The generatrix of the cylinder is a particular geodesic ($\alpha = 0$ corresponding to $\varphi = \beta$), the parallels of the cylinder (corresponding to $\gamma = 0, z = \delta$) represent another special case and in both cases the A and B points belong to either a generatrix or to a parallel. But if A and B are any points of the cylinder, the nature of domain in which one wants to define globally the geodesic will be essential. If we limit ourselves to a simply connected domain* D of the cylinder as the one represented in Fig. 0.8 (by simply connected we mean that there exists a continuous deformation which reduces D to any of its internal points), then, there exists a helix and only one which goes through any couple (A, B) of its points. Let us consider a connected surface of the cylinder such as the portion of the cylinder lying between any two planes orthogonal to its generatrices. This is a connected domain but not a simply connected one. There are then infinitely many helices of different slopes connecting two arbitrary points A and B of this domain (see Fig. 0.8b).

On a sphere S^2 the geodesics are by Euclidean geometry arcs of circles of maximal radius: there are two such arcs going through two points A and B. One of them is of length smaller than πR and this is the shortest distance between A and B. The other one is of length larger than πR. These two simple examples show indeed that the determination of geodesics becomes

---

*A domain is connected, if for any couple of its points P and P' one can find a path belonging to the domain and going through P and P'.

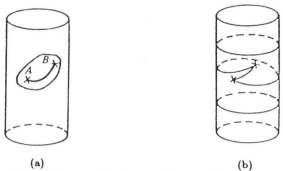

(a)

Simply connected domain on a cylinder.

(b)

Connected domain on a cylinder.

Fig. 0.8.

a real complex problem as soon as one wants to study global properties of geodesics and not only their local properties.

We finally notice that Eq. (4.8) of the geodesics may be taken as an intuitive definition of the Γ symbols. Unfortunately, it does not concern the Riemann tensor.

5. *Generalization of the concept of tangent and of tangent plane to a surface*

Let us consider a curve C on a surface $S : \mathbf{x} = \mathbf{x}(u,v)$; the equation of C can be written in parametric form

$$\mathbf{x} = \mathbf{x}(t) = \mathbf{x}(u(t), v(t)) \ , \tag{5.1}$$

where $u(t)$ and $v(t)$ are honest functions of t. The vector $\boldsymbol{\tau}_s^M(x(t)) = d\mathbf{x}(t)/dt$ is then the tangent vector at $M(\mathbf{x}) \in S$ and we may notice one of its properties: if we draw on S infinitely many curves C_M going through M, the set of the corresponding $\boldsymbol{\tau}_S^M$ constitutes the tangent plane T_S^M to M at S. This is a classical elementary concept.

Consider the set of all curves C_M going through M: the curves having the same tangent $\boldsymbol{\tau}_S^M$ build up a class and the C_M set is then divided into two equivalence classes. The element of the curve C_M in a small neighbourhood of M is called a **germ**.*

---

*The problem is more complex when S is given implicitly by $F(x) = F(x,y,z) = 0$. The

Let us now consider a fixed point $M_0(\mathbf{x}_0) \in S$ and let us choose $M(\mathbf{x})$ in a small neighbourhood of M_0. Let C_{M_0} be a curve going through M_0 and let $F(\mathbf{x}(t))$ be a function (which may for instance be taken C^∞) defined in a small neighbourhood of M_0. To every $\boldsymbol{\tau}_S^{M_0}$ let us associate

$$\tau^{M_0}(\tau_S^M, F) = \boldsymbol{\tau}_S^{M_0} \cdot \boldsymbol{\nabla} F(\mathbf{x})\big|_{\mathbf{x}=\mathbf{x}_0} \tag{5.2}$$

and the map $\mathcal{M}_0 C^\infty \to \mathbb{R}$ such that

$$\tau^{M_0}(\boldsymbol{\tau}_S^M, \cdot) = \boldsymbol{\tau}_S^{M_0} \cdot \boldsymbol{\nabla}_0 \ . \tag{5.3}$$

For a fixed M_0 and with $\tau_S^{M_0}$ variable the set of all differential operators $\tau^{M_0}(\tau_S^{M_0}, \cdot)$ can easily be given a structure of vector space. Consequently it is a map of the classical tangent plane into this vector space. One can also show that this map is bijective.

More generally if $\mathbf{V}(M)$ is a given vector field, we can consider $\mathbf{V}(M_0)$ at M_0 and as in (5.3), we define

$$\tau^{M_0}(\mathbf{V}(M), \cdot) = \mathbf{V}(M_0), \boldsymbol{\nabla}_0$$

which is directional derivative along $\mathbf{V}(M_0)$. This new vector space of differential operators is, in the language of differential geometry, the tangent vector space to S at M_0; its elements are tangent vectors, called simply vectors.

These are fundamental concepts of the intrinsic approach of differential geometry – they will be fully developed in the paragraphs 3 and 4 of Level 2. We have caught once again the mathematician giving the same name "vector" to two mathematical objects that are completely distinct: the first one is the vector in the Euclidean sense, the second one is the vector just introduced as a differential operator of first order. There is a danger of misunderstanding for the physicist here!

In the above considerations we remained at a level of the humblest everyday experience! The point, the straight line and the plane are objects of this experience: two parallel lines are lines that never meet, a bi-point vector is a directed segment, a trihedral frame is constituted by one of

---

curve C is given by 3 functions f, g, h such that $x = f(t)$, $y = g(t)$, $z = h(t)$; however these 3 functions are not independent, since for all t one must have $F(f(t), g(t), h(t)) = 0$, $\forall t$.

the corners of the laboratory, the reference planes being the floor and the two walls of the chosen corner. The points of the ambient space $\mathcal{E}_3$ are parametrized by three real numbers (x, y, z) or (x^1, x^2, x^3) that are 3 coordinates; infinitesimals are very small quantities,...

The aim of Levels 1 and 2 will be to elaborate these notions, to make them sufficiently abstract and general so that they can be described in a reasonably rigorous presentation.

A last word to stress that paragraph 3 provides, in an admittedly incomplete form, all the machinery that is essential for the development of general relativity. One could imagine that a reader in a great hurry would want to proceed directly from this paragraph 3 to the study of Einstein's equations and their solutions.

Complements, Exercises and Applications to Physics

Contents

Space Curves

$\boxed{Problem\ 0.1.}$ *Contravariant and covariant coordinates of a vector* $\mathbf{u} \in E_2$

Let $\{\mathbf{e}_1, \mathbf{e}_2\}$ be a basis of E_2, a vector $\mathbf{u}$

$$\mathbf{u} = x^1\mathbf{e}_1 + x^2\mathbf{e}_2$$

and a bilinear form

$$\begin{aligned}
\|\mathbf{u}\|^2 &= \langle x^1\mathbf{e}_i, x^j\mathbf{e}_j \rangle = g_{ij}x^i x^j \\
&= g_{11}(x^1)^2 + 2g_{12}x^1 x^2 + g_{22}(x^2)^2 \ ,
\end{aligned}$$

with

$$g_{ij} = \langle \mathbf{e}_i, \mathbf{e}_j \rangle = g_{ji} \ .$$

In particular if the basis is orthonormal

$$\|\mathbf{u}\|^2 = (x^1)^2 + (x^2)^2 \ .$$

Show that if we draw an orthonormal frame $\{\mathbf{i}, \mathbf{j}\}$ and a skew one $\{\mathbf{e}_1, \mathbf{e}_2\}$, the covariant component of $\mathbf{u}$, say $x_1 = g_{ij}x^j$ is the orthogonal projection of $\mathbf{u}$ on the frame $\{\mathbf{e}_1, \mathbf{e}_2\}$.

Hint:

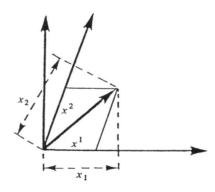

All functions we shall meet in the following exercises are supposed to be as many times derivable as needed (smooth or honest functions).

Problem 0.2 *Serret-Frenet formulas and applications*

(a) for a space curve given in parametric form, prove Serret-Frenet formulas

$$
\begin{pmatrix} dt/ds \\ dn/ds \\ db/ds \end{pmatrix} = \begin{pmatrix} 0 & 1/R & 0 \\ -1/R & 0 & 1/T \\ 0 & -1/T & 0 \end{pmatrix} \begin{pmatrix} \mathbf{t} \\ \mathbf{n} \\ \mathbf{b} \end{pmatrix}
$$

T is the torsion of the curve. Application to point kinematics.
Show that

(b) if all the tangents of a curve go through a point, the curve is a straight line.

(c) if all the osculating planes go through a point, the curve is planar.

(d) if all the tangents of a curve cut a given straight line Δ, the curve is planar and Δ belongs to its osculating plane (which is unique).

(e) if the tangent at any point M of a curve is orthogonal to $\mathbf{OM}$ (O origin), the curve is a spherical curve belonging to a sphere of center O.

(f) a conical curve (belonging to a cone) orthogonal to the generatrices of the cone, is spherical.

(g) a space curve is defined up to a displacement by its curvature and its torsion as functions of s.

Hints: (a) It is easier to deal separately with each of the equations of (a): notice first

$$
\mathbf{t} \cdot dt/ds = \mathbf{n} \cdot dx/ds = \mathbf{b} \cdot db/ds = o.
$$

In order to prove the first equation remark that $\mathbf{n}$ *belongs to the plane* $(\mathbf{t}, \mathbf{t} + dt/ds)$. *The proof of the third one starts from* $\mathbf{b} = \mathbf{t} \wedge \mathbf{n}$ *and its derivative, use then geometrical considerations. In the same way, the proof of the second equation starts* $\mathbf{n} = \mathbf{b} \wedge \mathbf{t}$. *For all the other questions, consult any textbook on analytical geometry. The question (g) is less straightforward than the other questions.*

Surfaces

| Problem 0.3. | *Geometry of surfaces (algebraic aspects)*

The theory of surfaces as developed in the 19th century and early 20th century dealt uniquely with bidimensional manifolds of E_3. Its fundamental concepts are important for a good comprehension of differential geometry. We consider a surface S immersed in E_3 and we look for local properties at $M \in S$ and in its infinitesimal neighbourhood using properties of curves C drawn on S and going through M ($M \in C \subset S$).

(a) $M \in S, C$ is a curve of S going through M, $\mathbf{N}$ the unit normal at S in M and $\{\mathbf{t}(M), \mathbf{n}(M), \mathbf{b}(M)\}$, the Serret-Frenet frame at M. Define for the curve C the function

$$\frac{1}{\rho_c(M)} = \mathbf{N} \cdot \frac{d\mathbf{t}}{ds} = \mathbf{N} \cdot \frac{d^2\mathbf{x}}{ds^2} \tag{I}$$

and use formulae (2.8) and (2.9) of Level 0 to bring $1/\rho_c(M)$ into the form

$$\frac{1}{\rho_c(M)} = \frac{\mathbf{x}_u \wedge \mathbf{x}_v}{\sqrt{EG - F^2}} \cdot \frac{d^2\mathbf{x}}{ds^2} . \tag{II}$$

Show that

$$d^2\mathbf{x} = \mathbf{x}_{uu} du^2 + 2\mathbf{x}_{uv} du dv + \mathbf{x}_{vv} dv^2 \tag{III}$$

and write Eq. (II) as follows:

$$\frac{1}{\rho_c(M)} = \frac{L du^2 + 2M du dv + N dv^2}{E du^2 + 2F du dv + G dv^2} , \tag{IV}$$

where

$$L = \frac{1}{\sqrt{EG - F^2}} \mathbf{x}_u \wedge \mathbf{x}_v \cdot \mathbf{x}_{uu} ,$$

$$M = \frac{1}{\sqrt{EG - F^2}} \mathbf{x}_u \wedge \mathbf{x}_v \cdot \mathbf{x}_{uv} ,$$

$$N = \frac{1}{\sqrt{EG - F^2}} \mathbf{x}_u \wedge \mathbf{x}_v \cdot \mathbf{x}_{vv} . \tag{V}$$

$L du^2 + 2M du dv + N dv^2$ is the **second form of Gauss**.

(b) Consider the function $f(x)/g(x)$ and let x_0 be the value of x for which the function is extremum, then show that

$$\text{Extr.} \left\{ \frac{f(x)}{g(x)} \right\} = \frac{f(x_0)}{g(x_0)} = \frac{f'(x_0)}{g'(x_0)} .$$

(c) Let $v = v(u)$ be a curve of S going through M, IV gives $\rho_c(M)$ as a function of dv/du, using (b) show that the extremum of $1/\rho_c(M)$ is given by

$$\left.\frac{1}{\rho_c(M)}\right|_{\text{extr.}} = \frac{M + N\overline{m}}{F + G\overline{m}} \,,$$

where $\overline{m}$ is the value of dv/du which gives to $1/\rho_c(M)$ its extremal value. In the same way, consider $dv/du = m'$ and show that

$$\left.\frac{1}{\rho_c(M)}\right|_{\text{extr.}} = \frac{L\overline{m}' + M}{E\overline{m}' + F} \,,$$

where $\overline{m}'$ has a similar meaning to m. Noting that $\overline{m}' = 1/\overline{m}$ and denoting $\psi(M) = \rho_c(M)/_{\text{ext}}$, show that ψ satisfies the second order equation:

$$(F - M\psi)^2 - (E - L\psi)(G - N\psi) = 0 \,. \qquad \text{(VI)}$$

(d) ψ_1 and ψ_2 being the roots of VI, show that the Gauss curvature is defined as

$$K(u, v) = \frac{1}{\psi_1\psi_2} = \frac{LN - M^2}{EG - F^2} \,.$$

One considers also the **mean curvature**

$$k(u, v) = \frac{1}{2}\left(\frac{1}{\psi_1} + \frac{1}{\psi_2}\right) \,;$$

Show that

$$k(u, v) = \frac{EN - 2FM + GL}{2(EG - F^2)} \,.$$

An application of the mean curvature will be found in Problem 0.14.

Remark : One denotes often

$$EG - F^2 = H^2 \,.$$

The results concerning the first and second forms of Gauss have been given following the traditional presentation of all texts on classical theory of surfaces. The interested reader will find in [Bibl. 12] Vol. 1 , p. 86 and following a more general presentation based on the theory of quadratic forms.

Hints: (c) Equation VI results from the elimination of $\overline{m}$ from the two equations

$$(M + N\overline{m})\psi = F + GM \ , \quad (L + M\overline{m})\psi = E + F\overline{m}.$$

(d) Recall that the sum and the product of the roots of $ax^2 + bx + c = 0$ are respectively $-\dfrac{b}{a}, \dfrac{c}{a}$.

| **Problem 0.4.** | *Geometry of surfaces (geometric aspects)* |

We define in the preceding exercise two functions: the first one $1/\rho_c(M)$ depends on $M \in S$ and the curve $C \subset S$, the second one is the Gauss curvature $K(M)$ which depends on M only. We want now to look more closely to the geometric significance of these functions.

(a) Formula I of the preceding exercise can also be brought into the following form

$$\frac{1}{\rho_c(M)} = \frac{\mathbf{N} \cdot \mathbf{dt}}{ds} = \mathbf{N}(M) \cdot \frac{\mathbf{n}(M)}{R_c(M)} = \frac{\cos \beta}{R_c(M)} \tag{I}$$

where R_c is the radius of curvature of C and β the angle between $\mathbf{n}$ and $\mathbf{N}$.

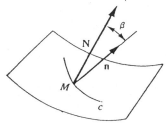

(b) Let a normal section γ of S at M be defined as a section of S by a plane containing $\mathbf{N}(M)$, the former expression takes the form

$$\frac{1}{\rho_\gamma(M)} = \pm \frac{1}{R_\gamma(M)} = \frac{L\,du^2 + 2M\,du\,dv + N\,dv^2}{E\,du^2 + 2F\,du\,dv + G\,dv^2} \tag{II}$$

where $v = v(u)$ is the equation of γ. Modify slightly the definition of $\rho_\gamma(M)$ by including the sign $\pm$ in front of R_γ, then $\rho_\gamma(M)$ will be positive if $\mathbf{n}$ and $\mathbf{N}$ point the same way (otherwise negative) and

$$\rho_\gamma(M) = R_\gamma(M) = \frac{E + 2F\frac{dv}{du} + G\left(\frac{dv}{du}\right)^2}{L + 2M\frac{dv}{du} + N\left(\frac{dv}{du}\right)^2} \ , \tag{III}$$

$\rho_\gamma(M)$ has now a clear-cut geometric meaning and there are two directions called **principal directions** for which R_γ takes its extremal values; ($\rho_c(M)$ takes then its extremal values ψ_1 and ψ_2, see question (c) of the preceding exercise).

(c) We want to obtain from Eq. III the value $\overline{m}$ of $m = dv/du$ corresponding to the extremas of $\rho_\gamma(M)$.

(i) First show that if $v = f(u)$ and $v = g(u)$ are two curves on S and $m = df/du$, $m' = dg/du$, then the angle α between the tangents at the two curves at $M(u,v)$ is given by

$$\cos \alpha = \frac{E + (m + m')F + mm'G}{\sqrt{E + 2Fm + Gm^2}\sqrt{E + 2Fm' + Gm'^2}} \ . \tag{IV}$$

(ii) Then for any curve $C : v = v(u)$ on S, show that $\overline{m}$ is a root of the equation

$$(FN - MG)\overline{m}^2 + (EN - GL)\overline{m} + EM - LF = 0 \tag{V}$$

and, using the formula III, that the two principal directions at any point of C are orthogonal.

(d) Draw from any fixed point O an equipollent vector to $\mathbf{N}(M)$. Let dS be an element of surface of S with C as a boundary; when M moves along C, the vector $\mathbf{N}(M)$ whose origin is O moves along a closed curve c drawn on the unit sphere S^2 with O as a center: c is called the spherical image of C. Call $d\sigma$ an element of the sphere S^2 with c as a boundary and show that the Gaussian curvature

$$K(M) = \frac{d\sigma}{dS} \ ,$$

a formula which displays the geometric content of the Gaussian curvature.

There are many other properties of the curves drawn on surfaces, a few more are quoted and proved hereafter.

(e) As is well-known, the equation of the tangent plane at $M(\mathbf{x}) \in S$ is $\mathbf{X} - \mathbf{x} \cdot \mathbf{N}(\mathbf{x}) = 0$. Show that it can be written as

$$\frac{\partial(y,z)}{\partial(u,v)}(X - x) + \frac{\partial(z,x)}{\partial(u,v)}(Y - y) + \frac{\partial(x,y)}{\partial(u,v)}(Z - z) = 0 \ .$$

As we already pointed out, every curve $C \subset S$, can be defined by a relation $v = v(u)$, show that the equation of the tangent to C at $M(\mathbf{x})$ is

$$(\mathbf{X} - \mathbf{x}) = \lambda \left(\mathbf{x}_u + \mathbf{x}_v \frac{dv}{du} \right) \ ,$$

x_u and x_v depend only on the point M, the value of dv/du is taken at M and characterizes the tangent in question.

(f) Let two curves C and C' go through M and have the same tangent at $M \in C \subset S$, show that $\rho_C(M) = \rho_{C'}(M)$ (see Eq. I of the preceding problem).

(g) Given any curve $C \subset S$ going through M, there corresponds a planar curve Γ with the same radius of curvature $R_c(M)$. This remark shows that one needs to study only the planar curves belonging to S.

(h) Given a planar curve Γ going through $M(x) \in S$ with $t(M)$ and $n(M)$ as tangent and also principal normal, and the normal section γ defined by $t(M)$ and $N(M)$, show that if one attributes to ρ_γ a sign as fixed in (b) one has

$$R_\Gamma(M) = \rho_\gamma(M) \cos \beta .$$

The geometric interpretation of that remark is known as **Meusnier's Theorem**.

We may now summarize the way we study the **local** behaviour of a surface around a point $M \in S$ using the properties of curves of S going through M. Given any curve C of S: its osculating plane at M determines a plane section γ of S. We notice that the curves C and γ have a common radius of curvature at M. Then we consider a normal section Γ at M which admits the same tangent at M than γ: the radius of curvature of γ is the orthogonal projection of the radius of curvature R_Γ on the principal normal $n(M)$ of γ. There is a question of sign to be determined as in this exercise (formulas II and III). Finally, the problem is reduced to the study of the normal sections Γ of S at M, the Gauss curvature $K(M)$ of S is then deduced.
For further information, the reader should consult any textbook on the theory of curves and surfaces, for instance [Bibl. 2] Vol. 2, Chap. 5.

Hints: (c) (i)the tangent vector to the curve $x = x(u, v(u))$ is $x_u + mx_v$. Considering the two curves $v = f(u)$ and $v = g(u)$, one obtains $\cos \alpha$ straightaway.
(ii) Consider then $1/\rho_\gamma$ in b as a function of m, then the value $\overline{m}$ for which $1/\rho_\gamma$ is extremum is a root of the equation

$$\frac{d}{dm}\left\{\frac{1}{\rho_\gamma}\right\} = 0 .$$ \hfill (VI)

On the other hand $\cos \alpha$ *is given by IV, use then the well-known formulas for the sum* $\overline{m} + \overline{m}'$ *and the product* $\overline{m}\,\overline{m}'$ *of the roots of Eq. VI to conclude that* $\cos \alpha = 0$.

(d) Following (2.10), Level 0, one has

$$d\delta = \|\mathbf{x}_u \wedge \mathbf{x}_v\| du\, dv = \sqrt{EG - F^2}\, du\, dv .$$

On the other hand

$$d\sigma = \|\mathbf{N}_u \wedge \mathbf{N}_v\| du\, dv .$$

Since $\mathbf{N}^2 = 1, \mathbf{N}_u.\mathbf{N} = \mathbf{N}_v.\mathbf{N} = 0$ *and* $\mathbf{N}_u, \mathbf{N}_v$ *are parallel to the tangent plane at* M *to which belong* $\mathbf{x}_u$ *and* $\mathbf{x}_v$ *one can write*

$$\mathbf{N}_u = \alpha \mathbf{x}_u + \beta \mathbf{x}_v , \quad \mathbf{N}_v = \alpha' \mathbf{x}_u + \beta' \mathbf{x}_v .$$

One can show by algebraic manipulation that

$$(\alpha\beta' - \alpha'\beta)(EG - F^2) = LN - M^2 .$$

Then from formula

$$\frac{d\sigma}{dS} = \frac{\|\mathbf{N}_u \wedge \mathbf{N}_v\|}{\sqrt{EG - F^2}} ,$$

results the final answer to (d).

(f) Since C *and* C' *have the same tangent at* M, *they correspond to a common value* dv/du *at* M.

(g) The space curve $C \subset S$ *has* $\mathbf{t}(M)$ *and* $\mathbf{n}(M)$ *as tangent vector and principal normal. Those two vectors define a plane, let* Γ *be the planar curve intersection of this plane with* S. *Since* C *and* Γ *have the same tangent, then* $\rho_C(M) = \rho_\Gamma(M)$ *but these curves have also the same principal normal, then* β *is a common angle for* C *and* Γ. *One concludes finally that* $R_c(M) = R_\Gamma(M)$.

(h) Γ *and* γ *have the same tangent, then* $\rho_\Gamma(M) = \rho_\gamma(M)$ *but* $\rho_\gamma(M) = 1/R_\gamma$ *(sign included !).*

| Problem 0.5. | *Lines of curvature and asymptotic lines*

(a) **A line of curvature** is defined as a curve on a surface such that its tangent at every of its points is a principal direction. Show that:

(i) at each point $M \in S$, there are two lines of curvature which are orthogonal: their mesh is then an orthogonal one.

(ii) the equation $v = v(u)$ or $u = u(v)$ of a line of curvature is a solution of the differential equation

$$\begin{vmatrix} E\,du + F\,dv & F\,du + G\,dv \\ L\,du + M\,dv & M\,du + N\,dv \end{vmatrix} = 0 \ .$$

(iii) show that the mesh of lines of curvature is the one corresponding to coordinate lines if and only if $F = M = 0$.

(b) An **asymptotic line** is a line whose osculating plane is tangent to the surfaces. Show that its equation is

$$\mathbf{N}(\mathbf{x}) \cdot d^2\mathbf{x} = 0 \ ,$$

or

$$L\,du^2 + 2M\,du\,dv + N\,dv^2 = 0 \ .$$

Hints: (a) (ii) Consider hint (c) of Ex. 0.3, eliminate ψ from the two linear equations and note that m is a convenient value of dv/du.

(a) (iii) Since the lines of curvature are orthogonal and also coordinate lines ($u = u_0$ for the one system and $v = v_0$ for the other), then $F = 0$. The conditions $dv/du = 0$ or $du/dv = 0$ lead to $GM = 0$ or $EM = 0$. On the other hand, $EG - F^2 = H^2$ is positive, then E and G are $\neq 0$, thus $M = 0$. The converse is also true.

(b) Notice that $d^2\mathbf{x}$ is in the osculating plane.

$\boxed{\textit{Problem 0.6}}$ *The surface S is given by an equation of the form $z = f(x,y)$* Let the points of the embedding space $\mathcal{E}_3$ be parametrized in an orthonormal frame and suppose S to be given by

$$z = f(x,y) \quad : \quad F(\mathbf{x}) = f(x,y) - z = 0 \ ,$$

its normal at M is

$$\mathbf{N}(M) = \frac{\boldsymbol{\nabla} \cdot F(M)}{\|\boldsymbol{\nabla} \cdot F(M)\|} = \frac{\mathrm{grad}\,F(M)}{\|\mathrm{grad}\,F(M)\|} \ .$$

One generally uses the notations

$$\frac{\partial f}{\partial x} = p \,, \frac{\partial f}{\partial y} = q \,, \frac{\partial^2 f}{\partial x^2} = r \,, \frac{\partial^2 f}{\partial x \partial y} = s \,, \frac{\partial^2 f}{\partial y^2} = t \ ,$$

then

$$\mathbf{N}(M) = \frac{1}{\sqrt{1+p^2+q^2}} \begin{pmatrix} p \\ q \\ -1 \end{pmatrix} .$$

One may also write the equation of S in a parametric form as follows:

$$\mathbf{x} = \mathbf{x}(u,v) = \begin{pmatrix} u \\ v \\ z(u,v) \end{pmatrix} ,$$

where $x = u, y = v$.

(a) Show that the equation of the tangent plane at the point $\{x, y, z\}$ of S is

$$Z - z = p(X - x) + q(Y - y)$$

and that

$$ds^2 = (1+p^2)dx^2 + 2pq\,dx\,dy + (1+q^2)dy^2 ,$$
$$H^2 = EG - F^2 = 1 + p^2 + q^2 .$$

(b) Calculate $\mathbf{x}_u, \mathbf{x}_v, \mathbf{x}_u \wedge \mathbf{x}_v$ and show that

$$L = \frac{r}{H}, \quad M = \frac{s}{H}, \quad N = \frac{t}{H},$$
$$K(x,y) = \frac{rt - s^2}{\left(1+p^2+q^2\right)^2} .$$

(c) Choose now a special system of coordinates with M as origin and the z-axis along $\mathbf{N}$. The components of $\mathbf{N}$ are $\{0, 0, 1\}$ and this implies that $p(M) = q(M) = 0$. Consider another point $P \in S$ such that $P \neq M$ with $\{\xi, \eta, \xi\}$ as coordinates and suppose that P belongs to an infinitesimal neighbourhood of M, then:

$$\zeta(\xi, \eta) = \frac{\xi^2}{2}r(M) + \xi\eta s(M) + \frac{\eta^2}{2}t + \cdots ,$$

where the coordinates of M are $\{0, 0, 0\}$.

Perform a rotation of the Mx, My axis around Mz such that in the new coordinate system $\{MX, MY, MZ\}$ one has

$$Z = z = \frac{X^2}{2\rho_1} + \frac{y^2}{2\rho_2} ,$$

where ρ_1 and ρ_2 are independent of X, Y, Z.

(d) The preceding formulas can also be rewritten in parametric form if one defines

$$\mathbf{X} = \mathbf{X}(u,v) : X = u, \quad Y = v, \quad Z = \frac{u^2}{2\rho_1} + \frac{v^2}{2\rho_2} .$$

Show that

$$ds^2 = \left(1 + \frac{u^2}{\rho_1^2}\right) du^2 + 2\frac{uv}{\rho_1\rho_2} du\,dv + \left(1 + \frac{v^2}{\rho_2^2}\right) dv^2$$

and

$$L = \frac{1}{\rho_1\sqrt{1 + \frac{u^2}{\rho_1^2} + \frac{v^2}{\rho_2^2}}}, \quad M = 0, \quad N = \frac{1}{\rho_2\sqrt{1 + \frac{u^2}{\rho_1^2} + \frac{v^2}{\rho_2^2}}},$$

$$K(u,v) = \frac{1}{\rho_1^2\rho_2^2\left(1 + \frac{u^2}{\rho_1^2} + \frac{v^2}{\rho_2^2}\right)^{\frac{3}{2}}} .$$

At the origin M, $K(u,v)$ is reduced to

$$K(M) = K(0,0) = \frac{1}{\rho_1^2\rho_2^2} .$$

(e) Introduce in the above expression of ds^2 the metric tensor g_{11}, g_{12}, g_{22} and using formulas (3.29),(3.31) and (3.32) of Level 2, show that the scalar curvature R (formulas (3.36) and (3.42)) is

$$R(M) = 2K(M) .$$

Hints: (b) One has

$$\mathbf{x}_u = \begin{pmatrix} \partial x/\partial u \\ \partial y/\partial u \\ \partial z/\partial u \end{pmatrix} = \begin{pmatrix} 1 \\ 0 \\ p \end{pmatrix}, \mathbf{x}_v = \begin{pmatrix} \partial x/\partial v \\ \partial y/\partial v \\ \partial z/\partial v \end{pmatrix} = \begin{pmatrix} o \\ 1 \\ q \end{pmatrix}, \dots$$

then

$$L = \frac{\mathbf{x}_u \wedge \mathbf{x}_v \cdot \mathbf{x}_{uu}}{H} = \frac{r}{H}, \dots$$

$$K(x,y) = \frac{EN - 2FM + GL}{2H^2} = \frac{rt - s^2}{(1 + p^2 + q^2)^2} .$$

(d) Same calculation as in (b).

Problem 0.7 *Curvature of a surface of revolution*

The space $\mathcal{E}_3$ is referred to as an orthonormal frame $\{Ox, Oy, Oz\}$ and we consider the curve C in the plane xOz

$$\mathbf{x} = \begin{pmatrix} x \\ y \\ z \end{pmatrix} = \begin{pmatrix} f(u) \\ o \\ h(u) \end{pmatrix}$$

This curve rotates around Oz and generates a surface of revolution S with Oz as axis; v is the angle between Ox and the plane of C, then any point M of S has for coordinates

$$\mathbf{x} = \begin{pmatrix} x \\ y \\ z \end{pmatrix} = \begin{pmatrix} f(u)\cos v \\ f(u)\sin v \\ h(u) \end{pmatrix} .$$

(a) Show that the surface S'

$$S' : x = \varphi(u)\cos v \;,\; y = \varphi(u)\sin v \;,\; z = u$$

represents also a surface of revolution with Oz as an axis and is obtained by the rotation of the planar curve

$$x = \varphi(u) \;,\; y = 0 \;,\; z = u$$

around the Oz axis.

(b) Nature of the curves $u = u_0$ and $v = v_0$: all points of the surface S which are common to the $z'z$ axis are singular for these representations, even if they are ordinary points of the surface.

(c) Show that

$$ds^2 = (f'^2 + h'^2)du^2 + f^2 dv^2 \ ,$$
$$H^2 = EG - F^2 = (f'f)^2 + (h'f)^2 \ ,$$

$$L = \frac{f}{H}(f'h'' - f''h') \ , \quad M = 0 \ , \quad N = \frac{h'}{H}f^2 \ ,$$
$$K(u,v) = \frac{h'(f'h'' - f''h')}{f(f'^2 + h'^2)^2} \ .$$

(d) Determine the mesh of lines of curvature of a surface of revolution.

(e) Same questions for the representation S'.

(f) The rotating curve around $z'z$ is given in the form $F(x,z) = 0$. Equation of the torus.

Hints: The answers to questions (a)(b)(c)(d)(e) are straightforward enough.

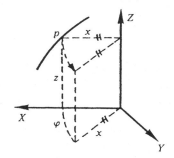

(f) The point $P(x, 0, z)$ of the curve $F(x,z) = 0$ is brought at $P(x,y,z)$ by the rotation φ around $z'z$. The equation of the corresponding meridian is now $F(r,z) = 0$ where $r = \sqrt{x^2 + y^2}$.

The equation of the corresponding surface is then

$$F\left(\sqrt{x^2 + y^2}, z\right) = 0 \ .$$

We apply this method to the torus obtained by rotating the circle

$$(x - d)^2 + z^2 = R^2$$

around $z'z : d$ is the distance of its center to $z'z$ and R its radius. Then in cylindrical coordinates the equation of a torus is

$$(r - d)^2 + z^2 - R^2 = 0$$

and after some simple transformations, one obtains its equations in rectangular coordinates.

$$\left(x^2 + y^2 + z^2 + d^2 - R^2\right)^2 = 4d^2\left(x^2 + y^2\right) .$$

| Problem 0.8 | *Surfaces of positive Gaussian curvature*

(a) Sphere S^2, radius a, center O.

$$x^2 + y^2 + z^2 = a^2.$$

Show that one has a parametrization

$$\mathbf{x} = \begin{pmatrix} x \\ y \\ z \end{pmatrix} = \begin{pmatrix} u \\ v \\ \pm\sqrt{a^2 - u^2 - v^2} \end{pmatrix} ,$$

with

$$ds^2 = \left(1 + \frac{a^2}{a^2 - u^2 - v^2}\right)du^2 + \frac{2uv}{a^2 - u^2 - v^2}dudv + \left(1 + \frac{v^2}{a^2 - u^2 - v^2}\right)dv^2 .$$

A parametrization in spherical coordinates is

$$\mathbf{x} = \begin{pmatrix} x \\ y \\ z \end{pmatrix} = \begin{pmatrix} a \sin u \cos v \\ a \sin u \sin v \\ a \cos v \end{pmatrix} ,$$

with

$$ds^2 = a^2 du^2 + a^2 \sin^2 u\, dv^2.$$

Calculate the Gaussian curvature to find

$$K(u, v) = \frac{1}{a^2} ,$$

S^2 is a surface with constant positive curvature.

Determine its lines of curvature. Show by drawing a certain triangle that the sum of its internal angles $\Sigma\alpha_j$ is $> \pi$.

(b) Ellipsoid of revolution around the $z'z$ axis:

$$\frac{x^2}{a^2} + \frac{y^2}{a^2} + \frac{z^2}{c^2} = 1 .$$

Show that

$$ds^2 = (a^2 \cos^2 u + c^2 \sin^2 u)du^2 + a^2 \sin^2 u \, dv^2 \ ,$$
$$H^2 = a^2 \sin^2 u (a^2 \cos^2 u + c^2 \sin^2 u) \ ,$$

$$L = -\frac{ac}{\sqrt{a^2 \cos^2 u + c^2 \sin^2 u}} \ , \quad M = 0 \ , \quad N = -\frac{ac \sin^2 u}{\sqrt{a^2 \cos^2 u + c^2 \sin^2 u}}$$

$$K(u,v) = \frac{c^2}{\sqrt{a^2 \cos^2 u + c^2 \sin^2 u}} \cdot$$

The ellipsoid has a positive Gaussian curvature which is not constant (as for S^2). Lines of Curvature.

Hint: A straightforward application of Ex. 0.7.

$\boxed{\text{Problem 0.9}}$ *Surfaces of negative Gaussian curvature*

    (a) Hyperbolic paraboloid

$$z = \frac{x^2}{2a^2} - \frac{y^2}{2b^2} \ ,$$

show that

$$K(x,y) = -\frac{1}{a^2 b^2} \frac{1}{\left(1 + \frac{x^2}{a^4} + \frac{y^2}{b^4}\right)^2} \cdot$$

This is an example of a surface of negative curvature function of the points of the surface.

    (b) The pseudosphere

(i) Consider the planar curve called tractrix

$$x = e^u \ , \quad z = \int_0^u \sqrt{1 - e^{2t}} dt \ ;$$

draw the corresponding curve and show that the portion of the tangent between the $z'z$ axis and the point of tangency has 1 as length.

(ii) Show that

$$K = -1 \ ,$$

the pseudosphere is a surface of constant negative Gaussian curvature.

Hints: (a) Application of (b), Ex. 0.6.

(b)(i) Equation of the tangent at the point $M(x, z)$

$$(Z - z) = \pm \sqrt{\frac{1 - x^2}{x^2}}(X - x)$$

coordinates of T

$$X_0 = 0 \quad Z_0 = z \pm \sqrt{1 - x^2}$$

$$MT = 1$$

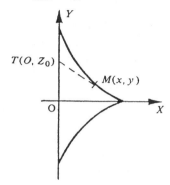

(ii) One has

$$\mathbf{x} = \begin{pmatrix} e^u \cos v \\ e^u \sin v \\ \int_0^u \sqrt{1 - e^{2t}}\, dt \end{pmatrix} ,$$

a straightforward application of Ex. 0.7 shows that $K = -1$.

$\boxed{\textbf{Problem 0.10}}$ *Surface of zero Gaussian curvature – Developable surfaces*

One considers

(a) the plane

$$z = ax + by + c .$$

Show that $K(x, y) = 0$.

(b) the cone of second degree with the origin as vertex

$$\frac{x^2}{a^2} + \frac{y^2}{b^2} + \frac{z^2}{c^2} = 0 .$$

Solve with respect to z and show that $K(x, y) = 0$.

(c) developable surfaces: in an orthonormal frame, consider the family of planes

$$A(\lambda)x + B(\lambda)y + C(\lambda)z + D(\lambda) = 0 , \qquad (I)$$

depending on the real parameter λ and also the family

$$A'(\lambda)x + B'(\lambda)y + C'(\lambda)z + D'(\lambda) = 0 , \qquad (II)$$

deduced from the first by taking its derivative with respect to λ. The elimination of λ — whenever possible — leads to a surface called **developable surface or envelope of a family of planes** depending on a single parameter λ. The intersection of the planes (I) and (II) is a family of straight lines depending on a single parameter λ: these are the **generatrices** of the developable surface. One also says that a developable surface is a **ruled surface** (see next exercise) with a constant tangent plane along a generatrix.

(i) Show that the generatrices of a developable surface are tangent to a space curve C (called their **envelop**).

(ii) Conversely, one has the following definition of a developable surface: given a space curve $C : \boldsymbol{\xi} = \boldsymbol{\xi}(u)$ where u is the length of an arc of C and $\mathbf{t}(u)$ its tangent, the equation

$$\mathbf{x}(u,v) = \boldsymbol{\xi}(u) + (v - u)\mathbf{t}(u)$$

represents following (a) a developable surface. Show that

$$d\mathbf{x} = (v - u)\frac{\mathbf{n}}{R}du + \mathbf{t}dv ,$$
$$ds^2 = \frac{(v - u)^2}{R^2}du^2 + dv^2 ,$$

$\mathbf{t}(u), \mathbf{n}, (u), R(u)$ are relative to the curve C defined intrinsically.

(iii) Show that all developable surfaces are of Gaussian curvature 0.

Hints: (c) The tangent plane

$$(Z - z) = p(X - x) + q(Y - y)$$

must depend on one parameter λ only, thus p and q are functions of λ and there exists a relation between p and q. One has $\partial(p,q)/\partial(x,y) = 0$. The answers to the other questions are simple enough.

| Problem 0.11 | *Ruled surfaces, generalized cylinder*

We consider the curve $C : \boldsymbol{\xi} = \boldsymbol{\xi}(u)$ (the parameter u is not necessarily its arc) and a given vector $\boldsymbol{\tau}(u)$. We define a ruled surface in parametric representation by

$$\mathbf{x}(u,v) = \boldsymbol{\xi}(u) + v\boldsymbol{\tau}(u) .$$

For $u = u_0 : \mathbf{x} = \mathbf{x}(u_0, v)$ represents a straight line going through the point $\boldsymbol{\xi}_0 = \boldsymbol{\xi}(u_0)$ and directed along the vector $\boldsymbol{\tau}(u)$. We may then conclude that a ruled surface can be generated by a straight line with $\boldsymbol{\tau}(u)$ as a direction and cutting a given curve C. Verify then any developable surface is a ruled surface, but the converse is not true.

(a) Show that any ruled surface is generated by a family of straight lines

$$x = a(u)z + \alpha(u) \; , y = b(u)z + \beta(u) \; ,$$

depending on a parameter u.

(b) If $\boldsymbol{\tau}(u)$ is a constant vector $\boldsymbol{\tau}_0$, the ruled surface is called a **generalized cylinder**. Show that for such a surface $M = N = 0$, thus $K(u,v) = 0$.

(c) Define the envelop of a family of curves $C_{(\lambda)}$ as a curve Γ which is tangent at each of its points to a curve $C(\lambda)$. Suppose then that the $C_{(\lambda)}$ family is given by $\mathbf{x} = \mathbf{x}(u, \lambda)$, show that a necessary condition for the existence of Γ is

$$\mathbf{x}_u \wedge \mathbf{x}_\lambda = 0 .$$

(d) Apply this condition to the set of equations in (a) to show that it is reduced to

$$\frac{da(\lambda)}{d\lambda}\frac{d\beta(\lambda)}{d\lambda} - \frac{db(\lambda)}{d\lambda}\frac{d\alpha(\lambda)}{d\lambda} = 0 .$$

Hints: (c) Consider the curve $\mathbf{x} = \mathbf{x}(u(\lambda), \lambda)$ *and look for the function* $u = u(\lambda)$ *such that*

$$\frac{d\mathbf{x}}{d\lambda} = \mathbf{x}_u \frac{du}{d\lambda} + \mathbf{x}_\lambda$$

is collinear to the tangent $\mathbf{x}_u$ *of the particular curve* $C_{(\lambda)}$ *we are interested in.*
The solution of the other questions is simple.

[*Problem 0.12*]   *Gaussian curvature and the metric*

We gave a solution of the problem of curvature along two different ways: the first one, given in the text (Sec.3, Level 0), started from the metric $ds^2 = g_{ij}dx^i dx^j$ and arrived at the expression of the total curvature R by means of the Christoffel symbols. The second one started from the equation of the surface in parametric form, either by $z = f(x, y)$ or $F(x, y, z) = 0$ and expressed the same metric ds^2 in the form $Edu^2 + 2Fdudv + Gdv^2$. The expression of the Gaussian curvature K was then given through the coefficients E, F, G, L, M, N. The connection between the two concepts apparently distinct was investigated in the Ex. 0.6(b). We want now to prove Gauss theorem by showing that the coefficients L, M, N can be expressed as functions of E, F, G: we thus reveal clearly how the two curvatures are connected with each other.

(a) Show, by using the rule of multiplication of determinants, that M^2 and LN are given by the following two determinants:

$$M^2 = \frac{1}{H^2}\left(\mathbf{x}_u \wedge \mathbf{x}_v \cdot \mathbf{x}_{uv}\right)^2 = \begin{vmatrix} E & F & \frac{1}{2}\frac{\partial E}{\partial v} \\ F & G & \frac{1}{2}\frac{\partial G}{\partial u} \\ \frac{1}{2}\frac{\partial E}{\partial v} & \frac{1}{2}\frac{\partial G}{\partial u} & \mathbf{x}_{uv}^2 \end{vmatrix},$$

$$LN = \frac{1}{H^2}\left(\mathbf{x}_u \wedge \mathbf{x}_v \cdot \mathbf{x}_{u^2}\right)\left(\mathbf{x}_u \wedge \mathbf{x}_v \cdot \mathbf{x}_{v^2}\right)$$

$$= \begin{vmatrix} E & F & \mathbf{x}_u \cdot \mathbf{x}_{v^2} \\ F & G & \frac{1}{2}\frac{\partial G}{\partial v} \\ \frac{1}{2}\frac{\partial E}{\partial u} & \mathbf{x}_v \cdot \mathbf{x}_{u^2} & \mathbf{x}_{u^2} \cdot \mathbf{x}_{v^2} \end{vmatrix}.$$

(b) In the former determinants we note the following four elements:

$$\mathbf{x}_{uv}^2, \quad \mathbf{x}_u \cdot \mathbf{x}_{v^2}, \quad \mathbf{x}v \cdot \mathbf{x}_{u^2}, \quad \mathbf{x}_{u^2} \cdot \mathbf{x}_{v^2}$$

and we want to prove that they also are functions of E, F, G and their derivatives with respect to u and v.

(i) Show that the expression $LN - M^2$ (numerator of $K(\mathbf{x})$) is given by the second and the third of the elements above (question (b)) and that the first and the last one form the following combination

$$\mathbf{x}_{uv}^2 - \mathbf{x}_{u^2} \cdot \mathbf{x}_{v^2} = \tau(u, v).$$

(ii) Show that

$$\mathbf{x}_u \cdot \mathbf{x}_{v^2} = \frac{\partial F}{\partial v} - \frac{1}{2}\frac{\partial G}{\partial u}$$
$$\mathbf{x}_v \cdot \mathbf{x}_{u^2} = \frac{\partial F}{\partial u} - \frac{1}{2}\frac{\partial E}{\partial v}$$

and finally,

$$\tau(u,v) = \frac{1}{2}\frac{\partial^2 G}{\partial u^2} + \frac{1}{2}\frac{\partial^2 E}{\partial v^2} - \frac{\partial^2 F}{\partial u \partial v}.$$

It is thus proved, following Gauss' own method, that the second form of Gauss and also the curvature K depend only on the local behaviour of ds^2. We noted in the text (Sec. 2, Level 0) that some predictions about global behaviour can also be made using the function $K(u,v)$, and finally it was proved in Ex. 0.6(e) that the total curvature $R(M)$ is twice the Gaussian curvature $K(M)$.

Hints: (b)(i) Develop each of the determinants along their last line.
(ii) Calculate

$$\frac{\partial F}{\partial v} = \mathbf{x}_u \cdot \mathbf{x}_{v^2} + \mathbf{x}_v \cdot \mathbf{x}_{uv}\,,$$
$$\frac{\partial G}{\partial u} = 2\mathbf{x}_v \cdot \mathbf{x}_{uv}\,,$$

to establish the first two relations of question (b)(ii).
Calculate also:

$$\frac{1}{2}\frac{\partial^2 G}{\partial u^2} = \mathbf{x}_{uv}^2 + \mathbf{x}_v \cdot \mathbf{x}_{u^2 v}\,,$$
$$\frac{1}{2}\frac{\partial^2 E}{\partial v^2} = \mathbf{x}_{uv}^2 + \mathbf{x}_u \cdot \mathbf{x}_{uv^2}\,,$$
$$\frac{\partial^2 F}{\partial u \partial v} = \mathbf{x}_{u^2} \cdot \mathbf{x}_{v^2} + \mathbf{x}_u \cdot \mathbf{x}_{uv^2} + \mathbf{x}_v \cdot \mathbf{x}_{u^2 v} + \mathbf{x}_{uv}^2\,,$$

in order to obtain the third relation. There are some lengthy calculations in this problem, but they require only some care and attention.

$\boxed{Problem\ 0.13}$ *Surfaces of constant curvature*

Let S be any surface and choose its coordinate curves as the geodesic lines and their orthogonal trajectories (see Ex. 0.5), then its ds^2 takes the form

$$ds^2 = du^2 + G(u,v)dv^2 .$$

(a) Using the determinant of the preceding exercise, show that

$$M^2 = \mathbf{x}_{uv}^2 - \frac{1}{4G}\left(\frac{\partial G}{\partial u}\right)^2 ,$$
$$LN = \mathbf{x}_{u^2} \cdot \mathbf{x}_{v^2} .$$

(b) Still using the results of the preceding exercise, show that

$$K(u,v) = -\frac{1}{2G}\frac{\partial^2 G}{\partial u^2} + \frac{1}{4G^2}\left(\frac{\partial G}{\partial u}\right)^2 .$$

(c) This last equation can be thought as a differential equation for the unknown function G. Show that by a change $G \to \sqrt{G}$ this equation takes the form

$$\frac{\partial^2}{\partial u^2}\sqrt{G} + K\sqrt{G} = 0$$

and compare it with the Eq. (3.38), Level 0.

(d) Suppose now that K is a constant independent of (u,v), and that one singles out 3 solutions:

$$K = 0 : \quad G = u^2$$
$$K = a^2 : \quad G = \frac{1}{a^2}\sin^2(au)$$
$$K = -a^2 : \quad G = \frac{1}{a^2}\sinh^2(au) .$$

The first two correspond to the plane and the sphere of radius a, the third one to a pseudosphere (see Ex. 0.9). Under strict differentiability conditions, one also shows that a surface of null curvature, positive or negative curvature is isometric to a plane, a sphere or a pseudosphere. Furthermore, a surface of constant curvature is isometric to itself – one therefore can define displacements on such a surface which depend on three parameters as in Euclid's space.

Hints: (a) $H^2 = G$.

(b) Notice that in the expression of $LN - M^2$, one finds the function $\tau(u, v)$: its expression is given in Ex. 0.12, question (b)(ii).

Problem 0.14   *Geometric significance of the mean curvature*

Let S be a surface and dS an element of surface

$$dS = \sqrt{EG - F^2}\, dudv = H\, dudv.$$

Under the following transformation depending on a function

$$M(\mathbf{x}) \in S \to M'(\mathbf{x}') \in S' : \mathbf{x}'(u, v) = \mathbf{x}(u, v) + \delta l(u, v)\mathbf{N}(u, v) , \quad \text{(I)}$$

the surface S is transformed into S' and dS into δdS. We shall suppose $\delta l(u, v)$ and its derivatives to be infinitesimal, then denoting by δE the variation of the coefficient E, one has

$$\delta E = E' - E : \mathbf{x}'^2(u) - \mathbf{x}_u^2 = \left(\mathbf{x}_u + \delta l_u \mathbf{N} + \delta l\, \mathbf{N}_u\right)^2 - \mathbf{x}_u^2 .$$

(a) Remarking that $\mathbf{N}.\mathbf{x}_u\, du = 0$, show that

$$\delta E = 2\delta l \mathbf{x}_u \cdot \mathbf{N}_u$$

and finally

$$\delta E = 2L\delta l , \quad \delta F = -2M\delta l , \quad \delta G = -2N\delta l .$$

(b) Consider the variation δdS under the preceding transformation I

$$\delta dS = \delta\{\sqrt{EG - F^2}\}\, dudv = \delta H\, dudv$$
$$= \frac{1}{2H}\left(\delta E \frac{\partial H^2}{\partial E} + \partial F \frac{\partial H^2}{\partial F} + \delta G \frac{\partial H^2}{\partial G}\right) dudv$$

and show that

$$\delta dS = -\frac{EN - 2FM + GL}{H^2} H\, dudv \delta l$$
$$= -2k(u, v)dS\delta l ,$$

where k is the mean curvature (see end of Ex. 0.3). This last formula contains all the geometric significance of k.

Hints: (a) Since $\mathbf{x}_u$ belongs to the tangent plane, then $\mathbf{x}_u \cdot \mathbf{N} = 0$. Develop to the first order to show that

$$\delta E = 2\delta l \; \mathbf{x}_u \cdot \mathbf{N}_u \; .$$

On the other hand

$$\frac{\partial}{\partial u}\{\mathbf{x}_u \cdot \mathbf{N}\} = 0 \; ,$$

develop it to obtain the final expression of δE. The infinitesimals $\delta F, \delta G$ are obtained by similar calculations.

$\boxed{\text{Problem 0.15.}}$ *Complements on surfaces with diagonal metric (see Sec. 3)*

Consider a surface with diagonal metric

$$ds^2 = g_{11}\left(du^1\right)^2 + g_{22}\left(du^2\right)^2 \; .$$

(a) Show that

$$g^{11} = \frac{1}{g_{11}} \; , \quad g^{22} = \frac{1}{g_{22}} \; .$$

(b) Let S be a surface given in parametric form

$$\mathbf{x} = \mathbf{x}\left(u^1, u^2\right)$$

and define

$$\mathbf{e}_1(M) = \frac{\partial \mathbf{x}}{\partial u^1} \; , \quad \mathbf{e}_2(M) = \frac{\partial \mathbf{x}}{\partial u^2} \; .$$

Starting from the definition 3.16, Level 0, of the Christoffel symbols

$$\partial_j \mathbf{e}_i = \Gamma^k{}_{ij}(\mathbf{x})\mathbf{e}_k \; , \quad i, j, k = 1, 2,$$

calculate for the special cases of the plane in $\mathcal{E}_3$ ((3.1) and (3.2), Level 0) and the sphere S^2 ((3.4) and (3.5), Level 0) the Christoffel symbols.

(c) Proceed as in formulas (3.32), (3.36), (3.37), (3.39) of Level 2 to calculate the scalar curvature of the plane and the sphere.

Variational Problems, Geodesic Lines

Problem 0.16   *Integral of energy*

When the Lagrangian L does not contain the variable t, there is a simple constant of motion, which for a mechanical system is the energy of the system. Show that

$$L(q, \dot{q}) - \sum_{1}^{N} \dot{q}^k \frac{\partial L}{\partial \dot{q}^k} = C$$

is a constant of motion of the system. Show that if some of the q^k variables — say q^a — are missing from the expression of L, one can define **constants of motion**

$$\frac{\partial L}{\partial \dot{q}^a} = C \ ,$$

or **first integrals** of the system of differential equations. One calls q^a a **cyclic variable**.

Hint: For a Lagrangian which does not depend on t explicitly

$$\frac{dL}{dt} = \frac{\partial L}{\partial q^k} \dot{q}^k + \frac{\partial L}{\partial \dot{q}^k} \ddot{q}^k = \frac{d}{dt} \{ \frac{\partial L}{\partial \dot{q}^k} \} \dot{q}^k + \frac{\partial L}{\partial \dot{q}^k} \ddot{q}^k$$

or,

$$\frac{d}{dt} \{ L(q, \dot{q}) - \frac{\partial L}{\partial \dot{q}^k} \dot{q}^k \} = 0 \, .$$

One denotes

$$p_k = \partial L / \partial \dot{q}^k \ ;$$

then the conservation law reads

$$\frac{d}{dt} \{ L(q, \dot{q}) - p_k \dot{q}^k \} = \frac{d}{dt} H(p, q) = 0 \, ,$$

the Hamiltonian H is a first integral when the time t does not enter explicitly in the expression of L. (See Problem 2.5.)

Problem 0.17　*Minimal surface of revolution*

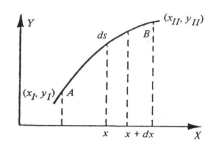

Let $\overset{\frown}{AB}$ be a certain arc of curve: when it rotates around OX, it generates a surface of revolution. Show that if $\overset{\frown}{AB}$ is an arc of catenary, then the generated surface is minimal.

Hint: An element of surface is $dS = 2\pi y\,ds$ and the total surface

$$S = \int_{x_I}^{x_{II}} 2\pi y\sqrt{1 + y'^2}\,dx\,,$$

then

$$L(y, y') = 2\pi y\sqrt{1 + y'^2}$$

and the corresponding constant of motion

$$y\sqrt{1 + y'^2} - \frac{yy'^2}{\sqrt{1 + y'^2}} = C\,,$$

or

$$\frac{y}{\sqrt{1 + y'^2}} = C$$

can be brought into the form

$$dx = \pm C\frac{dy}{\sqrt{y^2 - C}}\,.$$

A simple integration gives

$$x - x_0 = C\cosh^{-1}\frac{y}{C}$$

and finally

$$y = C \cosh \frac{x - x_0}{C},$$

the constants x_0 and C are fixed by taking into account the conditions at A and B.

Problem 0.18   *The brachistochrone*

Let a massive point M fall along a prescribed trajectory C going from A to B. Our aim is to determine C such that the time required by M to travel from A to B is minimum.

(a) Show that the energy integral of the corresponding Euler-Lagrange equation takes the form

$$\sqrt{\frac{1 + y'^2}{y - y_0}} - \frac{y'^2}{\sqrt{(y - y_0)(1 + y'^2)}} = C.$$

(b) Solve the differential equation representing the energy integral to show that

$$x - x_0 = \int \sqrt{\frac{y - y_0}{2a - (y - y_0)}} \, dy.$$

Express the parameter 2a as a function of C and of the coordinates (x_0, y_0) of the point A.

(c) Perform the integration to show that the brachistochronous curve is a cycloid.

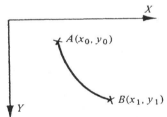

Hints: (a) Let v be the velocity of M, the time needed by M to describe an element of curve ds is ds/v and the total time from A to B

$$\tau = \int_A^B \frac{ds}{v} = \int_{x_0}^{x_1} \frac{\sqrt{1 + y'^2}}{v} \, dx.$$

Since M is submitted to gravity only, then

$$v = \sqrt{2g(y - y_0)} \,,$$

thus one has

$$L = \sqrt{2g(y - y_0)(1 + y'^2)}$$

and the energy integral takes indeed the form given in (a).
(b) One has

$$\sqrt{\frac{1 + y'^2}{y - y_0}} = C \,,$$

one gets the formula in (b) with

$$2a = \frac{1 - C^2 + C^2 y_0}{C^2} \,.$$

(c) One introduces in the integral in (b) the new variable

$$y - y_0 = 2a \, \sin^2 \frac{\theta}{2} \,,$$

then

$$x - x_0 = a(\theta - \sin \theta)$$

and one obtains the equation of the cycloid

$$x = x_0 + a(\theta - \sin \theta)$$
$$y = y_0 + a(1 - \cos \theta) \,.$$

$\boxed{\text{Problem 0.19}}$ *Extremal value of a generalized integral*

Several problems like the minimal revolution surface, the brachis-tochronous curve and some others lead to the study of the extremal value of the following integral,

$$\int_A^B \frac{ds}{y^\alpha} \,, \quad \alpha \in R \,,$$

to which corresponds the Lagrangian

$$L(y, y') = \frac{\sqrt{1 + y'^2}}{y^\alpha} \ .$$

(a) Show that the energy integral leads to the differential equation

$$y^\alpha \sqrt{1 + y'^2} = C$$

and that the value $\alpha = 0$ corresponds to the geodesics of Euclid's plane.

(b) Suppose $\alpha \neq 0$ and set

$$y' = \cotg t \ ,$$

then show that

$$x - x_0 = C\beta \int \sin \beta t dt \ , \quad y = C \sin \beta t \ ,$$

where $\beta = 1/\alpha$ and C is a constant.

(c) Study the following cases:

$\alpha = 1, \beta = 1$: the extremal curves are circles of center (x_0, a) and radius $C > 0$;

$\alpha = 1/2, \beta = 2$: one finds a brachistochronous curve (cycloid);

$\alpha = -1, \beta = -1$: this case corresponds to the equation of a minimal surface of revolution (catenary);

$\alpha = -1/2, \beta = -2$ the extremal curves are parabolas, possible trajectories of a massive point submitted to gravity only.

Hints: All calculations are straightforward.

| **Problem 0.20** | *Fermat's principle* |

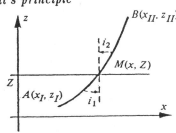

Following the figure, let I and II be two homogeneous and isotropic media separated by a parallel to $Ox : z = Z, Z$ constant. Fermat's principle implies that:

1. the light path (photon trajectory) connecting two points belonging to the same medium is a straight line and the velocity of the photon along this line is constant.

2. the time required by a photon travelling from (x_I, z_I) to (x, Z) or from (x, Z) to (x_{II}, z_{II}) is minimum.

(a) Apply Fermat's principle to the special case given in the figure and derive the Snell-Descartes law.

(b) Consider now an inhomogeneous but isotropic medium where the velocity of the photon is $v = v(z)$, a function of z only. State Fermat's principle for the optical path and derive the expression of Snell-Descartes law. Special case of

$$v(z) = \sqrt{2g(z - z_0)} \, ,$$

physical interpretation.

Hints: (a) If v_I and v_{II} are the photon velocities respectively in medium I and II, the time required to describe AMB is

$$T(x) = \frac{\sqrt{(x - x_I)^2 + (Z - z_I)^2}}{v_I} + \frac{\sqrt{(x - x_{II})^2 + (Z - z_{II})^2}}{v_{II}}$$

and this time should be minimum

$$\frac{dT}{dx} = \frac{x - x_I}{v_I \sqrt{(x - x_I)^2 + (z_I - Z)^2}} + \frac{x - x_{II}}{v_{II} \sqrt{(x - x_{II})^2 + (Z - z_{II})^2}}$$

$$= \frac{x - x_I}{v_I \|\mathbf{AM}\|} + \frac{x - x_{II}}{v_{II} \|\mathbf{BM}\|} = 0 \, .$$

i.e.

$$\frac{\sin i_I}{v_I} = \frac{\sin i_{II}}{v_{II}} \, ,$$

we may then introduce the index of refraction $n = c/v$ and obtain the Snell-Descartes law

$$n_I \sin i_I = n_{II} \sin i_{II} \, .$$

(b) For a medium similar to the one described in (b), Fermat's principles reads "the light path from A to B is the extremum of the integral".

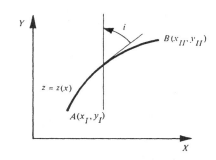

$$\mathcal{L} = \int_A^B \frac{ds}{v(M)} = \int_{x_I}^{x_{II}} \frac{\sqrt{1 + z'^2}}{v(z)} dx$$

with

$$L = \frac{\sqrt{1 + z'^2}}{v(z)}$$

the integral of energy is

$$z' \frac{\partial L}{\partial z'} - L(z, z') = \frac{z'^2}{v(z)\sqrt{1 + z'^2}} - \frac{\sqrt{1 + z'^2}}{v(z)} = C \,,$$

i.e.

$$\frac{1}{v(z)\sqrt{1 + z'^2}} = C \,.$$

But also $Z' = \cot i$, then the preceding formula reads

$$\frac{\sin i}{v(z)} = n(z)\sin i = C$$

and represents the expression of Snell-Descartes law. When

$$v(z) = \sqrt{2g(z - z_0)}$$

where g and z_0 are constants, one obtains the brachistochrone.

Problem 0.21. Geodesic lines and geodesic coordinates

Using once more Gauss notations, let $v = v(u)$ be the equation of a geodesic, extremal curve of

$$\int_A^B \sqrt{E + 2Fv' + Gv'^2} du \,.$$

(a) Show that the Euler-Lagrange equations are

$$\frac{\partial}{\partial v}\sqrt{E + 2Fv' + Gv'^2} = \frac{d}{du}\frac{F + Gv'}{\sqrt{E + 2Fv' + Gv'^2}} \; ; \qquad (I)$$

does the energy integral make any sense ?

(b) Consider a family of curves $\mathcal{C}$ such that through any point $M \in S$ goes one and only one of them. We now look for a geodesic going through M and orthogonal to the corresponding curve $C \subset \mathcal{C}$: these two conditions determine a unique solution of the second order differential Eq. I. We finally choose as coordinate curves these geodesic curves and their orthogonal trajectories, $v = v_0$ is one of the coordinate curves, thus Eq. I has the particular solution $v' = 0$. Since we defined an orthogonal mesh, then $F = 0$ and I is reduced to the condition $\partial\sqrt{E}/\partial v = 0$. E becomes a function of $u : E = E(u)$ and the ds^2 takes the form

$$ds^2 = E(u)du^2 + G(u, v)dv^2 .$$

Changing the variable $u \to \overline{u} = \int E(u)du$, we reduce the former ds^2 to the geodesic form

$$ds^2 = d\overline{u}^2 + G(\overline{u}, v)dv^2$$

which has been used in Ex. 0.13.

(c) As an application of the Eq. I, we look for surfaces of revolution. Using the notations of Ex. 0.7, we adopt the representation S'

$$x = \varphi(u)\cos v , \quad y = \varphi(u)\sin v , \quad z = u ,$$

a surface which is generated by the curve: $x = \varphi(z), y = 0$ around the $z'z$ axis. Show that

$$ds^2 = (1 + \varphi'^2)du^2 + \varphi^2 dv^2 ,$$

where the index prime represents a derivation with respect to u and the corresponding Lagrangian

$$L(v, v') = (1 + \varphi'^2) + \varphi^2 v'^2 .$$

Since $\partial L/\partial v = 0$, then $v(u)$ is a solution of

$$\frac{dv}{du}\varphi^2(u) = C \quad C : \text{constant} ,$$

and

$$v - v_0 = C \int \frac{du}{\varphi^2(u)} \ .$$

The geometric meaning of v is simple: it represents the azimuthal angle φ (not to be confused with the function $\varphi(u)$!).

As special cases, we consider:

1. the cone with the origin as a vertex

$$x = au \cos v \ , y = au \sin v \ , z = u \ ,$$

where a is the constant angle θ built up by the generatrix of the cone with Oz. The geodesics are

$$v - v_0 = -\frac{C}{a^2} \frac{1}{u} \ ;$$

2. the cylinder with generatrices parallel to Oz and a circular base of radius a in the plane xOy

$$x = a \cos v \ , y = a \sin v \ , z = u \ ,$$

whose geodesics are helices.

Hints: (a) Since E, F, G depend explicitly on (u, v), the energy integral does not make sense.

The other questions are simple enough to answer.

| Problem 0.22. | Variation of a multiple integral

This is a fundamental problem for all field theories and is consequently treated in all concerned textbooks. It can be stated in the following terms: let there be given several functions $f, g, \ldots$ of several variables $\{x^1 \ldots x^l\}$ symbolized by x, their derivatives $\partial f / \partial x^i, \partial g / \partial x^i$ symbolized by $\partial f, \partial g$ and an integral — called **action** — extended to a finite domain D of the variable x:

$$\mathcal{A} = \int dx^1 \cdots \int dx^l L(f(x), g(x), \partial f, \partial g)$$

where $\mathcal{L}$ — the **Lagrangian density** — is a known derivable function of $f, g, \ldots \partial f, \partial g, \ldots$. We want to find out which functions $f, g, \ldots$, make $\mathcal{A}$ stationary. Let us first be systematic with the notation: the functions $f, g, \ldots$ are denoted by the set of functions $\{Q_A(x)\}$ which in their turn are symbolized by $Q(x)$.

The index A refers to a single index as in the electromagnetic potential $\Phi^0(x), \Phi^1(x), \Phi^2(x), \Phi^3(x)$ or to two indices as in the 10 components of the gravitational potential $g_{\mu\nu}(x)$. Suppose the dimension of space given by (1) to be 4 (there is no difficulty in defining l equal to 2,3 or more than 4). Then the action is defined by the integral

$$A = \int dx^0 \ldots \int dx^3 L(Q, \partial Q) = \int_D L(Q, \partial Q) \mathrm{d}_4 x \,.$$

Submit each of the fields $Q(x)$ to a form variation* $\delta Q(x)$ such that

$$Q(x) \to Q'(x) = Q(x) + \delta Q(x)$$

and use the common notation

$$\partial_\mu Q_A(x) = Q_{A,\mu}(x) \,,$$

the resulting variation of $\mathcal{A}$ is then

$$\delta A = \int_D \left(\frac{\partial L}{\partial Q_A} \delta Q_A + \frac{\partial L}{\partial Q_{A,\mu}} \delta Q_{A,\mu} \right) d_4 x \,,$$

where the involved summations interest both A and μ. It results from the very definition of $\delta Q_{A,\mu}$ that

$$\delta Q_{A,\mu} = \partial_\mu Q'_A(x) - \partial_\mu Q_A(x) = \partial \mu \delta Q_A(x)$$

and the second term of the integral representing δA takes the form

$$\frac{\partial L}{\partial Q_{A,\mu}} \delta Q_{A,\mu} = \frac{\partial L}{\partial Q_{A,\mu}} \partial_\mu \delta Q_A$$

$$= \partial_\mu \left\{ \frac{\partial L}{\partial Q_{A,\mu}} \delta Q_A \right\} - \left(\partial_\mu \frac{\partial L}{\partial Q_{A,\mu}} \right) \delta Q_A \,.$$

Reverting finally to δA, one has

$$\delta A = \int_D \left(\frac{\partial L}{\partial Q_A} - \partial_\mu \frac{\partial L}{\partial Q_{A,\mu}} \right) \delta Q_A d_\mu x + \int_D \partial_\mu \left\{ \frac{\partial L}{\partial Q_{A,\mu}} \delta Q_A \right\} d_4 x \,,$$

---

*See also Sec. 11, Level 1 on Lie derivative.

with summation on both indices A and μ. At this point we impose specific conditions on $\delta Q_A(x)$: when the point x reaches the boundary ∂D of the domain of integration $D(x \in \partial D)$ then δQ_A vanishes and it is then easy to see that by performing the integrations in the last term of $\delta \mathcal{A}$, this term vanishes also. The final form of this variation is then

$$\delta \mathcal{A} = \int_D \Big(\frac{\partial L}{\partial Q_A} - \partial_\mu \frac{\partial L}{\partial Q_{A,\mu}} \Big) \delta Q_A \, d_4 x = 0$$

and since the δQ_A remain arbitrary, one gets the Euler-Lagrange equations;

$$\frac{\partial L}{\partial Q_A} - \partial_\mu \frac{\partial L}{\partial Q_{A,\mu}} = 0 \ ,$$

only one summation on μ is now involved in the second term and we have as many equations as $A = 1 \ldots N$.

(a) As an application, consider the electromagnetic field $A_\alpha(x), \alpha = 0, 1, 2, 3$ (denoted Φ^a above); $x = \{x^0, x^1, x^2, x^3\}$ with the Lorentz metric $\eta_{\mu\nu} : \{+1, -1, -1, -1\}$. One defines the antisymmetric tensor field $F_{\mu\nu}$ **(the field strength)**

$$F_{\mu\nu} = \partial_\mu A_\nu - \partial_\nu A_\mu \ .$$

Show that starting from the Lagrangian density

$$L(A, \partial A) = \frac{\varepsilon_0}{2} F_{\mu\nu} F^{\mu\nu} + j_\mu A^\mu \ ,$$

where j_μ are four given functions of x representing the **current**, one deduces the Euler-Lagrange equations

$$\varepsilon_0 \partial_\mu F^{\mu\lambda} = j^\lambda$$

(the system of units is the IS system).

| Problem 0.23. | Plateau's problem: surface of vanishing mean curvature

Consider a surface

$$z = f(x, y)$$

limited by a closed contour C whose projection in the xOy plane is γ. Let

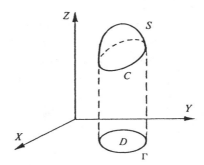

D be a domain in xOy delimited by Γ, the area of S is given by

$$S = \int_D \sqrt{EG - F^2}\ dxdy = \int_D \sqrt{1 + p^2 + q^2}dxdy\ .$$

Show that the integral is stationary for an extremal function f and that it defines a surface with zero mean curvature.

Hint: The density of Lagrangian is

$$L(f, \partial f) = \sqrt{1 + f_x^2 + f_y^2}\ .$$

Written with the notations p, q, r, s, t, the Euler-Lagrange equation is

$$0 = r(1 + q^2) - 2pqs + t(1 + p^2)\ ;$$

its right-hand side represents the numerator of the mean curvature $k(x, y)$ (see Exs. 0.3 and 0.14).

CHAPTER II — LEVEL 1
The Taxonomic Approach

The majority of works on general relativity written by physicists start with a rapid description of tensor analysis. The physicist trying to get rapidly to the heart of this technical apparatus uses the taxonomic method, i.e. the definition of scalar and other fields through their transformation laws, their variance under a change of reference frame. This method is not elegant, because of the avalanche of indices that it produces; it is also, in some sense, dangerous because it hides the intrinsic character of the mathematical objects which are defined; however, it has the advantage of being an easy access to the techniques of tensor analysis. Furthermore, this method is quite widespread; it has been used by physicists for a long time and is still used in very recent work (see for instance [Bibl. 8]). In order not to make the physicist reader feel that he is away from home, we have to dedicate a chapter to this method. However, we shall start by modifying this program a little bit by introducing an intrinsic approach at the beginning Sec. 2. The reader in a hurry may omit Part A of that paragraph and go directly to Part B. The intrinsic presentation of differential geometry will be the subject of the following chapter.

A. Manifolds – Tensor Fields – Covariant Differentiation

1. *Some Reminders*

The notions that interest us here are common knowledge, and we shall restrict ourselves to a very brief summary.

Let us consider sets V and V', the elements of which are called points $M, P, \ldots$ and $M', P' \ldots$ respectively. Consider a map φ that establishes a correspondence between points $M \in V$ and $M' \in V'$: we say that φ is an **injection** if two different points $M \neq P$ of V are mapped into two points $\varphi(M) \neq \varphi(P)$ of V' which are also distinct. One sees that if v' is the image of V, one has

$$\varphi(V) = v' \subset V',$$

consequently φ^{-1} exists if it is restricted to the set v':

$$\varphi^{-1}(v') = V.$$

One says then that one has a **bijection** between V and v'. The map φ is a **surjection** from V to V' if

$$\varphi(V) = V'.$$

This does not exclude the possibility that a single point of V' can be the image of several points of V. In this context a map $V \to V'$ which is both injective and surjective is a **bijection**. As an illustration, let us remind the reader that in Euclidean geometry the orthogonal projection of a plane onto a straight line is surjective while the projection of a straight line to another straight line is bijective. We shall also suppose that the reader is familiar with the concept of **open subset** of a topological space.

We can then define a **separated topological space V** as a set of points which satisfies the two following properties:

— every point $M \in V$ admits a neighbourhood $U_V(M)$ which is defined as an open set of V containing M^*;

— two distinct points M and P of $V (M \neq P)$ have separated neighbourhoods $U_V(M) \neq U_V(P)$ that do not intersect each other.

Coming back to V' let us consider the map φ, defined above, $V \xrightarrow{\varphi} V'$; one says that φ is continuous in M if and only if to a given neighbourhood $U_{V'}(M) \subset V$ we may associate another one such that $U_V(M) \subset U_{V'}(M')$.

---

*It may be an arbitrary set containing an open set containing M.

If in particular φ^{-1} exists, the map φ is continuous in M, if and only if $\varphi^{-1}(U_{V'}(M'))$ is a neighbourhood of $M \in V$.

The notion of **homeomorphism** plays a capital role in Differential Geometry: it can be defined as a bicontinuous bijection (bicontinuous means continuous in both directions). Two topological spaces are said to be **equivalent** if they are homeomorphic, and a property of V is said to be **topological** if it is preserved by a homeomorphism $V \to V'$. The continuous deformations which we introduced at Level 0 (Sec. 2) are homeomorphisms. All these reminders will be used in Level 2, but it is useful to keep them in mind even now.

2. *Tangent Vector Spaces and Contravariant Vector Fields*

This section is divided into two parts: in Part A we sketch an intrinsic definition of a contravariant vector field while in Part B we give a taxonomic definition of the same vector field which is familiar to physicists. The reader who is in a hurry can skip Part A and start his study in B which is self-sufficient.

Part A: A manifold $\mathcal{M}$ is defined as a separated topological vector space (a notion defined in the preceding paragraph) which is supposed to be topologically locally equivalent to an affine space $\mathcal{E}_N{}^*$. This means that there exists a homeomorphism ψ mapping neighbourhoods of $\mathcal{M}$ into $\mathcal{E}_N$; the dimension of $\mathcal{M}$ is defined as the dimension N of $\mathcal{E}_N$ and the space will be denoted by $\mathcal{M}_N$.

1. We shall first be content with generalizing the model of Level 0: the manifold $\mathcal{M}_2$ (i.e. the surface S) is immersed in the affine space $\mathcal{E}_3$. We shall therefore consider a manifold $\mathcal{M}_N$ such that its points $M \subset \mathcal{M}_N$ are parameterized in the basis $\{0, i_1 \ldots i_{N+1}\}$ of an affine space $\mathcal{E}_{N+1}$: the point M is thus parametrized by its coordinates X^k,

$$\mathbf{OM} = X^j \mathbf{i}_j \,, \tag{2.1}$$

as shown in the figure 1.1.

In order to express the fact that M belongs to $\mathcal{M}_N$, we shall suppose that every single coordinate X^k is a function of N real parameters $\{x^k, k = 1 \ldots N\}$. These parameters are the analogues of the parameters

---

*One should point out that only a local homeomorphism is assumed (not a global one), otherwise the definition will be too restrictive.

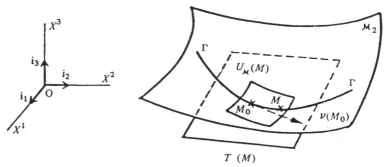

Fig. 1.1a. $\mathcal{M}_2$ manifold in $\mathcal{E}_3$ parameterized by the frame $\{0, i_1, i_2, i_3\}$.

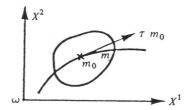

Fig. 1.1b. Corresponding configuration space $\omega = \Psi(0)$; $m_0 = \Psi(M_0)$; $m = \Psi(m)$.

$\{u, v\}$ which we met at Level 0: they are called **local or curvilinear coordinates**. They correspond to the **generalized coordinates** of a mechanical system with N degrees of freedom, and the set of all such coordinates is the **configuration space**. More precisely, we shall say at Level 2 that an open set of this space is a **chart** of $\mathcal{M}_N$. We shall make out of this space an affine space $\mathcal{E}_N$ (as a matter of fact, at Level 2, we shall be satisfied with a vector space $\mathbb{R}^N$) with ω as an origin. We see, consequently, that three spaces are involved: first an affine space $\mathcal{E}_{N+1}$ into which $\mathcal{M}_N$, a topologically separated space is immersed (all points of $\mathcal{M}_N$ are points of $\mathcal{E}_{N+1}$) and finally an affine space $\mathcal{E}_N$ which is the image of $\mathcal{M}_N$ under the homeomorphism ψ.

Let us now be more specific and stress that all our considerations are purely local and refer to a neighbourhood $U_\mathcal{M}(M_0) \subset \mathcal{M}_N$ and $M \in U_\mathcal{M}(M_0)$. The homeomorphism ψ gives an image $U_\mathcal{E}(m_0)$ such that

$$U_\mathcal{E}(m_0) = \Psi(U_\mathcal{M}(M_0)) , \qquad m_0 = \Psi(M_0) ,$$
$$M \in U_\mathcal{M}(M_0) , \qquad (m = \Psi(M)) \in U_\mathcal{E}(m_0) .$$

Any point $M \in \mathcal{M}_N$ is consequently considered either as a point of $\mathcal{E}_{N+1}$, and is parametrized by its coordinates $(X^1, \ldots X^{N+1})$ or as a point of $\mathcal{M}_N$

and its image m (through the diffeomorphism ψ) is parameterized by its coordinates $\{x^1 \ldots x^N\}$. We suppose then that each X^k is a function of $(x^1 \ldots x^N)$ and the vector **OM** can be written as

$$\mathbf{OM} = X^k(x^1 \ldots , x^N)i_k \quad k = 1, \ldots N+1 . \tag{2.2}$$

Having clarified these points, we consider now a curve $\Gamma \subset \mathcal{M}_N$ going through M_0 and M; to obtain Γ we suppose that the parameters x^k are C^∞ functions of the real parameter λ and are defined in the neighbourhood $U_{\mathcal{M}}(M_0)$. In the parametric Eq.* (2.2) of $\mathcal{M}_N$ we introduced the functions $x^k(\lambda)$ to get the relation

$$\mathbf{OM}(\lambda) = X^k(x^1(\lambda) \ldots x^N(\lambda))i_k , \tag{2.3a}$$

which is equivalent to the system

$$X^1 = X^1(x^1(\lambda) \ldots x^N(\lambda)) \ldots X^{N+1} = X^{N+1}(x^1(\lambda) \ldots x^N(\lambda)) . \tag{2.3b}$$

We now have at our disposal all the tools necessary to define the vectors of $\mathcal{M}_N$: we therefore assert the following: "the specification of a contravariant vector at the point $M_0 \in \mathcal{M}_N$ is equivalent to the definition of a tangent at M_0 to a certain curve $\Gamma \subset \mathcal{M}_N$".* In $\mathcal{E}_{N+1}$ the tangent to Γ at $M_0(M = M(\lambda), M_0 = M(\lambda_0))$ is

$$\begin{aligned} v_{M_0} = \frac{d\mathbf{OM}}{d\lambda}\bigg|_{\lambda=\lambda_0} &= \frac{dX^k(\lambda)}{d\lambda}\bigg|_{\lambda=\lambda_0} i_k \\ &= \frac{\partial X^k(x)}{\partial x^j}\frac{dx^j}{d\lambda}\bigg|_{\lambda=\lambda_0} i_k , \end{aligned} \tag{2.4}$$

and consequently v_M is a sum of terms, each one of which is the product of two factors: the first factor $\partial X^k/\partial x^j$ depends only on the definition of $\mathcal{M}_N$ and is independent of the choice of the curve Γ, while the second factor $dx^k(\lambda)/d\lambda$ characterizes Γ.

Simultaneously with Γ we may consider the curve $C \subset \mathcal{E}_N$:

$$C : x = x(\lambda) \Rightarrow x^k = x^k(\lambda) , \tag{2.5}$$

---

*The elimination of the N parameters $x^1 \ldots x^N$ between the coordinates $x^1 \ldots x^{N+1}$ gives (when possible) an equation $F(X^1 \ldots X^{N+1}) = 0$ of the hypersurface $\mathcal{M}_N$ immersed in $\mathcal{E}_{N+1}$.

the curve C determines Γ and vice-versa (bijective map). C is thus the image of Γ with respect to the homeomorphism ψ and one can postulate that the vector $\tau(\lambda)$ defined in $\mathcal{E}_N$ and having

$$\tau^k(x(\lambda)) = \frac{dx^k}{d\lambda} = \tau^k(m) : m = \psi(M) \tag{2.6}$$

as components, is the image of v_M.

Since the knowledge of the manifold $\mathcal{M}_N$ given by its parametric Eqs. (2.2) and of the vector $\tau(\lambda)$ given by (2.6) determines univocally v_M through (2.4), we may proceed as one does in analytical dynamics and neglect the physical space $\mathcal{E}_{N+1}$, working only in the configuration space built up by the multiplets $\tau(\lambda)$.

By extending this technique we can define **the τ^k as components of V_M**. This shift in notation is essential, and to neglect it is to court lack of understanding. Let us now consider, still in the framework of the generalization of $\mathcal{M}_2$ immersed in $\mathcal{E}_3$, the infinitesimal vector

$$d\mathbf{OM} = \frac{\partial \mathbf{OM}}{\partial x^i} dx^i \ . \tag{2.7}$$

This is an element of the affine N-dimensional space tangent to $\mathcal{M}_N$ at M, and corresponding to an arbitrary choice $\{dx^1 \ldots dx^N\}$ of the N differentials dx^j. The associated vector space $T_\mathcal{M}(M)$ is the **tangent vector space** to $\mathcal{M}_N$ at the point M and has a basis consisting of the vectors $\partial \mathbf{OM}/\partial x^i, i = 1 \ldots N$ (see (2.7)).

The next problem is now to describe the change of parametrization in $\mathcal{M}_N$: let us introduce another local set of coordinates $x' = \{x'^1 \ldots x'^N\}$ instead of the original coordinates $x = \{x^1, \ldots, x^N\}$ and let us assume that the transformation is given by*

$$x' = f(x) \Rightarrow \begin{pmatrix} x'^1 \\ \vdots \\ x'^N \end{pmatrix} = \begin{pmatrix} f'(x^1 \ldots x^N) \\ \vdots \\ f^N(x^1 \ldots x^N) \end{pmatrix} = \begin{pmatrix} x'^1(x) \\ \vdots \\ x'^N(x) \end{pmatrix} \ , \tag{2.8}$$

---

*The meaning of a point transformation is double. The first one considers such a transformation as a change of coordinates: the points of $\mathcal{M}_N$ are defined in the neighbourhood of a point M with respect to a certain local basis and a point transformation performs a change of basis. This is the point of view fundamental for the intrinsic approach in Level 2 (Sec. 1). But a point transformation may also be considered as follows: the local basis is kept fixed at M, but the manifold itself is deformed in the neighbourhood of M. These two points of view (the passive and the active) will be more extensively commented in Sec. 11 and 12 of this level, and in Secs. 1, 6 and 7 of Level 2.

the local equality being a rewriting of the preceding one. We shall assume that this transformation is inversible in $U_{\mathcal{E}}(m_0)$, i.e. such that the Jacobian $\partial(x')/\partial x \neq 0$.

In the new system of curvilinear coordinates, the curve Γ has C' (described by the equation $x' = x'(\lambda)$) as its image. The vector v_M tangent to Γ has components $\tau'^k = dx'^k/d\lambda$. We further have

$$
\begin{aligned}
\tau'^k(x(\lambda)) &= \frac{dx'^k(\lambda)}{d\lambda} = \frac{d}{d\lambda} f^k(x^1(\lambda) \ldots x^N(\lambda)) \\
&= \frac{\partial f^k(x)}{\partial x^j} \frac{dx^j}{d\lambda} = \frac{\partial x'^k(x)}{\partial x^j} \tau^j(\lambda) \; .
\end{aligned}
\tag{2.9a}
$$

In this way we can define the variance of the τ^k, and since by definition we have introduced the τ^k as components of v_M, we define also the variance of the vector v_M. In order to stress this point, one writes, instead of the preceding relations,

$$
v'^k(x') = \frac{\partial x'^k}{\partial x^j} v^j(x) \; ,
\tag{2.9b}
$$

which will be the starting point of our considerations in Part B where $v^k(x)$ will define a **contravariant vector field**.

2. We have not yet answered the question: how to define a contravariant vector field in $\mathcal{M}_N$ when this manifold is considered without an ambient affine space $\mathcal{E}_{N+1}$ into which it is immersed. The satisfactory answer to this question will be given only at Level 2, but we have already some elements of the answer. It is clear that in this new context the formulae (2.1), (2.2), (2.3) and the consequences we drew from them cannot be valid any more. However the configuration space $\mathcal{E}_N$, homeomorphic to $\mathcal{M}_N$, still keeps its meaning. In particular, the concepts of a curve $C \subset \mathcal{E}_N$ and its image $\Gamma \subset \mathcal{M}_N$ are well defined by the map ψ^{-1} (inverse of the map $\mathcal{M}_N \xrightarrow{\psi} \mathcal{E}_N$); the N-vector $\tau(m)$, defined by (2.6), is the tangent to the curve C.

The return of $\mathcal{M}_N$ is then achieved simply by stating that the $\tau^k(m)$ are the components of v_M, and the variance of v_M is defined by the formula (2.9b). With the help of the tools that we have at our disposal now we cannot go any further. The intrinsic discussion of Level 2 will show a different way; certain signposts for this road have already been set up in Sec. 5 of Level 0 and in the formula (4.11) of the present chapter.

Part B: Let us summarize Part A of this section by stating that it makes no sense to speak about bipoint vectors in a topological space which is different from $\mathcal{E}_N$, we came around this difficulty by associating canonically to a manifold an affine space in which the bipoint vector can be defined.

Roughly speaking, we shall follow the same idea in B without trying to be overly precise: we shall say that a contravariant field will be described in $\mathcal{M}_N$ by its set of N components $A^i(x^1 \ldots x^N)$ which are functions of curvilinear (or local) coordinates $\{x^1 \ldots x^N\}$ denoted by x.

We can substitute another system $\{x'^1 \ldots x'^N\}$ denoted by x' such that between x and x' the formulae (2.7) are still valid, i.e.

$$x \to x' = f(x) \Leftrightarrow x'^i = x'^i(x^1 \ldots x^N) \, , \qquad (2.10)$$

in a domain where the Jacobian $\partial(x')/\partial(x) \neq 0$. Let then $A'^i(x'^1 \ldots x'^N)$ be the new components of the controvariant field $A(M)$.

We shall say that $A(M)$ is a contravariant vector field if $A^i(x)$ and $A'^i(x')$ are related by the relationships (2.9), i.e.

$$A'^i(x') = \frac{\partial x'^i}{\partial x^j} A^j(x) \, . \qquad (2.11)$$

Our procedure is that of the "founding fathers" of tensor analysis and it is the favorite way of a large majority of physicists. It is really taxonomic since formula (2.11) allows us to define the class of N-uplets of functions of x, components of a contravariant vector field. An advantage of the taxonomic approach is that it displays a different aspect of the definition of a tensor field: any N-uplet of differentiable functions of the N variables $\{x^1 \ldots x^N\}$ can be considered, in a given reference frame, as the set of components of a contravariant field, as long as one imposes (2.11) as the law of variance under a point transformation. The same remarks hold for any tensor field.

This approach is furthermore in close correspondence with the intellectual habits of a physicist who is accustomed to handling transformation groups. Let us illustrate this point by considering Lorentz transformations which are unavoidable as soon as one considers beams of particles with velocities comparable to c (which happens in high-energy accelerators): a Lorentz transformation connects the coordinates of the same space-time point viewed from two different frames; if we call x the quadruplet $\{x^0 = ct, x^1, x^2, x^3\}$ and x' another quadruplet of the same kind, one has

$$x \to x' = Lx \Leftrightarrow x'^\alpha = L^\alpha_\nu x^\nu : \alpha, \nu = 0, 1, 2, 3 \qquad (2.12)$$

and a further condition on the 4 by 4 matrix L is obtained by requiring that it should not change the quadratic form $(x^0)^2 - (x^1)^2 - (x^2)^2 - (x^3)^2$.

Let us now consider an electromagnetic field described by the four-vector potential $A(x)$ and let $A^\alpha(x)$ and $A'^\alpha(x')$ be its components before and after a Lorentz transformation. Since Maxwell's equations keep their form when they undergo a Lorentz transformation, we see that the four-potential has the transformation law

$$A'^\alpha(x') = \frac{\partial x'^\alpha}{\partial x^\nu} A^\nu(x) \ .$$

(2.13)

This can be written, because of (2.12), in a form identical to (2.11). Consequently the physicist is on familiar ground here.

Remark:

One should insist on a point of vocabulary. We call, in fact, the N-uplet of numbers $\{A'_k(x)\}$ a vector field, in other words we are using the same expression for a function and for its value at a point (the same confusion also occurs between a vector and its components).

This notation where the argument x is displayed corresponds to another need: one has to distinguish the tensor field $A(x)$, the components of which $\{A^1 \dots A^N\}$ are a set of functions, and the tensor $A(x)$ with components $\{A^1(x) \dots A^1(x)\}$ which are a set of real numbers. The first object belongs to tensor analysis while the second one belongs to tensor calculus.

The correct statement would be to describe $A(x)$ as a tensor bound to a point x of coordinates $\{x^k\}$ while the object A is the tensor field; we shall carefully observe this distinction at Level 2. Note also that the word "field" has two different meanings: for a physicist, a vector field is an N-uplet of functions $\{A^k(x)\}$ defined at the point of coordinates $\{x^k\}$ of $M \in \mathcal{M}_N$; for a mathematician a field is just a map, as we shall see at Level 2 (and also in the chapter on fibre bundles). This point may cause misunderstandings.

3. *Jacobian, Jacobian Matrix and their Applications*

We come back to the point transformation defined by formula (2.7)

$$x'^i = x'^i(x^1 \dots x^N) \quad i = 1 \dots N \ .$$

(3.1a)

If one differentiates this relation, one obtains

$$dx'^i = \frac{\partial x'^i}{\partial x^j} dx^j \ ,$$

(3.1b)

which can be written in matrix form as

$$
\begin{pmatrix} dx'^1 \\ \vdots \\ dx'^N \end{pmatrix} = \begin{pmatrix} \dfrac{\partial x'^1}{\partial x^1} & \dfrac{\partial x'^1}{\partial x^2} & \cdots & \dfrac{\partial x'^1}{\partial x^N} \\ \cdots & & & \\ \dfrac{\partial x'^N}{\partial x^1} & \dfrac{\partial x'^N}{\partial x^2} & & \dfrac{\partial x'^N}{\partial x^N} \end{pmatrix} \begin{pmatrix} dx^1 \\ \vdots \\ dx^N \end{pmatrix} \ . \qquad (3.2)
$$

Let us denote the two column matrices by dx' and dx and the N by N matrix which also goes by the name of **Jacobian matrix** by $\mathbf{D}(x')/\mathbf{D}(x)$; we can then write (3.2) in a condensed form

$$
dx' = \frac{\mathbf{D}(x')}{\mathbf{D}(x)} dx \ , \qquad (3.3)
$$

with the following chain rule

$$
\frac{\mathbf{D}(x')}{\mathbf{D}(x)} = \frac{\mathbf{D}(x')}{\mathbf{D}(x'')} \frac{\mathbf{D}(x'')}{\mathbf{D}(x)} \ . \qquad (3.4)
$$

The **Jacobian** can be defined as the determinant of the Jacobian matrix:

$$
\frac{\partial x'}{\partial x} = \frac{\partial(x'^1 \ldots x'^N)}{\partial(x^1 \ldots x^N)} = \det \frac{\mathbf{D}(x')}{\mathbf{D}(x)} \ , \qquad (3.5)
$$

if $\partial(x')/\partial(x) \neq 0$, then the Jacobian matrix is invertible. This conclusion remains valid for (3.1b) and, by a classical theorem of analysis, it is also true for (3.1a).

The Jacobian matrix was introduced here as a square matrix, but we shall see at Level 2 that one also can consider Jacobians that are rectangular matrices. The concept of Jacobians is important for the physicist as a generalization of the gradient concept. Let us remember that this last notion has been introduced by considering a function $\varphi(x)$ (or if one prefers, a scalar field φ), at two infinitesimally close points x and $\xi = x + dx$. The problem to solve is to express $\varphi(\xi)$ in terms of $\varphi(x)$; the solution is immediate:

$$
\varphi(\xi) = \varphi(x) + \partial_k \varphi(x) dx^k \ .
$$

This is generally written for the space $\mathcal{E}_3$ as

$$
\varphi(\boldsymbol{\xi}) = \varphi(\mathbf{x}) + \boldsymbol{\nabla}\varphi \cdot \mathbf{dx} \ . \qquad (3.6)
$$

The following step consists in considering the same problem for a vector field $X(x)$ (for instance contravariant) of components $X^i(\xi)$; the answer is again immediate:

$$X^i(\xi) = X^i(x) + \frac{\partial X^i(x)}{\partial x^j} dx^j \ . \tag{3.7}$$

If we denote by $X(x)$ a column matrix with elements $X^k(x)$ one has

$$X(\xi) = X(\mathbf{x}) + \frac{\mathbf{D}(X)}{\mathbf{D}(x)} dx \ . \tag{3.8}$$

Examples of the use of the Jacobian in physics are numerous: one can quote for instance the concept of **gradient-vector** in the theory of deformable media. We can point out other applications: In order to define the **directional-derivative** along a vector $Y(x)$ in a space $\mathcal{E}_N$, one sets

$$\frac{d}{dY(x)} = Y^k(x) \frac{\partial}{\partial x^k} \ , \tag{3.9}$$

so that the infinitesimal change of a scalar field, when the point x is displaced at $\xi = x + Y(x)$, can be expressed as

$$\xi = x + dx = x + Y(x) : \varphi(\xi) = \varphi(x) + dx^k \frac{\partial \varphi(x)}{\partial x^k} = \varphi(x) + \frac{d\varphi(x)}{dY(x)} \ . \tag{3.10}$$

From this notion comes, in the theory of potential and for a space $\mathcal{E}_3$, the notion of a normal derivative to a surface S^*. In a point $\mathbf{x} \in S$ where the normal is $\mathbf{N}(\mathbf{x})$ we define the normal derivative by

$$\frac{d}{d\mathbf{N}(\mathbf{x})} = \mathbf{N}(\mathbf{x}) \cdot \boldsymbol{\nabla} \ . \tag{3.11}$$

Let us also notice another compact formula that can simplify some calculations: the variance of a vector has been defined by formula (2.11). Denote by $A'(x)$ the elements of the column matrix $A^k(x')$ and by $A(x)$ the one with elements $A^k(x)$; one can then write in compact notation

$$A'(x') = \frac{\mathbf{D}(x')}{\mathbf{D}(x)} A(x) \ . \tag{3.12a}$$

---

* $d/dY(x)$ will be denoted at Level 2 by v_M, a differential operator or vector on S.

A similar formula can also be given for a covariant tensor (see next paragraph). Let $\underline{A}(x)$ be the row matrix

$$\underline{A}(x) = (A_1(x) \ldots A_N(x))$$

then anticipating on formula (4.2), we may write

$$\underline{A}'(x') = \underline{A}(x)\frac{\mathbf{D}(x)}{\mathbf{D}(x')} = \underline{A}(x)\left(\frac{\mathbf{D}(x')}{\mathbf{D}(x)}\right)^{-1} , \qquad (3.12b)$$

or

$$\underline{A}(x) = \underline{A}'(x')\frac{\mathbf{D}(x')}{\mathbf{D}(x)} . \qquad (3.12c)$$

Finally, consider the infinitesimal form of the Jacobian when

$$x'^k = x^k + \delta x^k(x^1 \ldots x^N) \quad k \; 1 \ldots N , \qquad (3.13)$$

δx^k being an infinitely small, differentiable function of the multiplet $x = (x^1 \ldots x^N)$, one obtains the following expression of $\mathbf{D}(x')/D(x)$,

$$
\begin{aligned}
\frac{\mathbf{D}(x')}{\mathbf{D}(x)} &= \begin{pmatrix} 1 + \dfrac{\partial \delta x^1}{\partial x^1} & \dfrac{\partial \delta x^1}{\partial x^2} & \cdots & \dfrac{\partial \delta x^1}{\partial x^N} \\ \dfrac{\partial \delta x^2}{\partial x^1} & 1 + \dfrac{\partial \delta x^2}{\partial x^2} & \cdots & \dfrac{\partial \delta x^2}{\partial x^N} \\ \cdots & & & \end{pmatrix} \\
&= \begin{pmatrix} 1 & 0 & 0 & \cdots \\ 0 & 1 & 0 & \cdots \\ \cdots & & & \end{pmatrix} + \begin{pmatrix} \dfrac{\partial \delta x^1}{\partial x^1} & \cdots & \dfrac{\partial \delta x^1}{\partial x^N} \\ \dfrac{\partial \delta x^2}{\partial x^2} & \cdots & \dfrac{\partial \delta x^2}{\partial x^N} \\ \cdots & & \end{pmatrix} \qquad (3.14) \\
&= I + \frac{\mathbf{D}(\delta x)}{\mathbf{D}(x)} = I + \Delta(\delta x) ,
\end{aligned}
$$

with the obvious definition of $\Delta(\delta x)$. The inverse Jacobian takes the form

$$\left(\frac{\mathbf{D}(x')}{\mathbf{D}(x)}\right)^{-1} = \frac{\mathbf{D}(x)}{\mathbf{D}(x')} = I - \Delta(\delta x) . \qquad (3.15)$$

Finally the calculation of the corresponding Jacobian is also simplified; it is easy to see that up to the first order

$$\frac{\partial(x')}{\partial(x)} = 1 + \frac{\partial \delta x^1}{\partial x^1} + \ldots \frac{\partial \delta x^N}{\partial x^N} = 1 + \partial_j \delta x^j . \qquad (3.16)$$

4. *Tensor Field*

A **scalar field F** (tensor of order **0**) corresponds to a real function of N variables x^k such that if x is changed into x'

$$x \to x' \Rightarrow F'(x') = F(x) , \qquad (4.1)$$

i.e. F is such that its value at a point does not depend on the choice of the coordinate system.

A **covariant vector field** is by definition given by its N components $A_i(x)$, such that for x going to x', one has

$$x \to x' \Rightarrow A'_i(x') = \frac{\partial x^j}{\partial x'^i} A_j(x) . \qquad (4.2)$$

Finally, a **contravariant vector field** is such that its variance is given by (2.11). By taking tensor products of covariant and contravariant vector fields, one can define a p times contravariant and q times covariant field through their variance

$$x \to x' \Rightarrow T'^{i_1 \ldots i_p}_{j_1 \ldots j_q}(x')$$
$$= \frac{\partial x'^{i_1}}{\partial x^{k_1}} \cdots \frac{\partial x'^{i_p}}{\partial x^{k_p}} \frac{\partial x^{l_1}}{\partial x'^{j_1}} \frac{\partial x^{l_q}}{\partial x'^{j_q}} T^{k_1 \ldots k_p}_{l_1 \ldots l_q}(x) . \qquad (4.3)$$

One also considers other operations that can be done on tensors: sums, products, contraction,...; they are a set of techniques that constitute tensor algebra. The reader will find the required information on these techniques in Problem 1.2 at the end of this chapter.

Let us finally note that a **tensor density** p times contravariant and q times covariant will be defined by its variance

$$\mathcal{E}'^{i_1 \ldots i_p}_{j_1 \ldots j_q}(x') = \left(\frac{\partial(x)}{\partial(x')} \right)^W \frac{\partial x'^{i_1}}{\partial x^{r_1}} \cdots \frac{\partial x'^{i_p}}{\partial x^{r_p}} \frac{\partial x^{s_1}}{\partial x'^{j_1}} \cdots \frac{\partial x^{s_q}}{\partial x'^{j_q}} \times \mathcal{E}^{r_1 \ldots r_p}_{s_1 \ldots s_q}(x) , \quad (4.4)$$

where the real number W is the **weight** of the density. The rules governing the algebra of tensor densities are simple indeed: a sum or a difference of tensor densities of the same weight yield a tensor density of the same weight, the direct product of densities of orders n and n' and weights W and W' is a density of order $n + n'$ and weight $W + W'$, and finally the contraction of two indices in a tensor density of order n and weight W yields to a density of the same weight and of order $n - 2$. The proof of these three rules is very simple indeed.

The typical example of a tensor density in $\mathcal{M}_4$ is the the completely antisymmetric tensor $\varepsilon^{\lambda\mu\nu\rho}$ of the fourth order taking the values ± 1 according to the parity of the permutation $\binom{\lambda\mu\nu\rho}{0123}$ and 0 if two of its indices are identical.*

We studied above three types of field variances: the one of the scalar field, of the covariant vector field and finally the one of the contravariant vector field. These three transformation laws are called respectively transformation by invariance, by covariance and by contravariance; one can easily see that each family constitutes a group.

Let us conclude with a few remarks:

1. To a point $x \in \mathcal{E}_N$, affine space associated to $\mathcal{M}_N$ (see Part A, paragraph 2), let us associate a vector tangent in $M \in \mathcal{M}_N$ with components dx^k. We saw in (3.1b) when considering the point transformation $x \to x'$, that there exists the relation

$$dx'^k = \frac{\partial x'^k}{\partial x^j} dx^j \tag{4.5}$$

which shows that the vector field of components $dx^k(x)$ is a contravariant field.

2. Let $F(x)$ be a scalar field. From its definition $F'(x') = F(x)$, one gets

$$\frac{\partial F'(x')}{\partial x'^k} = \frac{\partial F(x)}{\partial x'^k} = \frac{\partial F(x)}{\partial x^j} \frac{\partial x^j}{\partial x'^k} . \tag{4.6}$$

Therefore the vector field of components $\partial_J F(x)$ is a covariant field, in other words the components of $\nabla. F(\mathbf{x})$ are those of a covariant field.

3. The following remark could have been made as well at the end of paragraph 2. Consider a contravariant vector field $A(x)$ and suppose that its components $A^i(x)$ are defined with respect to a certain basis $B = (e_1(x), \ldots e_N(x))$. It is certainly a major defect of the taxonomic approach that it neglects completely the intrinsic definition of a vector field; the introduction of the intrinsic point of view fully justifies the approach to that problem as studied at Level 2. But even within the present taxonomic approach, one can enlighten the reader on that point.

---

*See Exercise 1.7.

Indeed a point transformation $x \to x'$ amounts to defining vectors with respect to a new basis $B' = (e'_1(x'), \ldots e'_N(x'))$ (passive aspect of such a transformation): we mean by that the contravariant vetor field $A(M)$ has two sets of components $A^i(x)$ and $A'^i(x)$ such that

$$A(M) = A^i(x)e_i(x) = A'^i(x)e'_i(x') . \tag{4.7}$$

Considering the variance (2.11) of the A^k, we may write

$$A'^k(x')e'_k(x') = \frac{\partial x'^k}{\partial x^j}A_j(x)e'_k(x') , \tag{4.8}$$

taking (4.7) into account; and since formula (4.8) is valid for any field A, it follows that

$$e_i(x) = \frac{\partial x'^k}{\partial x^i}e'_k(x') , \tag{4.9}$$

or

$$e'_k(x') = \frac{\partial x^k}{\partial x'^j}e_j(x) . \tag{4.10}$$

The transformation $x \to x'$ induces on the vectors of the basis B a contragredient law of transformation (following H. Weyl).

Coming back to (4.6), we can see that the variance of $\partial/\partial x_k$ is the same as that of $e_k(x)$:

$$\frac{\partial}{\partial x'^k} = \frac{\partial x^j}{\partial x'^k}\frac{\partial}{\partial x^j} . \tag{4.11}$$

This remark, which seems fortuitous here, will help, after an appropriate elaboration, to define contravariant tensor fields in an intrinsic formulation (Level 2, Sec. 2).

5. *Covariant Derivatives*

The notion of curvature, introduced in Sec. 3 of Level 0, appeared there as a metric concept. Historically, that was indeed the method adopted by the founders of differential geometry. We will now see that it is convenient to go through an intermediate step and define the **covariant derivative** of any tensor field. But this will be possible only if we introduce a new structure on the manifold, an **affine connection** prior to any metric. In this way the concept of curvature will lose its metric appearance. From these considerations, there will result a new type of derivative: the **covariant derivative** which will retain all the algebraic properties of the common derivative and behave under a point transformation like a covariant vector field. Its introduction starts from a very simple property of the scalar field $F(x)$: consider its derivative $\partial_i F(x)$. A point transformation which transforms $F(x)$ into $F'(x') = F(x)$ transforms $\partial_i F(x)$ as follows (see (4.6)):

$$\frac{\partial F'(x')}{\partial x'^i} = \frac{\partial F(x)}{\partial x'^i} = (\partial_j F(x))\frac{\partial x^j}{\partial x'^i} \; ; \tag{5.1}$$

we may then conclude that $\partial_i F(x)$ is a covariant vector field.

We now ask the simple question: $A^i(x)$ being a contravariant vector field, does $\partial_k A^i(x)$ behave as a mixed tensor field, once covariant and once contravariant? To investigate this point, we start from formula (2.11) which we write as

$$A'^i(x') = \frac{\partial x'^i}{\partial x^j} A^j(x) \tag{5.2}$$

and calculate

$$\begin{aligned}
\frac{\partial}{\partial x'^k} A'^i(x') &= \left(\frac{\partial}{\partial x'^k}\frac{\partial x'^i}{\partial x^j}\right) A^j(x) + \frac{\partial x'^i}{\partial x^j}\frac{\partial}{\partial x'^k} A^j(x) \\
&= \frac{\partial^l x}{\partial x'^k}\frac{\partial^2 x^i}{\partial x^i x^j} A^j(x) + \frac{\partial x'^i}{\partial x^j}\frac{\partial x^l}{\partial x'^k}\partial_l A^j(x) \; .
\end{aligned} \tag{5.3}$$

The second term of the last line has indeed the expected kind of variance, but the first, generally different from 0 (unless $x \to x'$ is a linear transformation), shows that $\partial_k A^i(x)$ is not a tensor field.

The operation ∇_i that we shall now introduce transforms a tensor field into another, has all the algebraic properties of differentiation:

$$\nabla_i\{A^k(x) + B^k(x)\} = \nabla_i A^k(x) + \nabla_i B^k(x) \tag{5.4a}$$

and for any differentiable function $\alpha(x)$ gives

$$\nabla_i\{\alpha(x)A^k(x)\} = (\nabla_i\alpha(x))A^k(x) + \alpha(x)\nabla_i A^k(x) \; . \tag{5.4b}$$

Consider a covariant vector field $A_k(x)$ and define $\nabla_i A_k(x)$ as follows

$$\nabla_i A^k(x) = \partial_i A_k(x) + \Gamma^s{}_{ki}(x)A_s(x) \; , \tag{5.5}$$

where the $\Gamma^s{}_{ki}(x)$ are three-index functions of $(x^1 \ldots x^N)$. They describe an **affine connection field** whose components $\Gamma^\bullet{}_{\bullet\bullet}$ are called **affine connection coefficients, Christoffel symbols** or Γ**-symbols.** We want now to fix the transformation law of the Γ-symbols in such a way that $\nabla_i A_k(x)$ will have the right variance of a double covariant tensor field, with ∇_i satisfying (5.4).

The field $\nabla_i A_n$ will then transform as follows:

$$\nabla_i A_k(x) \rightarrow \nabla'_i A'_k(x') = \frac{\partial x^s}{\partial x'^i} \frac{\partial x^r}{\partial x'^k} \nabla_s A_r(x) \; . \tag{5.6}$$

To achieve this aim, we impose the following law of transformation to the Γ-symbols:

$$\partial'_i A^k(x') - \Gamma'^s_{ki}(x') A'_s(x') = \frac{\partial x^s}{\partial x'^i} \frac{\partial x^r}{\partial x'^k} \nabla_s A_r(x) \; . \tag{5.7}$$

On the other hand, we also have

$$\partial'_i A'_k(x') = \frac{\partial}{\partial x'^i} \left\{ \frac{\partial x^r}{\partial x'^k} A_r(x) \right\} = \frac{\partial^2 x^r}{\partial x'^i \partial x'^k} A_r(x) + \frac{\partial x^r}{\partial x'^k} \frac{\partial A_r(x)}{\partial x'^i} \; , \tag{5.8a}$$

where the last term of the right-hand side can be rewritten as

$$\frac{\partial x^r}{\partial x'^k} \frac{\partial x'^p}{\partial x'^i} \frac{\partial A_r(x)}{\partial x^p} \; , \tag{5.8b}$$

while the left-hand side of (5.7) takes the form

$$\frac{\partial^2 x^r}{\partial x'^i \partial x'^k} A_r(x) + \frac{\partial x^r}{\partial x'^k} \frac{\partial x^p}{\partial x'^i} \frac{\partial A_r(x)}{\partial x^p} - \Gamma'^s_{ki}(x') \frac{\partial x^p}{\partial x'^s} A_p(x) \; . \tag{5.9}$$

Let us consider (5.6) again and bring (5.5) into its right-hand side; after some calculations, one obtains

$$\frac{\partial^2 x^r}{\partial x'^i \partial x'^k} A_r(x) - \Gamma'^s_{ki}(x') \frac{\partial x^p}{\partial x'^s} A_p(x) \\ = -\frac{\partial x^p}{\partial x'^i} \frac{\partial x^q}{\partial x'^k} \Gamma^s_{pq}(x) A_s(x) \; . \tag{5.10}$$

The indices r and p in the two first terms of the left-hand side and the index s of the right-hand side are all dummy indices: let us denote all of them by l. We may also notice that this relation should remain valid independently of the choice of $A_l(x)$, then

$$\Gamma'^s_{ki}(x') \frac{\partial x^l}{\partial x'^s} = \frac{\partial^2 x^l}{\partial x'^i \partial x'^k} + \frac{\partial x^p}{\partial x'^i} \frac{\partial x^q}{\partial x'^k} \Gamma^l_{pq}(x) \; . \tag{5.11}$$

Multiplying both sides by $\partial x'^j/\partial x^l$, the law of transformation for the Γ-symbols becomes

$$\Gamma'^j{}_{ki}(x') = \frac{\partial x'^j}{\partial x^l}\frac{\partial x^p}{\partial x'^i}\frac{\partial x^q}{\partial x'^k}\Gamma^s{}_{pq}(x) + \frac{\partial x'^j}{\partial x^l}\frac{\partial^2 x^l}{\partial x'^i \partial x'^k} \; . \tag{5.12}$$

This shows that the Γ-symbols are not the components of a tensor field (because of the second term) unless the transformation $x \to x'$ is linear.

Finally, the verification of the relations (5.4) is immediate. Coming back to (5.1) and since we proved that $\partial_i F(x)$ is a vector field, we can write

$$\partial_i F(x) = \nabla_i F(x) \; . \tag{5.13}$$

We now consider the covariant derivation of a contravariant field with components $B^k(x)$ and calculate

$$\nabla_i\{A_k(x)B^k(x)\} = (\nabla_i A_k(x))B^k(x) + A_k(x)\nabla_i B^k(x)$$
$$= (\partial i A_k(x) - \Gamma^j{}_{ki}(x)A_j(x))B^k(x) + A_k(x)\nabla_i B^k(x) \; . \tag{5.14}$$

But, as $A_k B^k$ is a scalar field, one also has

$$\nabla_i\{A_k(x)B^k(x)\} = \partial_i\{A_k(x)B^k(x)\}$$
$$= (\partial_i A_k(x))B^k(x) + A_k(x)\partial_i B^k(x) \; . \tag{5.15}$$

Comparing (5.14) with (5.15) and noticing that these formulae are valid for any $A_k(x)$, one obtains the expression of the covariant derivative of a contravariant vector field

$$\nabla_i B^k(x) = \partial_i B^k(x) + \Gamma^k{}_{ji}(x)B^j(x) \; . \tag{5.16}$$

Two more remarks: we note first that the notations

$$\nabla_i A^k(x) = A^k_{ji}(x) \; ; \quad \nabla_i A_k(x) = A_{kji}(x) \tag{5.17}$$

are also in used, whereas for the ordinary derivative one writes

$$\partial_i A^k(x) = A^k{}_{,i}(x) \; ; \quad \partial_i A_k(x) = A_{k,i}(x) \; . \tag{5.18}$$

The second remark concerns the definition of the **covariant derivative along a vector v** as follows:[*]

$$\nabla_v = v^i(x)\nabla_i \; , \tag{5.19a}$$

---

[*]See also Ex. 1.18 for the covariant derivative along a given curve.

then

$$\nabla_{dx} = dx^i \nabla_i \ .$$ (5.19b)

The differentiation rules (5.5) and (5.16) can be generalized; for any higher tensor fields one has, for instance,

$$\nabla_\gamma T_{\alpha\beta} = \partial_\gamma T_{\alpha\beta} - \Gamma^\mu{}_{\alpha\gamma} T_{\mu\beta} - \Gamma^\mu{}_{\beta\gamma} T_{\alpha\mu} \ ,$$
$$\nabla_\gamma T^{\alpha\beta} = \partial_\gamma T^{\alpha\beta} + \Gamma^\alpha{}_{\mu\gamma} T^{\mu\beta} + \Gamma^\beta{}_{\mu\gamma} T^{\alpha\mu} \ .$$ (5.20)

For a proof, we consider successively $T_{jk} A^j A^k$ and $T^{jk} A_j A_k$ and apply the method used in formula (5.16).

We proceed along the same lines for any tensor field $T^{\bullet\bullet\bullet}_{\bullet\bullet\bullet}(x)$ and as a result it may be seen that its covariant derivative with respect to x^k is the sum of the term $\partial_k T^{\bullet\bullet\bullet}_{\bullet\bullet\bullet}$ and some other terms obtained through the application of the following rule: each contravariant index $A^{\bullet i \bullet}_{\bullet\bullet\bullet}$ corresponds to a term $\Gamma^i_{mk} A^{\bullet m \bullet}_{\bullet\bullet\bullet}$ and each covariant index $A^{\bullet\bullet\bullet}_{\bullet i \bullet}$ to a term $-\Gamma^m{}_{ik} A^{\bullet\bullet\bullet}_{\bullet m \bullet}$.

Finally, the expression of the covariant derivative of a tensor density field is given in Problem 1.21, question (e).

We are now ready for a comparison of the methods followed at Level 0 and the present level. At first sight, the two approaches may appear very different: at Level 0, we followed C.F. Gauss' point of view, at the present level we follow B. Riemann's. At Level 0, our starting point was the metric associated with a given surface, the notion of surface being considered as immediate and evident. We studied curves drawn on the surface we were interested in and the problem of the length of their arcs: all concepts were fundamentally metric even though it appeared later that one could draw topological consequences (see Sec. 2 of Level 0).

In the present chapter, the same theme is treated more abstractly. We started by defining a manifold $\mathcal{M}_N$ as a topological space with an homeomorphism $\mathcal{M}_N \to \mathbb{R}^N$. The concept of covariant derivative followed that of affine connection: this method is due to H. Weyl who tried hard to build up a unified theory of gravitation and electromagnetism. However, it is a surprise to see that a manifold with an affine connection structure is sufficiently "rich" to allow a geometric interpretation of the covariant differentiation. The notion of geodesic curve reappears then as an autoparallel curve (see the following paragraph), independently of any metric considerations, and so does the concept of curvature of a manifold. It is clear that one has to admit that the taxonomic approach of this chapter lacks elegance and conciseness; nonetheless it is true that the physicist needs analytic expressions that easily allow the prediction of quantitative laws and also numerical calculations on present-day computers. It is true that the avalanche of the many indices that appear darkens the understanding of the notions introduced, but one cannot see at present how the elegant presentation of the intrinsic approach could be transformed into a computing program. These are a few of the reasons that explain why physicists seem to prefer the taxonomic approach. But it must also be stressed that the intrinsic approach of the next chapter should not be neglected as some physicists often do, since it permits to draw into light the analogies between many concepts that might seem unrelated at first sight.

6. *Parallel Displacement and Self-Parallel Curves*

Let $\mathcal{E}_N$ be an affine space and consider two of its points, M located at x and M' located at $x' = x + dx$, and also a vector field $A(x)$. The parallel transport of the vector $A(x)$ from the point x to the point x' amounts to defining a vector $\hat{A}(x')$ at x' equipollent to $A(x)$

$$\hat{A}(x') = A(x) \ . \tag{6.1}$$

In particular

$$dA(x) = A(x') - A(x) = dx^i \partial_i A(x) \tag{6.2}$$

can be written as

$$dA(x) = A(x') - \hat{A}(x') \tag{6.3}$$

and for the $\mathcal{E}_N$ space, this new notation is without any further consequences. On the contrary, for a manifold $\mathcal{M}_N$ where there is no notion of equipollence anymore, one can take (6.3) as a pattern to define the displaced vector $\hat{A}(x')$ by replacing $dA(x)$ by $\nabla_{dx} A(x)$,

$$\nabla_{dx} A^k(x) = A^k(x') - \hat{A}^k(x') \tag{6.4}$$

whose right-hand side may be developed as follows:

$$\begin{aligned}
\nabla_{dx} A^k(x) &= dx^i \partial_i A^k(x) + \Gamma^k{}_{ij}(x) A^i(x) dx^j \\
&= A^k(x + dx) - A^k(x) + \Gamma^k{}_{ij}(x) A^i(x) dx^j \ ;
\end{aligned} \tag{6.5}$$

and compare these last two equations to get

$$\hat{A}^k(x') = A^k(x) - \Gamma^k{}_{ij}(x) A^i(x) dx^j \ . \tag{6.6}$$

This is an equation that defines the parallel displaced $\hat{A}_k(x')$ of the field $A_k(x)$, moved from the point $M(x)$ to the point $M'(x')$.

We shall say that a field $A_k(x)$ is uniform in a certain domain D if for each point $P(y) \in D$ one has

$$\hat{A}^k(y) = A^k(y) \ . \tag{6.7a}$$

In this case, there results from relation (6.6) an equation which the field should satisfy,

$$A^k(x') = A^k(x) - \Gamma^k_{ij}(x)A^i(x)dx^j$$
$$\Rightarrow dA^k(x) + \Gamma^k{}_{ij}(x)A^i(x)dx^j \Rightarrow \nabla_{dx}A^k(x) = 0 \ , \tag{6.7b}$$

valid for any dx and thus implying

$$\nabla_i A^k(x) = 0 \ , \tag{6.7c}$$

imposing a total of N^2 conditions on the N unknown functions $A^1(x) \dots A^N(x)$. Such a system of differential equations is not generally compatible. But if we impose on the field $A^k(x)$ to be uniform only along a certain curve $\Gamma \subset \mathcal{M}_N$ (with $C : x = x(\lambda)$ as its image), the N^2 conditions (6.7a) which can be written as

$$\nabla_{dx/d\lambda}A^k(x(\lambda)) = \frac{dx^j}{d\lambda}\nabla_j A^k(x(\lambda)) = 0 \ , \tag{6.8a}$$

or

$$\frac{dx^j}{d\lambda}(\partial j A^k(x) + \Gamma^k{}_{ij}(x)A^i(x)) = 0 \ , \tag{6.8b}$$

or

$$\frac{dA^k(x(\lambda))}{d\lambda} + \Gamma^k{}_{ij}(x(\lambda))A^i(x(\lambda))\frac{dx^j}{d\lambda} = 0 \ , \tag{6.8c}$$

are reduced to a system of N equations for N unknown functions $A^k(x)$, a system which is generally compatible.

Therefore, if we consider $A^k(x)$ to be given at a point $M_0(x_0^1 \dots x_0^N) \in \Gamma$, the theory of differential systems, provided that some classical conditions are fulfilled, allows to calculate $A^k(x)$ in a neighbourhood (belonging to C) of x_0.

We will generally say that the vector $A^k(x)$ has been deduced from the vector $A^k(x_0)$ by parallel displacement along Γ from the point M_0 to the point M. In the next paragraph (formulae (7.5) to (7.12)), we will define the curvature of a manifold using the notion of parallel displacement of a vector field along a closed curve: this is a major application of this concept.

Another important application is the definition of an autoparallel curve. Such a curve, $\Gamma \subset \mathcal{M}_N$, with local image C and equation $x = x(\lambda)$ defines a vector field $\tau(x) = dx/d\lambda$. If the field τ is uniform along C, then one can write

$$\frac{d\tau^k(x(\lambda))}{d\lambda} + \Gamma^k{}_{ij}(x(\lambda))\tau^i(x(\lambda))\tau^j(x(\lambda)) = 0 \; , \tag{6.9}$$

or

$$\frac{d^2x^k}{d\lambda^2} + \Gamma^k{}_{ij}(x(\lambda))\frac{dx^i}{d\lambda}\frac{dx^j}{d\lambda} = 0 \; . \tag{6.10}$$

There is a marked similarity between an auto-parallel curve and the geodesic Eq. (4.9), Level 0. But we shall come back to this point in the last paragraph of this chapter.

We may conclude with two last remarks: we notice first that if we consider two vectors tangent at the points x and x' of the autoparallel curve C, each vector is the displaced parallel of the other. We also remark that the parallel displacement of 2 vectors $A^k(x)$ and $B^k(x)$ along the same C curve leaves invariant their contracted product.

Let $\tau(x)$ be the vector field of the tangents to C and $A^k(x')(B^k(x'))$ the parallel displaced vectors of $A^k(x)(B_k(x))$, where x and x' are points of C. With the notation used in (6.8a) and (6.8b), we have

$$\begin{aligned}
\nabla_\tau A^k(x) &= \tau^i(x)\nabla_i A^k(x) = 0 \; , \\
\nabla_\tau B_k(x) &= \tau^i(x)\nabla_i B_k(x) = 0 .
\end{aligned} \tag{6.11}$$

Then multiplying respectively these equations by $B_k(x)$ and $A^k(x)$ and adding the results, we get

$$B_k(x)\nabla_\tau A^k(x) + A^k(x)\nabla_\tau B_k(x) = \nabla_\tau\{A^k(x)B_k(x)\} = 0 \; . \tag{6.12}$$

Now, the product $A^k(x)B_k(x)$ defines a scalar field so ∇_τ can be expressed through ordinary partial derivatives

$$\nabla_\tau = \tau^k(x)\partial_k = \frac{dx^k}{d\lambda}\partial_k = \frac{d}{d\lambda} \tag{6.13}$$

along C, so that (6.12) amounts to

$$d\{A^k(x)B_k(x)\} = 0 \; . \tag{6.14}$$

One deduces that the contracted product is indeed independent of x. One can finally remark that the geometric content of an autoparallel curve is very "rich": if one tries to establish a correspondence between properties of a manifold $\mathcal{M}_N$ and properties of an affine space $\mathcal{E}_N$, then the object corresponding to a geodesic line is a straight line.

7. *Curvature Tensor Field or Riemannian Tensor or Curvature Tensor*

In order to introduce the concept of **curvature**, let us define the **commutator**,

$$[\nabla_k, \nabla_l] = \nabla_k \nabla_l - \nabla_l \nabla_k \,, \tag{7.1}$$

and consider

$$[\nabla_k, \nabla_l] A^j(x) \,. \tag{7.2}$$

By a calculation more laborious than difficult one gets

$$[\nabla_k, \nabla_l] A^j(x) = R^j{}_{ikl}(x) A^i(x) + (\Gamma^i{}_{kl(x)} - \Gamma^i{}_{lk(x)}) \nabla_i A^j(x) \,, \tag{7.3}$$

where the object R^j_{ikl} (introduced at Level 0 under the name of Riemann tensor) has the expression

$$R^j{}_{ikl} = \partial_k \Gamma^j{}_{il} - \partial_l \Gamma^j{}_{ik} + \Gamma^j{}_{hk} \Gamma^k{}_{il} - \Gamma^j{}_{hl} \Gamma^h{}_{ik} \,, \tag{7.4}$$

each term of the right-hand side being a function of x.

The first point to note is that $R^j{}_{ikl}$ is a tensor field once contravariant and three times covariant; indeed, the left-hand side of (7.4) has clearly this property, on the other hand, according to the variance (5.12) of the Γ-symbol, the bracket $(\Gamma^i_{kl} - \Gamma^j_{kl})$, the last term in (7.3), is indeed a tensor field once contravariant and twice covariant, since it is easy to verify that the terms arising under a point transformation that could disturb the tensor character of the commutator in (7.3) are eliminated. Therefore, the right-hand of (7.3) is indeed a tensor field. We thus conclude that $R^j_{ikl}(x)$ deserves indeed the name of tensor curvature. We call $\Gamma^i_{kl} - \Gamma^i_{lk}$ the **torsion tensor**; the corresponding manifold is said to possess a torsion. All that being clarified, there is still an unanswered question: why call R^j_{ikl} the component of the curvature tensor?

To answer it, let us come back to the formula (6.6) and note that when $dx = 0$, i.e. when $x' = x$, then $\hat{A}^k(x) = A^k(x)$, so that this same formula (6.6) can be written as

$$A^k(x') - \hat{A}^k(x) = \delta \hat{A}^k(x) = -\Gamma^k{}_{ij}(x) A^i(x) dx^j \,. \tag{7.5}$$

Note the notation $\delta A^k(x)$ recalls that the last term of this relation is not usually a total differential. When one goes from a point $m(x)$ of $\mathcal{E}_N$ to a point p along a path C_1, one has

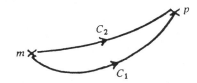

Fig. 1.2.

$$\hat{A}^k_{C_1}(p) = -\int_{C_1} \Gamma^k{}_{ij}(x)A^i(x)dx^j \ , \tag{7.6}$$

and if we choose the path C_2 going also from m to p we have

$$\hat{A}^k_{C_2}(p) = -\int_{C_2} \Gamma^k_{ij}(x)A^i(x)dx^j \ . \tag{7.7}$$

Subtract (7.6) from (7.7) and write

$$\Delta \hat{A}^k(p) = \hat{A}^k_{C_2}(p) - \hat{A}^k_{C_1}(p) \ . \tag{7.8}$$

By calling C_+ the closed curve built up by the curves C_1 and C_2 and by moving along C_+ anticlockwise, one deduces

$$\Delta \hat{A}^k(p) = -\oint_{C_+} \Gamma^k_{ij}(x)A^i(x)dx^j \ . \tag{7.9a}$$

For a covariant field, (7.9a) should be replaced by

$$\Delta \hat{A}_k(p) = \oint_C \Gamma^i_{kj}(x)A_i(x)dx^j \ . \tag{7.9b}$$

Stokes formula is quite easily generalized (see formula (16.3) of Level 2) to $\mathcal{E}_N$ and if S is a surface of $\mathcal{E}_N$ that has C_+ as edge we may write

$$\Delta \hat{A}_k(p) = \frac{1}{2}\int_S \left[\partial_l\{\Gamma^i_{km}(x)A_i(x)\} - \partial_m\{\Gamma^i_{kl}(x)A_i(x)\}\right]dS^{lm}$$
$$= \frac{1}{2}\int_S [(\partial_l\Gamma^i_{km}(x)A_i(x) - (\partial_m\Gamma^i_{kl}(x))A_i(x) + \Gamma^i_{km}(x)\partial_l \tag{7.10}$$
$$A_i(x) - \Gamma^i_{kl}(x)A_i(x)]dS^{lm} \ ,$$

where $dS^{lm} = dx^l \wedge dx^m$. Let us assume that the transport of the vector $A^k(x)$ along C_+ is a parallel displacement, then the condition (6.7) always allows us to write that along C_+,

$$\Delta A_k(p) = 0 \Rightarrow \partial_i A_k(x) = \Gamma^l_{ik}(x) A_l(x) . \tag{7.11}$$

Finally formula (7.10) takes the form

$$\Delta \hat{A}_k(p) = \frac{1}{2} \int_S [\partial_l \Gamma^i_{km}(x) - \partial_m \Gamma^i_{kl}(x) + \Gamma^j_{km}(x) \Gamma^i_{lj}(x)$$
$$- \Gamma^j_{kl}(x) \Gamma^i_{mj}(x)] A_i(x) dS^{lm} \tag{7.12}$$
$$= \frac{1}{2} \int_S R_{iklm} A^i(x) dS^{lm} .$$

In this way the curvature tensor reappears and may be given a geometric meaning. Indeed, let $A(x)$ be a vector field defined on the closed curve C_+, and let us start at the point $p \in C_+$ and transport the vector A along C_+ by means of a parallel displacement ($\nabla_i A_k(x) = 0$) coming back to the point p. The initial value of the field in p and its final value after this parallel transport at the same point p are generally different, and this difference is expressed by the surface integral of the Riemann tensor through formula (7.12). If this difference vanishes, whatever C_+ is in the neighbourhood of p, the space is locally flat; if it differs from 0, the manifold $\mathcal{M}_N$ together with the connection Γ is locally curved: this is the deep significance of the curvature tensor.

Let us make two remarks before exhibiting some algebraic properties of R^j_{ikl}. First, all the above considerations are local and all properties described are also local, valid only in the neighbourhood of $U_\mathcal{M}(P)$. A second trivial remark is related to the notion of a flat surface: a cylinder is flat, and so is any developable surface; on the other hand, the manifold $\mathcal{E}_2$ is a plane. Thus a flat surface is not necessarily a plane; each concept has a different geometric content.

One derives from the Riemann tensor several other tensor fields: let us consider R^j_{ikl}; by contracting the contravariant index with the second covariant index, one obtains Ricci's tensor:

$$R^l_{ilk} = R_{ik} = \partial_l \Gamma^l_{ik} - \partial_k \Gamma^l_{il} + \Gamma^l_{ik} \Gamma^p_{lp} - \Gamma^k_{il} \Gamma^l_{kp} . \tag{7.13}$$

Coming back to the definition (7.4) of the Riemannian tensor, we see that it is antisymmetric in its two last indices: therefore, instead of having N^4 components it only has $N^3(N-1)/2$ components (for $N = 4$ it has $(4)^3 \times 3/2 = 96$ components).*

B. Riemannian Geometry

8. *Riemannian Metric*

Let us consider again the $\mathcal{E}_N$ affine space homeomorphic to the $\mathcal{M}_N$ manifold and let us define at $x \in \mathcal{E}_N$ a field of N by N matrices $G(x)$ with elements $g_{ij}(x)$ that are real, symmetric: we shall say that the $\mathcal{M}_N$ **manifold is Riemannian** if in addition $g_{ij}(x)$ are the components of a doubly covariant tensor field.* Therefore, in a change of coordinates $M \to M'$, locally represented by the point transformation

$$x \to x' = x'(x) = x'(x^1 \dots x^N) \,, \tag{8.1}$$

(notation already used), we have

$$g_{rs}(x) \to g'_{rs}(x') = \frac{\partial x^i}{\partial x'^r} \frac{\partial x^j}{\partial x'^s} g_{ij}(x) \,. \tag{8.2}$$

Then, to all contravariant $A^k(x)$ vector fields we will be able to associate the **scalar**

$$g_{ij}(x) A^i(x) A^j(x) \,; \quad g_{ij} = g_{ji} \,. \tag{8.3}$$

Therefore, if we **define** the covariant tensor field by

$$A_i(x) = g_{ij}(x) A^j(x) \,, \tag{8.4}$$

we have

$$A^i(x) A_i(x) = g_{ij}(x) A^i(x) A^j(x) \,. \tag{8.5}$$

Consider now two tensor fields $A^k(x)$ and $B^k(x)$. We may define a scalar product

$$\langle A, B \rangle_x = g_{ij}(x) A^i(x) B^j(x) \,. \tag{8.6}$$

Indeed, this expression is bilinear in A and B and also symmetric so that

$$\langle A, B \rangle_x = \langle B, A \rangle_x \,. \tag{8.7}$$

---

*There is another interpretation of the curvature notion through the geodesic deviation, see Ex. 1.17.

*The $G(x)$ matrix has already been considered at Level 0, formulae (3.11) to (3.14).

Let us introduce the covariant field

$$B_i(x) = g_{ij}(x)B^j(x) \; , \qquad (8.8)$$

and the preceding scalar product can be written as

$$\langle A, B \rangle_x = A^i(x)B_i(x) = A_i(x)B^i(x) \; . \qquad (8.9)$$

Since no claim of positivity has been made for $\langle A, A \rangle_x$ it cannot represent a norm; however, we may call it the Riemannian length of the vector $A^i(x)$. It becomes a norm in all cases where (8.3) is definite positive.

In particular, we can consider in $\mathcal{E}_N$ the vector dx; usually an expression ds^2 is introduced which by definition is

$$ds^2(x) = g_{ij}(x)dx^i dx^j \; . \qquad (8.10)$$

Obviously, the notation is bad because ds^2 can have negative values; we have to wait until Level 2, where we will consider ds^2 as a differential 2-form. In any case where ds^2 is definite positive, the metric is said to be **proper Riemannian**.

The notation ds^2 goes back to Gauss who introduced for a surface the curve element ds^2 which is then positive definite. We have developed this point at Level 0, formula (2.7) and following.

At this same Level 0 we have defined the contravariant components $g^{ij}(x)$ of the metric (formula (3.12)) with the general property

$$g^{ir}g_{rj} = g_{ir}g^{rj} = \delta^i_j = \delta^j_i = g^i_j = g^j_i \; , \qquad (8.11)$$

and we introduced without any justification the law of indices. We can now justify it by showing that the $g_{ij}A^j$ is indeed a covariant vector field that we can denote by $A_i(x)$. As a matter of fact, after a point transformation,

$$
\begin{aligned}
A'_i(x') &= g'_{ij}(x')A'^j(x') = \frac{\partial x^r}{\partial x'^i}\frac{\partial x^s}{\partial x'^j}g_{rs}(x)\frac{\partial x'^j}{\partial x^m}A^m(x) \\
&= \frac{\partial x^r}{\partial x'^i}g_{mr}(x)A^m(x) = \frac{\partial x^r}{\partial x'^i}A_r(x) \; .
\end{aligned} \qquad (8.12)
$$

However, it should be noted that without any metric, covariant, contravariant and mixed tensor fields were considered as different mathematical objects. But, once a metric has been chosen, we may look upon a tensor field $A(x)$ as a single mathematical object with covariant, contravariant or mixed components, since we know how to raise or lower indices. Then one should be careful about the order of an index (first, second, third ... in the array of all indices) which should not be changed in the raising or lowering of that index. We clarify this point with the following example. The two mixed tensor fields

$$A^i{}_j = g^{il} A_{lj} \; ; \quad A_j{}^i = g^{il} A_{jl} \tag{8.13}$$

are different (except if A_{ij} is symmetric). See Ex. 1.3.

Coming back to the formula (8.2), we note that if $G(x)$ is the matrix with elements $g_{ij}(x)$, we have

$$\det G'(x') = g'(x') = \left| \frac{\partial(x^1 \dots x^N)}{\partial(x'^1 \dots x'^N)} \right|^2 g(x) \; , \tag{8.14a}$$

where $g'(x')$ and $g(x)$ are the respective determinants of $G'(x')$ and $G(x)$, and following definition (4.4), the determinant $g(x)$ is a tensorial scalar density of weight 2; furthermore

$$\sqrt{|g'(x')|} = \left| \frac{\partial(x^1 \dots x^N)}{\partial(x'^1 \dots x'^N)} \right| \sqrt{|g(x)|} \; . \tag{8.14b}$$

Let us then consider the expression

$$\sqrt{|g(x)|} d_N(x) = \sqrt{|g(x)|} dx^1 \dots dx^N \; . \tag{8.15}$$

Taking into account the well-known transformation law of the volume element $d_N x$ induced by a point transformation, we have

$$\begin{aligned} \sqrt{|g'(x')|} d_N x' &= \left| \frac{\partial(x^1 \dots x^N)}{\partial(x'^1 \dots x'^N)} \right| \sqrt{|g(x)|} \left| \frac{\partial(x'^1 \dots x'^N)}{\partial(x^1 \dots x^N)} \right| dx \\ &= \sqrt{|g(x)|} d_N x \; , \end{aligned} \tag{8.16}$$

according to the Jacobian chain rule. We therefore conclude that $\sqrt{|g(x)|}$ $d_N x$ is a scalar for any point transformation.

We finally consider the Riemannian length $g_{ij} A^i A^j$ of the field vector A at a point $x \in \mathcal{M}_N$. It is known from the elementary theory of quadratic forms that one can choose a suitable basis such that

$$\langle A, A \rangle_x = g_{ij} A^i A^j = \sum_{i=1}^{m} (A'^i)^2 - \sum_{j=m+1}^{N} (A'^j)^2 \; . \tag{8.17}$$

The number m of the positive squares is called the **index** of the form and is characteristic of the Riemannian manifold (law of inertia for quadratic forms). If $N = 4$ and $m = 1$, the manifold is called **hyperbolic** and may be taken as the manifold of general relativity. One usually labels the coordinates of $M \in \mathcal{M}_4$ as (x^0, x^1, x^2, x^3): they are the general (not necessarily orthogonal!) coordinates of M. A basis (e_α) of M_4 will be orthogonal iff

$$\langle e_\alpha, e_\beta \rangle_x = 0 \quad \text{if} \quad \alpha \neq \beta ; \quad \langle e_0, e_0 \rangle_x = +1 ; \langle e_j, e_j \rangle_x = -1 , \qquad (8.18)$$

the index i running from 1 to 3 and α from 0 to 3. We define the null-cone at x as the set of vectors orthogonal to themselves:

$$\langle B, B \rangle_x = g_{\mu\nu}(x) B^\mu(x) B^\nu(x) . \qquad (8.19)$$

A vector field $A(x)$ such that $\langle A, A \rangle_x < 0$ is space-like; a time-like vector field is such that $\langle A, A \rangle_x > 0$.

9. *Riemannian Connection*

The introduction to the Riemannian Metric will lead to the introduction of a connection without torsion, called **Riemannian connection.** Its Christoffel symbols will be completely determined by ds^2, provided we suppose that the manifold we are considering is of zero torsion (Sec. 7). The first thing to note is that the covariant derivative of the metric tensor vanishes:

$$\nabla_m g_{ij}(x) = 0 . \qquad (9.1)$$

Indeed, $\nabla_{dx} = dx^i \nabla_i$ behaves as a scalar under any point transformation; it does not participate in the raising or lowering of the indices. Consequently the only covariant vector that one can associate with $\nabla_{dx} A^i(x)$ (where A^i is an arbitrary contravariant field) is $\nabla_{dx} A_i(x)$. On the other hand, for reasons of internal coherence,

$$\nabla_{dx} A_i(x) = g_{ik}(x) \nabla_{dx} A^k(x) , \qquad (9.2)$$

which has to be equal to

$$\nabla_{dx} \{g_{ik}(x) A^k(x)\} = (\nabla_{dx} g_{ik}(x)) A^k(x) + g_{ik(x)} \nabla_{dx} A^k(x) .$$

The comparison between these two relations gives

$$(\nabla_{dx} g_{ik}(x)) A^k(x) = 0 \tag{9.3}$$

for any dx and $A^k(x)$, then Eq. (9.1) follows immediately.

It is easy to obtain an expression of the Γ as a function of the g_{ij}: let us return to (9.1) which we write as

$$
\begin{aligned}
\nabla_l g_{ik}(x) &= \partial_l g_{ik}(x) - g_{rk}(x) \Gamma^r_{il}(x) - g_{ir}(x) \Gamma^r_{kl}(x) \\
&= \partial_l g_{ik}(x) - \Gamma_{kil}(x) - \Gamma_{ikl}(x) = 0 ,
\end{aligned}
\tag{9.4}
$$

where

$$\Gamma_{ijk} = g_{ir} \Gamma^r_{jk} . \tag{9.5}$$

At this point, no further hypothesis has been made on the Γ-symbols (except for their derivability). Suppose now that they are **symmetric** with respect to their two last indices, then the Riemannian manifold is said to be without torsion, following Sec. 7. We want to show that for such a manifold the affine connection coefficients can be expressed by the g_{ij} only.

Returning to (9.4) let us write two other analogous equations obtained by a circular permutation of the indices l, i, k:

$$
\begin{aligned}
\partial_i g_{kl} - \Gamma_{lki} - \Gamma_{kli} &= 0 , \\
\partial_k g_{li} - \Gamma_{ilk} - \Gamma_{lik} &= 0 .
\end{aligned}
\tag{9.6}
$$

Let us add (9.4) to the first of Eq. (9.6) and let us subtract from the result the second. By the symmetry assumption that we just made, we obtain (9.7):

$$\Gamma_{ikl} = \frac{1}{2} (\partial_l g_{ki} + \partial_k g_{il} - \partial_i g_{lk}) . \tag{9.7}$$

This last equation shows that if $\mathcal{M}_N$ is endowed with an affine connection without torsion, then the Christoffel symbols are uniquely determined as functions of the g_{ij}'s. Finally, from (9.5) and (9.7), one obtains

$$\Gamma^i_{kl} = g^{ij} \Gamma_{jkl} = \frac{1}{2} g^{ij} (\partial_l g_{kj} + \partial_k g_{jl} - \partial_j g_{kl}) . \tag{9.8}$$

10. *Riemannian Tensor*

Formula (9.7) allows us to express the Riemann tensor as a function of the g_{ij}. In addition, new symmetries appear because of the Riemannian character of the manifold. For instance its completely covariant form can be written as[*]

$$R_{iklm} = g_{in} R^n_{\ klm} = \frac{1}{2}(\partial_k \partial_l g_{im} + \partial_i \partial_m g_{kl} - \partial_k \partial_m g_{il} - \partial_i \partial_l g_{km})$$
$$+ g_{np}(\Gamma^n_{\ kl}\Gamma^p_{\ im} - \Gamma^n_{\ km}\Gamma^p_{\ il}) \ . \qquad (10.1)$$

We can also note that the second parenthesis takes the form

$$g^{np}(\Gamma_{nkl}\Gamma_{pim} - \Gamma_{nkm}\Gamma_{pil}) \ , \qquad (10.2)$$

and in all these formulae the argument x is understood.

We remark further that (7.3) can be simplified to

$$[\nabla_k, \nabla_l]A^i = R^i_{\ nkl}A^n \ , \qquad (10.3)$$

since the torsion term disappears, and as a first consequence of (10.3) one has

$$R^i_{\ klm} + R^i_{\ mlk} + R^i_{\ mkl} = 0 \ . \qquad (10.4a)$$

As another consequence of (10.3), one verifies that

$$[\nabla_k, \nabla_l]A_i = -R^n_{\ ikl}A_n \ , \qquad (10.4b)$$

a formula which is valid for any tensor field, for instance,

$$[\nabla_k, \nabla_l]A_{ij} = R^n_{\ jkl}A_{in} - R^n_{\ ilk}A_{nj} \ . \qquad (10.4c)$$

We have already seen (see text following (7.13)) that R is antisymmetric with respect to its two last indices.

$$R^i_{\ klm} = -R^i_{\ kml} \ . \qquad (10.5)$$

Finally, (10.1) allows us to verify the following symmetries:

— antisymmetry with respect to the two first indices;

---

[*]The calculations which lead from $R^i_{\ klm}$ to R_{iklm} are lengthy; they use the following relation which is easy to prove:

$$g_{ij}\partial_k g^{il} = -g^{jl}\left(\Gamma^m_{\ ki}g_{mj} + \Gamma^m_{\ kj}g_{mi}\right) \ .$$

— antisymmetry with respect to the two last indices;

— symmetry with respect to the interchange of the couple consisting of the first two indices and the couple consisting of the last two indices.

It follows that $R_{iklm} = 0$ if $i = k$ or $l = m$.

Regarding other tensor fields which can be constructed from $R^i{}_{klm}$ we have mentioned in (7.13) the Ricci tensor which can also be written as

$$R_{ik} = g^{lm} R_{limk} = R_{ki} \; . \tag{10.6}$$

The symmetry of R_{ik} results from the symmetry just mentioned.*

Together with this tensor, one can define also the **scalar curvature**

$$R = g^{ik} R_{ik} \; . \tag{10.7}$$

Let us consider a two-dimensional surface S with a line element ds^2 already introduced at Level 0 (formulae (3.10)):

$$ds^2 = (dx^1)^2 + \chi^2(x^1, x^2)(dx^2)^2 \; . \tag{10.8}$$

It has been shown that between the Gauss curvatures $K(x)$ and $R(x)$ there exists the following relation proved in (9.40) of Level 0 (see also Problems 1.1 and 1.16):

$$K(x) = \frac{R(x)}{2} = -\frac{2}{\chi} \frac{\partial^2 \chi}{(\partial x^1)^2} \; . \tag{10.9}$$

Let us finally mention the Bianchi identity from which one deduces an important compatibility condition on Einstein equations (see Problem 1.11):

$$\nabla_i \left\{ R^i{}_k - \frac{1}{2} g^i_k R \right\} = 0 \; . \tag{10.10}$$

C. Miscellaneous Questions
11. *Deformation of Manifolds — Lie Derivatives*

A point transformation $x^k = x'^k(x^1 \ldots x^N)$ can be interpreted in two ways:

(a) It can be considered as a passive transformation, and this is the point of view which we have taken in paragraph 1 of this chapter. The point

---

*Note that the contraction of the first indices i and l or the two last ones m and k gives a vanishing result.

$M \in \mathcal{M}_N$ is described in two different local frames. Its local coordinates are x in the first and $x' = x'(x)$ in the second frame. Such a relation defines then the relationship between these two coordinate systems.

Analogously, geometric objects in $\mathcal{M}_N$ allow two representations, each of which is relative to a frame. The relationship between those two pictures defines the transformation law of the object through the properties of the group built up by such transformations (Sec. 4, formula (4.5) and following as well as Level 2, end of Sec. 5).

(b) One can also give to a point transformation the meaning of an active transformation: the point $M \in \mathcal{M}_N$ is mapped into a point $M' \in \mathcal{M}_N$, where $\mathcal{M}_N$ is the locally **deformed manifold**. Then $x' = x'(x)$ defines a local deformation of $\mathcal{M}_N$, x and x' being the local coordinates of those points in the same local coordinate frame.

We shall return to these questions in Sec. 7 of Level 2. The physicist reader will then realize that the concepts of homeomorphism and of chart elucidate considerably these questions. Here, we shall study the active local transformations when $\mathcal{M}_N = \mathcal{M}'_N$; let us consider an infinitesimal transformation $x \rightarrow x + \delta x$ where the N-uplet $\delta x = (\delta x^1 \ldots, \delta x^N)$ is infinitesimal. Let us put

$$\delta x(x) = \xi(x)d\tau : x' = x + \xi(x)d\tau , \qquad (11.1a)$$

where $\xi(x)$ is the N-uplet $(\xi^1(x), \ldots \xi^N(x))$, called the **generator** of the infinitesimal transformation.

We also notice that (11.1a) has yet another interpretation: consider the family of deformation with τ as a parameter

$$x'^k = x'^k(x^1 \ldots x^N {}_j \tau) : x'^k(x^1 \ldots x^N, 0) = x^k . \qquad (11.1b)$$

The infinitesimal transformation in a neighbourhood of $\tau = 0$,

$$x'^k = x^k + \xi^k(x)d\tau : \xi^k(x) = \left.\frac{\partial x'^k}{\partial \tau}\right|_{\tau=0} \qquad (11.1c)$$

is of the form (11.1a). Let $A(x)$ now stand for the set of components $A^{i_1 \ldots i_p}{}_{j_1 \ldots j_q}$ of a tensor field; we suppose that the active transformation (11.1a) induces on $A(x)$ a change identical to the change due to the passive transformation (4.3) which defines its variance or transformation law.

By definition the **Lie derivative L_ξ (A) of a field A in the direction**
ξ (*x*) is

$$L_\xi(A) = -\frac{A'(x) - A(x)}{\delta\tau} = -\frac{\delta A(x)}{\delta\tau} \; , \tag{11.2}$$

valid for each of the components of $A(x)$.

The variation

$$\delta A(x) = A'(x) - A(x)$$

is the **form variation of A(x)**. But it should be stressed that the two
fields A and A' must be defined at the same point $x \in \mathbb{R}_N$, because only
the difference of two tensor fields of the same transformation law at the
same point x has a transformation law common to the two fields. We can
conclude that $L_\xi A(x)$ is a tensor field having the same variance as $A(x)$.

On the other hand we define the **total variation**

$$\Delta A(x) = A'(x') - A(x) \; , \tag{11.3a}$$

which, contrary to $\delta A(x)$, is not necessarily a tensor field since the two
tensor fields A' and A have different points x and x' arguments.

The expression of $\Delta A(x)$ can also be written as

$$\begin{aligned}
\Delta A(x) &= A'(x') - A'(x) + A'(x) - A(x) \\
&= A'(x) - A'(x) + \delta A(x) \; . \tag{11.3b}
\end{aligned}$$

We notice also that a limited expansion of $\Delta A(x)$ gives

$$\begin{aligned}
\Delta A(x) &= A'(x') - A(x) = A'(x + \delta x) - A'(x) + A'(x) - A(x) \\
&= \delta x^k \partial_k A'(x) + \delta A(x) \; ,
\end{aligned}$$

from which one deduces

$$\delta A(x) = \Delta A(x) - \delta x^k \partial_k A'(x) \; , \tag{11.4}$$

but $A'(x) - A(x) = \delta A(x)$ is of first order in $d\tau$; one can therefore replace the
last term by $\delta x^k \partial_k A(x)$, up to second order infinitesimally small corrections.
Therefore one can bring $L_\xi A(x)$ into the form

$$L_\xi A(x) = -\frac{\delta A(x)}{\delta\tau} = -\frac{\Delta A(x)}{\delta\tau} + \frac{\delta x^k}{\delta\tau}\partial_k A(x) = -\frac{\Delta A(x)}{\delta\tau} + \xi^k \partial_k A(x) \; .$$

$$\tag{11.5}$$

Two examples will illustrate this method.

1. The tensor field $A(x)$ is the scalar field $F(x)$; so that $F'(x') = F(x)$: formula (11.5) allows us to write

$$L_\xi F(x) = \xi^k(x) \partial_k F(x) \ . \tag{11.6}$$

The Lie derivative of a function F is its directional derivative along $\xi(x)$.

2. The tensor field $A(x)$ is the contravariant vector field $X(x)$. In this case, $L_\xi(X)$ is also contravariant and we shall denote $(L_\xi X(x))^i = L_\xi X^i(x)$. The field $X'^i(x)$ is deduced from $X^i(x')$ according to the usual procedure; taking into account (11.1) and (11.5):

$$X'^i(x') = \frac{\partial x'^i}{\partial x^k} X^k(x) = \left(\frac{\partial x^i}{\partial x^k} + \frac{\partial \xi^i}{\partial x^k} d\tau \right) X^k(x) = X^i + \frac{\partial \xi^i}{\partial x^k} X^k d\tau \ ,$$

since we have $\partial_k x^i = \delta_k^i$. Finally,

$$L_\xi X^i(x) = \xi^k \partial_k X^i - X^k \partial_k, \xi^i \ , \tag{11.7a}$$

where the right-hand side is also called the **Lie bracket**[*] and may be written as

$$[\xi(x), X(x)]^i = L_\xi X^i(x) \ . \tag{11.7b}$$

This bracket is antisymmetric with respect to its factors and satisfies the **Jacobi identity** for 3 contravariant vector fields

$$[X^1, [X^2, X^3]] + [X^2, [X^3, X^1]] + [X^3, [X^1, X^2]] = 0 \ .$$

One obtains so an algebra called **Lie algebra**. We shall come back to this point at Level 2.

Let us notice that in formulae (11.6) and (11.7) one can replace the ordinary derivative by the covariant one: it is obvious for (11.6); for (11.7) one verifies that

$$\xi^i \nabla_i X^k - X^i \partial_i \xi^k = \xi^i (\partial_i X^k - \Gamma_{ji}^k X^j) - X^i (\partial_i \xi^k - \Gamma_{ji}^k X^j) \ ,$$

where the terms including Γ-symbols cancel each other. One can therefore write instead of (11.7a)

$$L_\xi X^k(x) = \xi^l \nabla_i X^k - X^i \nabla_i \xi^k \ . \tag{11.8}$$

---

[*]We shall see that in the intrinsic formulation (Level 2, formula (4.21)) the Lie bracket is indeed a commutator.

The method we just used can also be applied to a covariant vector:

$$L_\xi X_k = \xi^i \partial_i X_k + X_i \partial_k \xi^i \ , \tag{11.9a}$$

and again ordinary derivatives can be replaced by covariant ones:

$$L_\xi X_k = \xi^i \nabla_i X_k + X_i \nabla_k \xi^i \ . \tag{11.9b}$$

Let us give three other formulae whose proof is left to the reader:

$$L_\xi A_{kl}(x) = (\nabla_k \xi^i) A_{il}(x) + (\nabla_l \xi^i) A_{ki}(x) + \xi^i \nabla_i A_{kl}(x) \ , \tag{11.10a}$$

$$L_\xi A^{kl}(x) = -(\nabla_i \xi^k) A^{il}(x) - (\nabla_i \xi^l) A^{li}(x) + \xi^i \nabla_i A^{kl}(x) \ , \tag{11.10b}$$

$$L_\xi A^k_l(x) = (\nabla_l \xi^i) A^k_i(x) - (\nabla_i \xi^k) A^i_l(x) + \xi^i \nabla_i A^k_l(x) \ . \tag{11.10c}$$

In the next paragraph, we shall study the problem of isometry which is an immediate application of the Lie derivative concept.

12. *Isometries*

Let us consider again the manifold $\mathcal{M}_N$ that we will deform continuously. Consider in $\mathbb{R}^N$ the couple $(x, x + dx)$ to which we associate by diffeomorphism the couple (M, M') of $\mathcal{M}_N$, this being done before the transformation. After deformation, to the same couple $(x, x + dx)$ we associate (P, P') of $\mathcal{M}_N$, the deformed manifold obtained from $\mathcal{M}_N$. Let us suppose $\mathcal{M}_N$ to be proper Riemannian with a metric g_{ij} before deformation which becomes g'_{ij} after deformation; the elementary arc $\overparen{MM'}$ has a length $g_{ij} dx^i dx^j$ and $g_{ij} dx^i dx^j$ is that of the arc PP'. The lengths of these two arc elements are generally different, but if $g_{ij}(x) = g'_{ij}(x)$ the form variation δg_{ij} of g_{ij} vanishes:

$$\delta g_{ij}(x) = g'_{ij}(x) - g_{ij}(x) = 0 \ , \tag{12.1}$$

which implies, according to the definition of Lie derivative,

$$L_\xi g_{ij}(x) = 0 \ . \tag{12.2}$$

We shall call **isometry** any deformation satisfying (12.1) and therefore (12.2). The above discussion exhibits the geometric content of the concept of isometry and gives a more precise meaning to the often-used expression "isometry preserves lengths". Condition (12.2) gives rise to the system of **Killing equations**: indeed if we replace in (11.9) the general tensor A_{kl} by g_{kl} and notice $\nabla_i g_{kl} = 0$, we obtain

$$
\begin{aligned}
L_\xi g_{ij}(x) &= (\nabla_i \xi^k)g_{kj} + (\nabla_j \xi^k)g_{ki} + \xi^k \nabla_k g_{ij} \\
&= \nabla_i \xi^j(x) + \nabla_j \xi^i(x) = -\frac{\delta g_{ij}(x)}{\delta \tau} \ .
\end{aligned}
\tag{12.3}
$$

The Killing equations can consequently be written as

$$
L_\xi g_{ij}(x) = \nabla_i \xi^j(x) + \nabla_j \xi^i(x) = 0 \ .
\tag{12.4}
$$

The solutions $\xi(x)$ of the system are the **Killing vectors**. Their study is of fundamental importance in the theory of the symmetries of manifolds. The properties exhibited allow a considerable simplification of the solutions of Einstein's equations. See Problem 1.16.

As a consequence of the Killing equations, we may point out that if we define a N by N matrix with elements $\nabla_i \xi_j$, then this matrix is skew-symmetric.

Let us consider as an example the Lorentz space of special relativity. In this case the covariant derivative can be replaced by the usual derivatives since the Γ-symbols vanish. One can verify that the generating vectors

$$
\xi_\mu(x) = \varepsilon_{\mu\rho} x^\rho = \eta_\mu \quad \mu, \rho = 0, 1, 2, 3 \ ,
$$

where the $\xi_{\mu\rho}$ and η_α are independent of x, generate a system of solutions of (12.4) provided that the tensor $\varepsilon_{\mu\rho}$ is antisymmetric. One obtains the **Poincaré group** of transformations (i.e., inhomogeneous Lorentz transformations). This group has 10 parameters: the four η_μ describing the translations and the tensor $\varepsilon_{\mu\rho}$ related to the pseudorotations of space-time (see Problem 1.18). We finally remark that the Killing equations may be brought into the form

$$
-\delta g_{ij}(x) = \nabla_i \delta x^j + \nabla_j \delta x^i = 0 \ ,
\tag{12.5}
$$

where $\delta x(x)$ has been defined in (11.1a).

13. *Geodesic and Self-Parallel Curves*

The theory of geodesic lines, which was sketched in Sec. 4 of Level 0 in a form valid for two curvilinear coordinates, can be extended to the case of N variables $\{x^1 \dots x^N\}$ almost without change. We set

$$s_{\widehat{AB}} = \int_{t_A}^{t_B} \sqrt{g_{ij}(x(t))\dot{x}^i \dot{x}^j}\, dt \tag{13.1}$$

everytime the metric g_{ij} is properly Riemannian and that implies that the square root in (13.1) is real. If this condition is satisfied, one can obtain the equations of geodesics by calculations analogous to the ones developed in Sec. 4, Level 0 (formulae (4.5) to (4.9)). In an arbitrary parametrization*, one has

$$\ddot{x}^n + \Gamma^n{}_{ij}(x(t))\dot{x}^i \dot{x}^j - \dot{x}^n \frac{d^2 s/dt^2}{ds/dt} = 0 \ . \tag{13.2}$$

If we use the intrinsic parameterization ($t = s$) then (13.2) simplifies to

$$\frac{d^2 x^n}{ds^2} + \Gamma^n{}_{ij}(x(t))\frac{dx^i}{ds}\frac{dx^j}{ds} = 0 \ . \tag{13.3}$$

We shall not go deeper into this theory: the difficulties mentioned at Level 0, Sec. 4 remain, and require very careful handling. Furthermore, the definition we just gave will be generalized in Sec. 14, Level 2.

Let us however notice an interesting analogy between geodesics and self-parallel curves: if we compare (6.10) to (13.3) we see that any self-parallel curve which solves the system (6.10) with the choice $\lambda = s$ is geodesic and vice versa.

We finally remark that with this intrinsic parameterization one can immediately derive a first integral of the system (13.3). Indeed,

$$ds^2 = g_{ij}(x(s))dx^i dx^j \tag{13.4}$$

gives immediately

$$1 = g_{ij}(x(s))\frac{dx^i}{ds}\frac{dx^j}{ds} \ , \tag{13.5}$$

which is indeed a first integral.

---

*The notations of (13.2) are different from the notations used at Level 0, formula (4.9): the parameters $(u^1 \dots u^N)$ of that formula are here denoted by $(x^1 \dots x^N)$.

The similarity between geodesics and autoparallel curves is to be used in the definition of null geodesics, i.e. geodesics corresponding to $ds^2 = 0$. In that case, neither Eq. (13.3) nor Eq. (13.5) has a meaning, but a null geodesic can still be defined as an autoparallel curve as defined by (6.10).

We shall use a parameter $\lambda \neq s$; then such a geodesic will be given by the equation

$$\frac{d^2 x^n}{d\lambda^2} + \Gamma^n{}_{ij}(x(\lambda))\frac{dx^i}{d\lambda}\frac{dx^j}{d\lambda} = 0 \; , \tag{13.6a}$$

with an integral

$$g_{ij}(x(\lambda))\frac{dx^i}{d\lambda}\frac{dx^j}{d\lambda} = 0 \; . \tag{13.6b}$$

Complements, Exercises and Applications to Physics

Contents

Differential Geometry

Applications to Physics

Differential Geometry

Problem 1.1. *Examples of manifolds: the sphere S^3, the hyperboloid H_3 and the hyperplane $\mathcal{E}_3$. Spaces of constant curvature*

In paragraph 3, Level 0, we introduced the sphere S^2 and the plane $\mathcal{E}_2$ as manifolds embedded into the affine space $\mathcal{E}_3$. We shall now look for a generalization of this method.

(a) Consider a $\mathcal{M}_4$ manifold with the following metric

$$ds^2 = dx^2 + dy^2 + dz^2 + dw^2 \ ,$$

in which we shall embed the following S^3 manifold:

$$x = a \sin \chi \sin \theta \cos \varphi \ , \qquad z = a \sin \chi \cos \varphi$$
$$y = a \sin \chi \sin \theta \sin \varphi \ , \qquad w = a \cos \chi \ ,$$

represented in generalized spherical coordinates in $\mathcal{M}_4$. Show that by elimination of χ, θ, φ in the above expressions of x, y, z, w, one obtains the following relation

$$x^2 + y^2 + z^2 + w^2 = a^2 \ ,$$

which, in an orthonormal frame, represents a hypersphere in $\mathcal{E}_4$. Show that the ds^2 previously given now becomes

$$ds^2 = a^2 d\chi^2 + a^2 \sin^2 \chi d\sigma^2 \ ,$$

with

$$d\sigma^2 = d\theta^2 + \sin^2 \theta d\varphi^2 \ .$$

Show that the intersection of S^3 by the plane $w = w_0$ (w_0 is a constant) is a sphere S^2 with radius a_0:

$$x = \pm a_0 \sin \theta \cos \varphi \ , \qquad z = \pm a_0 \cos \theta \ ,$$
$$y = \pm a_0 \sin \theta \sin \varphi \ , \qquad a_0 = \sqrt{a^2 - w_0^2} \ ,$$

the volume of that intersection is

$$\frac{4}{3} \pi |a^2 - w_0^2|^{3/2} = \frac{4}{3} \pi a_0^3 \ ,$$

and its surface area is

$$4\pi|a^2 - w_0{}^2| = 4\pi a_0{}^2 .$$

Note that if one puts the ds^2 in the standard form $g_{ij}dx^i dx^j$ with $x^1 = \theta, x^2 = \varphi, x^3 = \chi$, one has

$$g_{11} = a^2 \sin^2 \chi , \quad g_{22} = a^2 \sin^2 \chi \sin^2 \theta , \quad g_{33} = a^2 ,$$

and show that

$$\int \sqrt{|g|}dx^1 dx^2 dx^3 = 2\pi^2 a^3 ,$$

which is the 3-volume of S^3 (see Level 1, formula (8.16)). The 4-volume is then

$$\int_0^a 2\pi^2 r^2 dr = \frac{1}{2}\pi a^4$$

and the concepts of 3- and 4-volume correspond respectively to the elementary geometrical concepts of area and volume of the sphere.

Notice that for $\chi = 0 : x = y = z = 0$ and $w = +a$ whatever the values of θ and φ, and for $\chi = \pi : x = y = z = 0$ and $w = -a$ for all values of θ and φ. For the values 0 and π of θ, $x = y = 0$ for all values of χ and φ. All these values represent singular points of the spherical representation of S^3.

Show finally that one has another representation of S^3:

$$x = \frac{a}{A}t , \quad y = \frac{a}{A}u , \quad z = \frac{a}{A}v , \quad w = \pm\frac{a}{A} ,$$

with

$$A = \sqrt{1 + t^2 + u^2 + v^2} .$$

(b) We now look for a ds^2:

$$ds^2 = dx^2 + dy^2 + dz^2 - dw^2 ,$$

and the manifold H_3:

$$x = a\,\mathrm{sh}\,\chi \sin\theta \cos\varphi , \qquad z = a\,\mathrm{sh}\,\chi \cos\theta ,$$
$$y = a\,\mathrm{sh}\,\chi \sin\theta \sin\varphi , \qquad w = a\,\mathrm{ch}\,\chi ,$$

which is an hyperboloid in an orthonormal frame

$$x^2 + y^2 + z^2 - w^2 = -a^2 .$$

Show that the ds^2 takes the form

$$ds^2 = a^2 d\chi^2 + \alpha^2 \operatorname{sh}^2\chi d\sigma^2 \ ,$$
$$d\sigma^2 = d\theta^2 + \sin^2\theta d\varphi^2 \ .$$

The intersection of H_3 by the plane $w = w_0$ is a sphere S^2:

$$x^2 + y^2 + z^2 = a^2 + w_0{}^2 = a_0 \ ,$$

its volume being

$$\frac{4}{3}\pi a^3 |\operatorname{sh}^2\chi_0|^3 : \operatorname{sh}^2\chi = \frac{a^2 + w_0{}^2}{a^2} \ ,$$

and its area $4\pi a^2 \operatorname{sh}^2\chi_0$, both of which tend to ∞ with χ_0. The 3-volume of H^3 between the planes $w = w_0$ and $w = a \operatorname{ch}\chi_0$ is

$$a^3 \int_0^{\chi_0} d\chi \int_0^{2\pi} d\varphi \int_0^{\pi} d\theta \operatorname{sh}^2\chi \sin\theta = \pi a^3 (\operatorname{sh} 2\chi_0 - 2\chi_0) \ ,$$

which also tends to ∞ with χ_0.

(c) The spherical coordinates of a point M of an hyperplane $\mathcal{E}_3$ are r, θ, φ. In order to unify notations, put $r = aX$, where a is a constant. Cartesian coordinates x, y, z take the form

$$x = aX \sin\theta \cos\varphi \ , \quad y = aX \sin\theta \sin\varphi \ , \quad z = aX \cos\theta$$

and the ds^2 is given by

$$ds^2 = a^2 dX^2 + a^2 X^2 d\sigma^2 : \ d\sigma^2 = d\theta^2 + \sin^2\theta d\varphi^2 \ ,$$

or

$$ds^2 = dx^2 + dy^2 + dz^2 \ .$$

It may be noticed that the formula

$$ds^2 = a^2 dX^2 + a^2 \varpi^2(X^2) d\sigma^2 \ ,$$

may represent the ds^2 of S^3 ($\varpi^2 = \sin^2 X$), either the one relative to $H_3(\varpi^2 = \operatorname{sh} X)$ or the one of $\mathcal{E}_3(\varpi^2 = X^2)$.

The Riemannian $\mathcal{M}_4$ manifold can be generalized into a $\mathcal{M}_N$. Let $M \in \mathcal{M}_N$ with generalized (curvilinear) coordinates $(x^1 \ldots x^N)$, and consider the $\mathcal{M}_{N+1}$ Riemannian manifold with $(x^1 \ldots x^N, w)$ coordinates and $\hat{ds}^2$ defined as follows:

$$\hat{ds}^2 = c_{ij} dx^i dx^j + \frac{1}{K} dw^2 = \hat{g}_{mn} dx^m dx^n : \; m, n = 1 \ldots N+1 \; ,$$

where $\hat{g}_{ij} = c_{ij} = c_{ji}$, for $i, j \neq N+1$ and running from 1 to N; $\hat{g}_{N+1,N+1} = 1/K$, all other $\hat{g}$'s vanish and $x^{N+1} = w$. Define now a $\mathcal{M}_N$ manifold by the implicit equation

$$K c_{ij} x^i x^j + w^2 = 1 \; ,$$

which is said to be embedded in $\mathcal{M}_{N+1}$ (the embedding manifold).

(d) Show that the ds^2 of $\mathcal{M}_N$ can be brought into the form

$$ds^2 = \left(c_{ij} + K \frac{c_{ik} c_{jl} x^k x^l}{1 - K c_{kl} x^k x^l} \right) dx^i dx^j = g_{ij} dx^i dx^j \; ,$$

i.e.,

$$g_{ij} = c_{ij} + K \frac{c_{ik} c_{jl} x^k x^l}{1 - K c_{kl} x^k x^l} \; .$$

(e) Show that the affine coefficient is given by

$$\Gamma^p{}_{ij} = K x^p g_{ij} \; .$$

(f) Show that the equation of the geodesics of $\mathcal{M}_N$ is

$$\frac{d^2 x^p}{ds^2} + K x^p = 0 : \; p = 1 \ldots N$$

and discuss its solutions.

(g) Calculate R_{jikl} and R_{il} to show that

$$R_{jikl} = K(g_{jk} g_{il} - g_{jl} g_{ik}) \; ,$$
$$R_{il} = K(N-1) g_{il} \; .$$

(h) Suppose that there exists a special reference frame in which the matrix C with c_{ij} as elements is of the form $1/|K|I$ where I is the unit N by N matrix.

Show that the $\hat{ds}^2$ of the embedding manifold takes the form

$$\hat{ds}^2 = \frac{1}{|K|} \sum_1^N (dx^i)^2 + \frac{1}{K} dw^2$$

and is a flat space (see also Problem 1.15).

Put $K|K| = k$, and $r^2 = \Sigma(x^i)^2$, and the embedded manifold $\mathcal{M}_N$ can be represented by

$$kr^2 + w^2 = 0 \ .$$

For $k = 1$ and $k = -1$ we obtain generalization of the sphere and the hyperboloid of questions (a) and (b). Note that the coordinates we are using are not the "ordinary" orthogonal ones (see Ex. 1.9(h)).

Show that the ds^2 of the embedded space is then

$$ds^2 = \frac{1}{|K|} \left[\sum_1^n (dx^i)^2 + \frac{k}{1 - kr^2} \left(\sum_1^N x^i dx^i \right)^2 \right] \ .$$

Consider the special case $N = 3$, and perform the point transformation

$$x^1 = r \sin\theta \cos\varphi \ , \quad x^2 = r \sin\theta \sin\varphi \ , \quad x^3 = r \cos\theta \ ,$$

$(r, \theta, \varphi$ would have been spherical coordinates if we had an orthonormal frame). Show that the ds^2 takes the form

$$ds^2 = \frac{1}{|K|} \left(\frac{dr^2}{1 - kr^2} + r^2 d\sigma^2 \right) \ ,$$

where $d\sigma^2$ has been defined in (a). This ds^2 will be present in the Robertson-Walker solution of Einstein's equations, Problem 1.24. Show that one can arrive at the same result, starting from $ds^2 = a^2 d\chi^2 + a^2 \varpi^2(\chi) d\sigma^2$ (from question (c)) and performing the change of variable $u = \varpi(\chi)$. It should be noticed, however, that the proof of the fundamental formulae given in question (g) requires a lengthy calculation using (10.1), Level 1.

All the manifolds we have been studying belong to the class of spaces of constant curvature; they are also maximally symmetric spaces. See also Problem 1.16.

Solution:

(a) *For the equation of the intersection of S^3 with the plane $w = w_0$, it is enough to remark that*

$$\cos \chi_0 = \frac{w_0}{a} \ , \quad \sin^2 \chi_0 = \frac{a^2 - w_0{}^2}{a^2} \ .$$

The 3-volume of S^3 is given by

$$\int_0^\pi d\chi \int_0^{2\pi} d\varphi \int_0^\pi d\theta a^3 \sin^2 \chi \sin \theta = 2\pi^2 a^3 \ .$$

Questions (b) and (c): Use analogous considerations as in (a).

(d) *Differentiate the equation for $\mathcal{M}_N$, one gets*

$$K c_{ij} x^i dx^j + w dw = 0$$

and

$$dw^2 = \frac{K^2}{w^2} (c_{ij} x^i dx^j)^2 = \frac{K^2}{1 - K c_{kl} x^k x^l} (c_{ij} x^i dx^j)^2 \ .$$

Finally,

$$ds^2 = c_{ij} x^i dx^j + \frac{K}{1 - K c_{kl} x^k x^l} (c_{ik} c_{jl} x^k x^l) dx^i dx^j \ .$$

(e) *Calculations become involved if not systematically done; proceed as follows. From formula (9.7), Level 1*

$$\Gamma^p{}_{ij} = \frac{1}{2} g^{pm} \Gamma_{mij} = \frac{1}{2} g^{pm} (\partial_j g_{mi} + \partial_i g_{mj} - \partial_m g^{ij}) \ ,$$

take the first derivative, for instance,

$$\partial_j g_{mi} = \partial_j \ \frac{K c_{mk} c_{il} x^k x^l}{1 - K c_{pq} x^p x^q}$$

$$= \frac{K}{1 - K c_{pq} x^p x^q} (c_{mj} c_{il} x^l + c_{mk} c_{ij} x^k)$$

$$+ \frac{2K^2}{(1 - K c_{pq} x^p x^9)^2} c_{mk} c_{il} c_{jp} x^k x^l x^p \ ,$$

bring into $\Gamma^p{}_{ij}$ *this last expression of* $\partial_j g_{mi}$ *and the two other analogous derivatives. We end up with*

$$\Gamma^p{}_{ij} = g^{pm}\left[\frac{K}{1 - Kc_{pq}x^p x^q}c_{ij}c_{mk}x^k + \frac{K^2}{(1 - Kc_{pq}x^p x^q)^2}c_{mk}c_{il}c_{jp}x^k x^l x^p\right]$$

$$= g^{pm}\frac{K}{1 - Kc_{pq}x^p x^q}c_{mk}x^k\left(c_{ij} + \frac{K}{1 - Kc_{pq}x^p x^q}c_{il}c_{jp}x^l x^p\right)$$

$$= \frac{K}{1 - Kc_{pq}x^p x^q}g^{pm}c_{mk}x^k g_{ij} \ .$$

The factor $g^{pm}c_{mk}x^k$ *can also be simplified as follows;*

$$\delta_k^p x^k = g^{pm}g_{mk}x^k = g^{pm}\left(c_{mk} + K\frac{c_{mr}c_{kl}x^r x^l}{1 - Kc_{rs}x^r x^s}\right)x^k \ ,$$

a relation equivalent to

$$x^p = g^{pm}c_{mk}x^k + g^{pm}c_{mr}x^r\frac{Kc_{kl}x^k x^l}{1 - Kc_{rs}x^r x^s} \ ,$$

i.e.

$$x^p = g^{pm}c_{mk}x^k\frac{1}{1 - Kc_{rs}x^r x^s} \ .$$

We finally bring $g^{pm}c_{mk}x^k$ *into the expression of* $\Gamma^p{}_{ij}$ *to obtain the required result.*

(f) Introduce the expression e of $\Gamma^p{}_{ij}$ *into Eq. (13.3), Level 1:*

$$\frac{d^2 x^p}{ds^2} + \Gamma^p{}_{ij}\frac{dx^i}{ds}\frac{dx^j}{ds} = \frac{d^2 x^p}{ds^2} + Kx^p g_{ij}\frac{dx^i}{ds}\frac{dx^j}{ds} = 0 \ .$$

Since $(g_{ij}dx^i/ds)dx^j/ds = 1$ *(see Eq. (13.5), Level 1), the equation of geodesics becomes*

$$\frac{d^2 x^p}{ds^2} + Kx^p = 0 \ .$$

Its fundamental solutions are $\sin(\sqrt{k}s)$ *and* $\cos(\sqrt{k}s)$ *for* $K > 0$; *if* $K < 0$ *one has* $\sinh(\sqrt{-k}s)$ *and* $\cosh(\sqrt{-k}s)$; *and if* $k = 0, x^p = A^p s + B^p$, A^p *and* B^p *being the real components of the contravariant vectors A and B.*

(g) Consider the expression of $R^j{}_{ikl}$ *as given by (7.4), Level 1:*

$$R^j{}_{ikl} = \partial_k\Gamma^j{}_{il} - \partial_l\Gamma^j{}_{ik} + \Gamma^j{}_{hk}\Gamma^h{}_{il} - \Gamma^j{}_{hl}\Gamma^h{}_{ik}$$

*and perform the derivation of the first two terms using the expression of
the Γ-symbols as given in (e):*

$$R^j{}_{ikl} = K(\delta^j_k g_{il} - \delta^j_l g_{ik}) + K(x^k \partial_k g_{il} - x^j \partial_l g_{ik}) + \Gamma^j{}_{hk}\Gamma^h{}_{il} - \Gamma^j{}_{hl}\Gamma^h{}_{ik} \; .$$

*The first bracket is clearly a tensor field ($\delta^j{}_k = g^j{}_k$) since $R^j{}_{ikl}$ is one,
then the remaining terms build up a tensor field too. We now use geodesic
coordinates (see Ex. 1.10): in such a frame the g_{ij} are stationary ($\partial_k g_{ij} =
0$) and the Γ-symbols vanish, then*

$$R^j{}_{ikl} = K(g^j_k g_{il} - g^j_l g_{ik}) \; ,$$

*a tensorial equation valid in any frame. We finally deduce the expressions
of R^j_{ikl} and R_{il} as requested.*

*(h) The first questions are evident. As far as the change of variables is
concerned, it is sufficient to note*

$$\sum_1^3 x^i dx^i = r dr$$

and

$$ds^2 = \frac{1}{|K|}\left(dr^2 + \frac{kr^2}{1 - kr^2}dr^2\right) \; ,$$

which are the required result.

*The proof for the last comment will be given for S^3 corresponding to
$K = 1$, then*

$$ds^2 = a^2(d\chi^2 + \sin^2\chi d\sigma^2) \; .$$

*From $u = \sin\chi$, one gets $d\chi^2 = du^2/(1 - u^2)$. Back to ds^2 we obtain the
previous form by identifying a^2 and $1/|K|$.*

$\boxed{\text{Problem 1.2.}}$ *The terms "tensor" and "tensor fields" will be used interchangeably On the other hand, $\{x^1 \ldots x^N\}$ will denote curvilinear (or configuration space) coordinates*

Tensor algebra

(a) Show that a linear combination of tensors of the same type and order
whose coefficients are constants or scalars is a tensor of the same type and
order.

(b) Show that the multiplication of the components of any number of tensor, results in a tensor called the **product (or tensor product)**. Its order is obtained by adding respectively the number of covariant indices and contravariant indices.

(c) For any mixed tensor $A^{ij}{}_{rst}$, the expression $A^{ij}{}_{rsj}$, the sum of N components of this tensor, is a tensor of the third order. This process is general, it allows to obtain from a mixed tensor of order p a tensor $p-2$ and is called **contraction**.

(d) Using (b) and (c), show that to the tensors A_{ij} and B^{rst} corresponds a new tensor $A_{ij}B^{jst}$. This operation is sometimes called **inner product** of tensors or **contracted product**.

(e) Any set of functions in sufficient number can be taken as the components of a tensor of any type and order in one given coordinate system, provided that its components in any other system are defined by the previously given variance laws.

(f) Show that if the components of a tensor are zero at a point, they are zero at this point in every coordinate system.

(g) If A^i and B_i are the components of any contravariant and covariant vector respectively, $A^i B_i$ is a scalar. Conversely, if $A^i B_i$ is a scalar and either A^i or B_i are the components of an arbitrary vector, the other set are also components of a vector.

Remarks:

1. It is clear that all tensor fields which enter in the previously defined operations are to be considered at the same point $(x^1 \ldots x^N)$.

2. At Level 2 (intrinsic approach), the operation (a) will lead to the tensor $A \pm B$ provided that A and B are tensors of the same type and order and (b) will lead to the tensor product $A \otimes B$ for any A and B.

Hint: Apply the variance laws summarized by the formula (4.3) of Level 1.

| Problem 1.3. | *Symmetrization and antisymmetrization of tensors** |

(a) Show that any tensor of second order can be decomposed into a symmetric part ($A_{ij}^{(s)} = A_{ji}^{(s)}$) and a skew symmetric part ($A_{ij}^{(as)} = -A_{ji}^{(as)}$).

(b) Show that a symmetric tensor of second order has $N(N+1)/2$ components whereas a skew symmetric tensor has $N(N-1)/2$, their sum being N^2, the total number of components of a tensor of second order. If $N = 4$, a symmetric tensor has 10 components and a skew symmetric tensor 6 components only; for $N = 3$ these numbers are respectively 6 and 3: this last tensor is called a **pseudovector**.

(c) The symmetrization and antisymmetrization processes can be generalised for any tensor: for instance, the tensor $A_{ij}\ldots_{(lmn)}\ldots$ is symmetric with respect to its indices l, m, n if it does not change under any permutation of l, m, n; the tensor $A_{ij}\ldots_{(lmn)}\ldots$ is skew symmetric with respect to l, m, n if it keeps its sign or changes it when the permutation of l, m, n is respectively even or odd.

Completely symmetric tensor $A_{(ijkl)}$ is then obtained from A_{ijk} as follows:

$$A_{(ijk)} = \frac{1}{3!}(A_{ijk} + A_{jki} + A_{kij} + A_{jik} + A_{ikj} + A_{kji})$$

and the completely skew symmetric tensor $A_{[ijk]}$ takes the form

$$A_{[ijk]} = \frac{1}{3!}(A_{ijk} + A_{jki} + A_{kij} - A_{jik} - A_{ikj} - A_{kji}).$$

(d) Consider a symmetric tensor in all its indices covariant or contravariant, of order s, defined in an N-dimensional manifold. We therefore have N objects which are to be grouped into sets of s elements (indices). If $s > N$, some of the indices (among the s objects) should repeat themselves as indices of the components of the tensor. Then one has to consider repeated combinations of N objects classified in sets of $s(> N)$ elements: two sets are to be considered distinct, if they differ either by one object or by the number of times that the same object occurs. For instance, if $N = 3$, the

---

*See also Problem 1.7.

components T_{11133} and T_{12133} are distinct, and also T_{11133} and T_{11333}. The number of such sets,

$$B_s^N = \binom{s+N-1}{N} = \frac{(s+N-1)!}{s!(N-1)!} ,$$

gives the number of components of the symmetric tensor in all its indices of order s.

(e) Let A_i, B_i, C_i be the components of the three covariant vectors A, B, C and show that

$$T_{ijk} = \begin{vmatrix} A_i & B_i & C_i \\ A_j & B_j & C_j \\ A_k & B_k & C_k \end{vmatrix}$$

is a skew symmetric tensor $T_{[ijk]}$.

Hint: (a) $A_{ij} = \frac{1}{2}(A_{ij}+A_{ji})+\frac{1}{2}(A_{ij}-A_{ji}) = A_{ij}^{(s)}+A_{ij}^{(as)} = A_{(ij)}+A_{[ij]}$.

| Problem 1.4. | *Irreducible tensor fields*

Characterize a class c of tensor fields by a common given property: any tensor field of c which after a point transformation still belongs to c is called **irreducible**. An irreducible scalar tensor field is called an **invariant**. Examples follow:

(a) Consider the class of mixed tensor fields $A_{j_1 j_2 \cdots}^{i_1 i_2 \cdots}$ such that the tensor obtained by contraction of two of the indices among the i's and j's vanishes: such tensors are irreducible.

(b) Show that a symmetric tensor field $A^{(s)}{}_{ij}$ and a skew symmetric one $A^{[as]}{}_{ij}$ are irreducible.

(c) Show that the scalar $\operatorname{Tr} A = A^i{}_i$ is invariant and that the tensor field

$$\overline{A}{}^i{}_j = A^i{}_j - \frac{1}{N}\delta_j^i \operatorname{Tr} A$$

is irreducible with respect to the property $\operatorname{Tr} A = 0$.

(d) It should be noticed that for a Riemannian manifold, the rank of an index (first, second,... in the array of all indices) is significant (see (8.13),

Level 1). Thus, $A^i{}_i$ is to be replaced by the unambiguous A^i_i or A^i_i which are different scalars. Consider a symmetric field $A_{ij} = A_{ji}$ and show that $A^i{}_j = A_j{}^i$ and the concept of Tr A does not carry any ambiguity. For an antisymmetric field $F_{ij} = -F_{ji}$, show that $F^i{}_j = -F_j{}^i$.

The notion of irreducibility is particularly important in the theory of elementary particles since an irreducible tensor wave function represents particles of the same spin and not a mixture of particles with different spins. For instance, an irreducible wave function is associated with mesons of spin 1 and not a mixture of mesons of spins 1 and 0. The same remark applies to the graviton.

Hints: (a) Consider a tensor field $A^i{}_{jk}$ for instance: the contraction of two of its indices i and k leads to another tensor field $A^i{}_{ji}$. If $A^i{}_{ji} = 0$ in one frame, it vanishes in all frames.

(b) After a point transformation, the symmetric tensor field $A^{(s)}{}_{ij}$ is transformed into

$$\frac{\partial x^r}{\partial x'^i} \frac{\partial x^q}{\partial x'^j} A^{(s)}_{rq} ,$$

while $A^{(s)}{}_{ji}$ is transformed into

$$\frac{\partial x^r}{\partial x'j} \frac{\partial x^q}{\partial x'^l} A^{(s)}_{rq} ,$$

the two expressions being equal because of the symmetry of $A^{(s)}{}_{ij}$. The same kind of proof applies to $A^{(as)}{}_{ij}$.

$\boxed{\textit{Problem 1.5.}}$ *Diagonalization of the metric*

Let $\mathbf{G}(x)$ be a matrix with elements $g_{\mu\nu}(x)$ and $g(x) = \det\mathbf{G}(x)$. In matrix notation dx is a column matrix with dx^i as elements; the transpose being denoted by the index T, then dx^T is a line matrix with elements dx^i. The ds^2 of a Riemannian manifold can be put into the form

$$ds^2 = g_{\mu\nu}dx^\mu dx^\nu = dx^T \mathbf{G} dx : \mu, \nu = 1 \ldots N .$$

Since $\mathbf{G}(x)$ is a symmetric real matrix, there exists a unitary matrix (same dimension as $\mathbf{G}$!) such that

$$S^{-1}(x)\mathbf{G}(x)S(x) = X(x) \ ,$$

where $X(x)$ is a symmetric real matrix of elements $X_k(x)$.

(a) Show that

$$g(x) = \prod_1^N X_k(x) \ .$$

(b) Consider the N linear forms*

$$d\xi = S^{-1}(x)dx : d\xi^j = S^{-1}{}_k^j(x)dx^k : j, k = 1, \ldots N \ ,$$

then

$$ds^2 = d\xi^T X(x)d\xi \ .$$

Perform a change of scale on each of the $\delta\xi^j$ such that

$$ds^2 = \sum_1^N \varepsilon_i(\delta X^i)^2 : \varepsilon_i = \pm 1 \ .$$

It should be clearly said that the transformation $dx = S(x)d\xi$ is generally not a point transformation. Special integrability conditions on S_j^j are required for it to be so.

Hints: (a) One has

$$g(x) = \det\{S^{-1}\mathbf{G}S\} = \prod_1^N X^k(x) \ .$$

(b) One has

$$ds^2 = dx^T\mathbf{G}dx = d\xi^T S^T\mathbf{G}Sd\xi = d\xi^T X d\xi \ ,$$

since the unitary of S implies $S^{-1} = S^T$.

---

*We recall that the language used at the present Level 1 is that of elementary calculus.

We perform the change of scale

$$d\xi^i = \frac{1}{\sqrt{|X_i|}} dX^i$$

and obtain

$$ds^2 = d\xi^k \delta_{ik} X_i d\xi^k = \sum_1^N \frac{X_i}{|X_i|} (d\xi^i)^2 \ .$$

| Problem 1.6. | *Embedding of a manifold in an Euclidean space*

(a) Consider an Euclidean affine space $\mathcal{E}_N$ with a ds^2:

$$ds^2 = \sum_{j=1}^N \varepsilon_j (dX^j)^2 : \ \varepsilon_j = \pm 1 \tag{1}$$

and suppose each of the X^j to be a function of an n-uplet $(x^1 \ldots x^n)$:

$$X^j = X^j(x^1 \ldots x^n) : \ j = 1 \ldots N \ , \tag{2}$$

then the preceding ds^2 can be written as

$$ds^2 = g_{kl} dx^k dx^l \ . \tag{3}$$

Show that

$$g_{kl} = \sum_1^N \varepsilon_j \frac{\partial X^j}{\partial x^k} \frac{\partial X^j}{\partial x^l} \ . \tag{4}$$

(b) If $n < N$ and if the x^k may be eliminated among the N equations X^j, one is led to a certain number of relations involving the X^j only. For instance if $n = N - 1$, the elimination of the x^k (when possible) will lead to a single equation $F(X^1 \ldots X^N) = 0$. One says then that the manifold of $N - 1$ dimensions $(F = 0)$ is embedded in the $\mathcal{E}_N$ space and this has been used in Sec. 3, Level 0, to embed S^2 and H_2 in $\mathcal{E}_3$. If $n = N - 2$ we are generally led to two equations $F(X^1 \ldots X^N) = 0, G(X^1 \ldots X^N) = 0$ and if $n = 1$, one says that we embedded a curve in $\mathcal{E}_3$. No conclusion can be drawn if $n > N$.

(c) Conversely, we may look at the inverse problem: let the ds^2 be given in the form (3), is it possible to find a transformation (2) such that the ds^2 can be brought into the form (1), i.e., can we embed the manifold $\mathcal{M}_n$ into the affine space $\mathcal{E}_N$? Clearly, we have to satisfy (4) where the $g_{kl}(x)$ are known functions and the $X^j(x^1 \ldots x^n)$ the unknown functions for the system of $n(n+1)/2$ partial differential equations, $n(n+1)/2$ being the number of elements of an n by n symmetric matrix.

The problem is complex and we may point out two remarkable cases:

(1) $N = n(n+1)/2$ corresponds to the case where the number of the unknown functions X^j is equal to the number of equations in (4). Then the system may be resolved: one sees, for instance, that for $n = 2, N = 3$ (Sec. 3 of Level 0); for $n = 3, N = 6$ (see Ex. 1.1); for $n = 4$ one has $N = 10$, which is the case corresponding to general relativity.

(2) If $N < n(n+1)/2$, the number of unknown functions is smaller than the number of equations, and the system (4) is overdetermined and has no solution in general.

(c) Embedding of a sphere S^2 is an affine space $\mathcal{E}_3$ defined in an orthonormal reference system.

Hints: (a) From (2) one has

$$dX^j = (\partial_k X^j(x))dx^k \ ,$$

and the ds^2 becomes

$$ds^2 = \sum^N \varepsilon_j \left(\frac{\partial X^j}{\partial x^k} dx^k \right)^2 = \left(\sum^N \varepsilon_j \frac{\partial X^j}{\partial x^k} \frac{\partial X^j}{\partial x^l} \right) dx^k dx^l$$

$$= g_{kl} dx^k dx^l \ .$$

(b) The ds^2 of $\mathcal{E}_3$ is $ds^2 = dx^2 + dy + dz^2$; the equation of the sphere S^2 is

$$F(x, y, z) - a^2 = x^2 + y^2 + z^2 - a^2 = 0 \ .$$

For any point of $S^2 : xdx + ydy + zdz = 0$, then

$$ds^2 = dx^2 + dy^2 + \frac{(xdx + ydy)^2}{z^2}$$

$$= \left(1 + \frac{x^2}{z^2}\right) dx^2 + \left(1 + \frac{y^2}{z^2}\right) dy^2 + \frac{2xy}{z^2} dxdy \ .$$

| Problem 1.7. | *Numerical tensors and tensor densities*

(a) Show that the Kronecker deltas $\delta^i{}_j$ are the components of an invariant tensor.

(b) The permutation $\binom{i_1 \cdots i_N}{1 \cdots N}$ of N objects $1 \ldots N$, is said to be even or odd if one requires an even or odd number of transposition to go over from $(i_1 \ldots i_N)$ to $(1 \ldots N)$. The symbols $\varepsilon^{rs}, \varepsilon^{ijk}, \varepsilon_{\alpha\beta\gamma\delta}$ take the value ± 1 if

$$\begin{pmatrix} i & j & k \\ 1 & 2 & 3 \end{pmatrix}, \quad \begin{pmatrix} \alpha & \beta & \gamma & \rho \\ 0 & 1 & 2 & 3 \end{pmatrix},$$

are even or odd permutations, they take the value 0 if any two of their indices are equal. The respective range of variation of the indices $r, s; i, j, k; \lambda, \mu, \nu, \rho$ is then $1, 2; 1, 2, 3$ and $0, 1, 2, 3$. Show that

$$\varepsilon^{rs} a_r b_s = \begin{vmatrix} a_1 & b_1 \\ a_2 & b_2 \end{vmatrix} = \det(\mathbf{ab}) \ ,$$

$$\varepsilon^{ijk} a_i b_j c_k = \begin{vmatrix} a_1 & b_1 & c_1 \\ a_2 & b_2 & c_2 \\ a_3 & b_3 & c_3 \end{vmatrix} = \det(\mathbf{a}, \mathbf{b}, \mathbf{c}) \ ,$$

$$\varepsilon^{\lambda\mu\nu\rho} a_\lambda b_\mu c_\nu d_\rho = \begin{vmatrix} a_0 & b_0 & c_0 & d_0 \\ a_1 & b_1 & c_1 & d_1 \\ a_2 & b_2 & c_2 & d_2 \\ a_3 & b_3 & c_3 & d_3 \end{vmatrix} = \det(a, b, c, d) \ .$$

(c) Define a contravariant tensor density of weight 1 (Definition 4.5 of Level 1), the components of which are represented in a certain frame by the symbols $\varepsilon^{\alpha\beta\gamma\rho}$: it is generally called the Levi-Civita 4-tensor (see also question (f)). Show that after a point transformation one has $\varepsilon'^{\alpha\beta\gamma\rho} = \varepsilon^{\alpha\beta\gamma\rho}$, i.e. $\varepsilon^{\alpha\beta\gamma\rho}$ are the contravariant components of an invariant tensor density of weight 1.

(d) For Riemannian space, we consider the following set of components (see Problem 2.3):

$$-\frac{1}{g}g_{\alpha\lambda}g_{\beta\mu}g_{\gamma\nu}g_{\rho\sigma}\varepsilon^{\lambda\mu\nu\sigma}$$

and call them $\varepsilon_{\alpha\beta\gamma\rho}$. Show that $\varepsilon_{\alpha\beta\gamma\rho} = -1$ (or $+1$) if $\left(\begin{smallmatrix}\alpha & \beta & \gamma & \rho \\ 0 & 1 & 2 & 3\end{smallmatrix}\right)$ is even (or odd) and 0 if two or more of the indices $\alpha\beta\gamma\rho$ are equal. Considering a covariant invariant tensor density of weight -1 which is identical to $\varepsilon_{\alpha\beta\gamma\rho}$ in some frame, show that it is invariant. Such a tensor density is denoted by the same symbol $\varepsilon_{\alpha\beta\gamma\rho}$.

(e) How many independent components have a completely antisymmetric tensor of fourth order defined in a 4-dimensional manifold?

(f) Apply similar considerations and definitions to ε^{rs} and ε^{ijk} as those given in (b).

(g) Show that

$$\eta^{\alpha\beta\gamma\rho} = \frac{1}{\sqrt{\sqrt{|g|}}}\varepsilon^{\alpha\beta\gamma\rho} \ , \quad \eta_{\alpha\beta\gamma\rho} = \sqrt{|g|}\varepsilon_{\alpha\beta\gamma\rho}$$

are respectively contravariant and covariant tensors and that

$$\begin{aligned}
{}^*\phi^{\alpha\beta\gamma\rho} &= \varepsilon^{\alpha\beta\gamma\rho\varphi} \ , & {}^*T^{\alpha\beta\gamma} &= \varepsilon^{\alpha\beta\gamma\rho}X_\rho \ , \\
{}^*T^{\alpha\beta} &= \frac{1}{2!}\varepsilon^{\alpha\beta\gamma\rho}T_{\gamma\rho} \ , & {}^*X^\alpha &= \frac{1}{3!}\varepsilon^{\alpha\beta\gamma\rho}T_{\beta\gamma\rho} \ , \\
& & \varphi^* &= \frac{1}{4!}\varepsilon^{\lambda\mu\nu\rho}T_{\lambda\mu\nu\rho} \ ,
\end{aligned}$$

are tensor densities.[*]

(h) The Kronecker tensors δ^i_j can be generalized: we shall consider the

---

[*]See also Problem 2.3.

following antisymmetric symbols:

$$\varepsilon^{j_1 j_2}_{i_1 i_2} = \begin{vmatrix} \delta^{j_1}_{i_1} & \delta^{j_1}_{i_2} \\ \delta^{j_2}_{i_1} & \delta^{j_2}_{i_2} \end{vmatrix}$$

$$\varepsilon^{j_1 j_2 j_3}_{i_1 i_2 i_3} = \begin{vmatrix} \delta^{j_1}_{i_1} & \delta^{j_1}_{i_2} & \delta^{j_1}_{i_3} \\ \delta^{j_2}_{i_1} & \delta^{j_2}_{i_2} & \delta^{j_2}_{i_3} \\ \delta^{j_3}_{i_1} & \delta^{j_3}_{i_2} & \delta^{j_3}_{i_3} \end{vmatrix} \quad ,$$

and a similar expression for $\varepsilon^{j_1 \ldots j_p}_{i_1 \ldots i_p} (p \le N)$, the upper indices labelling the lines, the lower ones the columns. Show, using elementary properties of the determinants, that one can also give an equivalent definition

$$\varepsilon^{j_1 \ldots j_p}_{i_1 \ldots i_p} = \begin{cases} 0 & \text{if } j_1 \ldots j_p \text{ is not a permutation of } i_1 \ldots i_p \\ 1 & \text{if } j_1 \ldots j_p \text{ is an even permutation of } i_1 \ldots i_p \\ -1 & \text{if } j_1 \ldots j_p \text{ is an odd permutation of } i_1 \ldots i_p \end{cases}$$

and conclude that they are components of a mixed tensor. Show that T being any p-covariant tensor

$$T_{[i_1 \ldots i_p]} = \frac{1}{p!} \varepsilon^{j_1 \ldots j_p}_{i_1 \ldots i_p} T_{j_1 \ldots j_p} \quad ,$$

are components of an antisymmetric tensor.

Take the special case of 4-dimensional manifold with points M labelled by (x^0, x^1, x^2, x^3) to show that

$$\varepsilon^{\alpha\beta\gamma\rho}_{\lambda\mu\nu\rho} = \varepsilon^{\alpha\beta\gamma}_{\lambda\mu\nu} \quad , \qquad \varepsilon^{\alpha\beta\rho}_{\lambda\mu\rho} = 2\varepsilon^{\lambda\mu} \quad ,$$
$$\varepsilon^{\alpha\rho}_{\lambda\rho} = 3\delta^{\alpha}_{\lambda} \quad , \qquad \delta^{\rho}_{\rho} = 4 \quad .$$

(i) Some of the previous results can be stated in an intrinsic form: let S_p be the group of the permutations of the p first integers $1 \ldots p$ and consider any of its elements π which transform

$$(1 \ldots p) \to (\pi(1) \ldots \pi(p)) \quad .$$

Let T be any convariant p-tensor, then by definition

$$(\pi T)_{i_1 \ldots i_p} = T_{\pi(i_1) \ldots \pi(i_p)} \quad .$$

One says that T is a symmetric tensor if for any π

$$(\pi T)_{i_1 \ldots i_p} = T_{i_1 \ldots i_p} \ .$$

It is antisymmetric if

$$(\pi T)_{i_1 \ldots i_p} = (\text{sign } \pi) T_{i_1 \ldots i_p} \ ,$$

where sign $\pi = \pm 1$ according to the parity of the permutation π (even or odd): these concepts have already been introduced in question (b). Show then that

$$ST = \frac{1}{p!} \sum_\pi \pi T \ ,$$

$$A_s T = \frac{1}{p!} \sum_\pi (\text{sign} \pi) \pi T \ ,$$

are completely symmetric or antisymmetric tensor fields.

(j) Consider an $\mathbb{R}^2$ vector space with metric $ds^2 = (dx^1)^2 + (dx^2)^2 + (dx^3)^2$ and show that

$$\varepsilon_{ijk} \varepsilon_{lm}{}^k = \delta_{il} \delta_{jm} - \delta_{im} \delta_{jl} \ ,$$

$$\varepsilon_{ijk} \varepsilon_l{}^{jk} = 2\delta_{il} \ , \qquad \varepsilon_{ijk} \varepsilon^{ijk} = 6 \ .$$

Hints: (a) If δ^i_j are the components of a tensor in some frame, the variance law is

$$\delta''{}_j = \frac{\partial x'^i}{\partial x^k} \frac{\partial x^l}{\partial x'^j} \delta^l_k = \frac{\partial x'^i}{\partial x^k} \frac{\partial x^k}{\partial x'^j} = \delta^i_j \ .$$

(b) If $\varepsilon^{\alpha\beta\gamma\rho}$ are the components of a tensor density of weight l, its variance is by definition

$$\varepsilon'^{\alpha\beta\gamma\rho} = \frac{\partial(x)}{\partial(x')} \frac{\partial x'^\alpha}{\partial x^\lambda} \frac{\partial x'^\beta}{\partial x^\mu} \frac{\partial x'^\gamma}{\partial x^\nu} \frac{\partial x'^\rho}{\partial x^\sigma} \varepsilon^{\lambda\mu\nu\sigma} \ .$$

Denoting by $\partial x'^\alpha$ the quadruplet of functions,

$$\partial x'^\alpha = \left\{ \frac{\partial x'^\alpha}{\partial x^0} \ , \quad \frac{\partial x'^\alpha}{\partial x^1} \ , \quad \frac{\partial x'^\alpha}{\partial x^2} \ , \quad \frac{\partial x'^\alpha}{\partial x^3} \right\} \ ,$$

one has a consequence of (b):

$$\varepsilon'^{\alpha\beta\gamma\rho} = \frac{\partial(x)}{\partial(x')}\det(\partial x'^{\alpha}, \partial x'^{\beta}, \partial x'^{\gamma}, \partial x'^{\rho})$$

$$= \frac{\partial(x)}{\partial(x')}\varepsilon^{\alpha\beta\gamma\rho} \ .$$

The last equality requires some considerations which are left to the reader. One obtains finally the invariance of $\varepsilon^{\alpha\beta\gamma\rho}$.

(c) From the definition of $\varepsilon_{\alpha\beta\gamma\rho}$, *one has*

$$\varepsilon_{\alpha\beta\gamma\rho} = -\frac{1}{g}\det(g_{\alpha}g_{\beta}, g_{\gamma}, g_{\rho}) \ ,$$

where g_{α} *denotes the quadruplet of functions* $(g_{\alpha_0}, g_{\alpha_1}, g_{\alpha_2}, g_{\alpha_3})$. *It requires some considerations to show that the determinant is* g *if* $\left(\begin{smallmatrix} \alpha & \beta & \gamma & \rho \\ 0 & 1 & 2 & 3 \end{smallmatrix}\right)$ *is even,* $-g$ *if it is odd, and 0 if two or more of the indices are equal.*

The law of transformation of the density $\varepsilon_{\alpha\beta\gamma\rho}$ *is*

$$\varepsilon'_{\alpha\beta\gamma\rho} = \left(\frac{\partial(x)}{\partial(x')}\right)^{-1}\frac{\partial(x)^{\lambda}}{\partial x'^{\alpha}}\frac{\partial(x)^{\mu}}{\partial x'^{\beta}}\frac{\partial(x)^{\nu}}{\partial x'^{\gamma}}\frac{\partial(x)^{\sigma}}{\partial x'^{\rho}}\varepsilon_{\lambda\mu\nu\sigma}$$

$$= \frac{\partial(x')}{\partial(x)}\frac{\partial(x)}{\partial(x')}\varepsilon_{\alpha\beta\gamma\rho} = \varepsilon_{\alpha\beta\gamma\rho} \ .$$

The last equality requires some consideration.

(d) We denote the coordinates of the points of $\mathcal{M}_4$ *by* (x^0, x^1, x^2, x^3); *then only those contravariant components* $A^{\alpha\beta\gamma\rho}$ *of the antisymmetric tensor A which are different from 0 are those that correspond to all permutations of the four members 0, 1, 2, 3.* $A^{\alpha\beta\gamma\rho}$ *is then equal either to* $+A^{0123}$ *or* $-A^{0123}$ *depending upon whether* $\left(\begin{smallmatrix} \alpha & \beta & \gamma & \rho \\ 0 & 1 & 2 & 3 \end{smallmatrix}\right)$ *is even or odd. The tensor A has only one independent component. It is also called a pseudo-scalar and in* $\mathcal{M}_4$ *there are no skew-symmetric tensors of order higher than 4.*

(e) See formula (8.14), Level 1.

(f) One verifies easily the following formula:

$$\varepsilon_{ijl}\varepsilon^{lmn} = \begin{vmatrix} \delta_{il} & \delta_{im} & \delta_{in} \\ \delta_{jl} & \delta_{jm} & \delta_{jn} \\ \delta_{kl} & \delta_{km} & \delta_{kn} \end{vmatrix} \ .$$

Contract then any needed number of indices. Notice that since the space is Euclidean the raising or lowering of an index does not change the value of a tensor.

$\boxed{\textit{Problem 1.8.}}$ *Volume and surface elements in* $\mathcal{M}_4$

(a) Show that the antisymmetric numerical tensors defined in the previous problem can also be written as

$$\varepsilon^{\alpha\beta}_{\mu\nu} = \begin{vmatrix} \delta^\alpha_\mu & \delta^\beta_\mu \\ \delta^\alpha_\nu & \delta^\beta_\nu \end{vmatrix} \;, \quad \varepsilon^{\alpha\beta\gamma}_{\mu\nu\rho} = \begin{vmatrix} \delta^\alpha_\mu & \delta^\beta_\mu & \delta^\gamma_\mu \\ \delta^\alpha_\nu & \delta^\beta_\nu & \delta^\gamma_\nu \\ \delta^\alpha_\rho & \delta^\beta_\rho & \delta^\gamma_\rho \end{vmatrix} \;,$$

$$\varepsilon^{\alpha\beta\gamma\rho}_{\lambda\mu\nu\sigma} = \begin{vmatrix} \delta^\alpha_\lambda & \delta^\alpha_\mu & \delta^\alpha_\nu & \delta^\alpha_\sigma \\ \delta^\beta_\lambda & \delta^\beta_\mu & \delta^\beta_\nu & \delta^\beta_\sigma \\ \delta^\gamma_\lambda & \delta^\gamma_\mu & \delta^\gamma_\nu & \delta^\gamma_\sigma \\ \delta^\rho_\lambda & \delta^\rho_\mu & \delta^\rho_\nu & \delta^\rho_\sigma \end{vmatrix} \;;$$

show furthermore that

$$\varepsilon^{\alpha\beta\gamma\tau}_{\lambda\mu\nu\tau} = \varepsilon^{\alpha\beta\gamma}_{\lambda\mu\nu} \;, \quad \varepsilon^{\alpha\beta\tau}_{\mu\nu\tau} = 2\varepsilon^{\alpha\beta}_{\mu\nu} \;, \quad \varepsilon^{\alpha\tau}_{\mu\tau} = 3\delta^\alpha_\mu \;.$$

(b) Define the 2-dimensional elements of a surface in $\mathcal{M}_4$ spanned by two infinitesimal vectors dx and δx as

$$dS^{\alpha\beta} = \varepsilon^{\alpha\beta}_{\mu\nu} dx^\mu \delta x^\nu$$

and show that one has

$$dS^{\alpha\beta} = \begin{vmatrix} dx^\alpha & \delta x^\alpha \\ dx^\beta & \delta x^\beta \end{vmatrix}$$

and that one obtains the components of the exterior product

$$dS = dx \wedge \delta x \;.$$

Define the 3-dimensional elements of a volume in $\mathcal{M}_4$ spanned by three infinitesimal vectors $dx, \delta x, \Delta x$ as

$$dV^{\alpha\beta\gamma} = \begin{vmatrix} dx^\alpha & \delta x^\alpha & \Delta x^\alpha \\ dx^\beta & \delta x^\beta & \Delta x^\beta \\ dx^\gamma & \delta x^\gamma & \Delta x^\gamma \end{vmatrix}$$

and show that they are the components of the exterior product

$$dV = dx \wedge \delta x \wedge \Delta x \ .$$

(c) Show that

$$^*dS^{\alpha\beta} dS_{\alpha\beta} = 0 \ .$$

Hint: For the definitions of $dS^{\alpha\beta}$ and $^*dS_{\alpha\beta}$ see question (g) of the previous exercise. One then has

$$^*dS^{\alpha\beta} dS_{\alpha\beta} = \varepsilon^{\alpha\beta}_{\mu\nu} dx^\mu dx^\nu \frac{1}{2} \varepsilon_{\alpha\beta\rho\sigma} dx^\rho dx^\sigma$$

$$= \frac{1}{2} \varepsilon_{\alpha\beta\rho\sigma} (\delta^\alpha_\mu \delta^\beta_\nu - \delta^\beta_\mu \delta^\alpha_\nu) dx^\mu dx^\nu dx^\rho dx^\sigma \ .$$

Each of the terms of this expansion is a determinant with 0 as its value since it has two identical lines.

| Problem 1.9. | *Differential operators on manifolds $\mathcal{M}_N$*

$\mathbf{G}(x) = $ matrix of elements g_{ij} and $g(x) = \det \mathbf{G}(x)$.

(a) Show that $dg = g g^{ik} dg_{ik}$ and that one also has

$$\frac{\partial g}{\partial g_{ik}} = g g^{ik} \ , \qquad \frac{\partial g}{\partial g^{ik}} = -g g_{ik} \ ,$$

$$g^{ik} \partial_l g_{ik} = \frac{1}{g} \partial_l g = 2 \partial_l \{ \log |g|^{1/2} \} \ .$$

(b) Define the covariant components of the gradient of a function $\varphi(x^1 \ldots x^N)$ as

$$(\text{grad}_x \varphi)_i = \nabla_i \varphi(x) = \partial_i \varphi(x) : \ \partial_i = \frac{\partial}{\partial x^i} \ .$$

(c) Define the divergence of a vector field $X(x)$ by

$$\text{div}_x X = \nabla_i X^i = \partial_i X^i + \Gamma^i_{\ ik} X^k \ ,$$

and show that

$$\Gamma^i_{\ ik} = \partial_k \{ \log |g|^{1/2} \}$$

and

$$\text{div}_x X = \frac{1}{|g|^{1/2}} \partial_i \{|g|^{1/2} X^i\} \ .$$

What is the volume integral of $\text{div}_x X$?

(d) Define the covariant components of the curl of a covariant vector field $X(x)$ as

$$\text{curl}_x X = \nabla_i X_j - \nabla_j X_i$$

and show that

$$(\text{curl}_x X)_{ij} = \partial_i X_j - \partial_j X_i \ .$$

(e) For a function $\varphi(x^1 \dots x^N)$ one defines the two Beltrami parameters and the generalized d'Alambertian:

$$\Delta_1 \varphi = g^{ij} \nabla_i \varphi \nabla_j \varphi = g^{ij} \partial_i \varphi \partial_j \varphi \ ,$$
$$\Delta_2 \varphi = \text{div. grad.} \varphi = \nabla_i \{g^{ij} \nabla_j \varphi\} = \Box \varphi \ .$$

Show that one may write

$$\Delta_2 \varphi = g^{ij} \nabla_i \nabla_j \varphi = g^{ij} (\partial_i \partial_j \varphi - \Gamma^k{}_{ij} \partial_k \varphi) \ ,$$
$$\Box \varphi = \Delta_2 \varphi = \frac{1}{|g|^{1/2}} \partial_i \{|g|^{1/2} g^{ij} \partial_j \varphi\} \ .$$

(f) Suppose that $\mathcal{M}_N$ is $\mathcal{E}_3$ with an orthonormalized metric

$$ds^2 = g_{ij} dx^i dx^j \ : \ g_{ij} = \delta_{ij} \ .$$

Show that the former definitions become the well-known expressions of grad., div., curl and $\Delta \varphi$.

(g) Suppose that the system of coordinates in $\mathcal{E}_3$ is an orthogonal one (not necessarily normalized!):

$$ds^2 = g_{ij} du^i du^j \ ,$$

where

$$g_{ij} = U_i^2 \delta_{ij} \ , \quad g^{ij} = \frac{1}{U_i^2} \delta_{ij} \ .$$

Denoting $\partial_i = \partial/\partial u_i$, show that

$$(\mathrm{grad}_u\varphi)_i = \partial_i\varphi , \quad (\mathrm{curl}\,\mathbf{X})_{ij} = \partial_i X_j - \partial_j X_i ,$$

$$\mathrm{div}_u\mathbf{X} = \frac{1}{U_1 U_2 U_3}\partial_i\{U_1 U_2 U_3 X^i\} ,$$

$$\Delta_1\varphi = \sum U_1^2(\partial_i\varphi)^2 ,$$

$$\Delta_2\varphi = \frac{1}{U_1 U_2 U_3}\sum \partial_i\left\{U_1 U_2 U_3\frac{1}{U_i^2}\partial_i\varphi\right\} .$$

(h) With respect to any orthogonal frame, neither contravariant components nor covariant ones are the "ordinary" components of a vector, i.e., those that are used currently (in physics for instance!) and are obtained by the orthogonal projections of the given vector on the coordinate axis of an orthornomal frame.

If a_1, a_2, a_3 are the ordinary components of the vector $\mathbf{A}$ then $\|\mathbf{A}\|^2 = a_1^2 + a_2^2 + a_3^2$, the frame being orthogonal. If one writes

$$\|\mathbf{A}\|^2 = \sum^3 U_i^2(A^i)^2 ,$$

then

$$a_i = U_i A^i : \ A^i = \frac{a_i}{U_i} .$$

On the other hand

$$\|\mathbf{A}\|^2 = A_i A^i = \sum^3 A_i\frac{a_i}{U_i} = \sum^3 a_i^2 ,$$

$$\frac{A_i}{U_i} = a_i : \ A_i = U_i a_i .$$

Show then that the ordinary covariant components of $\mathrm{grad}_u\varphi$, $\mathrm{curl}_u\mathbf{A}$ are respectively

$$\frac{1}{U_i}\partial_i\varphi , \quad \frac{1}{U_i U_j}(\partial_i\{U_j a_j\} - \partial_j\{U_i a_i\}) ,$$

and that

$$\mathrm{div}_u\mathbf{A} = \frac{1}{U_1 U_2 U_3}\sum^3 \partial_i\left\{U_1 U_2 U_3\frac{a_i}{U_i}\right\} ,$$

$$\Delta_1\varphi = \sum^3 U_i^2(\partial_i\varphi)^2 ,$$

$$\Box\varphi = \Delta_2\varphi = \frac{1}{U_1 U_2 U_3}\sum^3 \partial_i\left\{U_1 U_2 U_3\frac{1}{U_i^2}\partial_i\varphi\right\} .$$

Hints: (a) Following the rule giving the derivative of a determinant, dg is obtained by the contracted product dg_{ik} by the minor G^{ik} corresponding to the element g_{ik}. But by definition $g^{ik} = G^{ik}/g$, thus

$$dg = G^{ik} dg_{ik} = g g^{ik} dg_{ik} ,$$

which is equivalent to the first formula of question (a) if one takes into account that

$$dg = \frac{\partial g}{\partial g_{ik}} dg_{ik} .$$

The second formula of question (a) is obtained by the use of $d\{g^{ij} g_{jk}\} = 0$. We now consider the third one: from $dg = \partial_l g \, dx^l$, one gets

$$\frac{dg}{g} = \frac{1}{g} \partial_l g \, dx^l : \quad g^{ik} dg_{ik} = \frac{1}{g} \partial_l g \, dx^l$$

$$\Rightarrow g^{ik} \partial_l g_{ik} = \frac{1}{g} \partial_l g = 2 \partial_l |g|^{1/2} .$$

(c) By the definition of $div_x X$ one has

$$div_x X = \partial_i X^i + \partial_i \{\log |g|^{1/2}\} X^i = \partial i X^i + \frac{1}{2|g|} (\partial_i |g|) X^i .$$

The last equation of question (c) reads also

$$div_x X = \frac{1}{|g|^{1/2}} [(\partial_i |g|^{1/2}) X^i + |g|^{1/2} \partial_i X^i] ,$$

with a result identical to the previous one.

A volume integral of $div_x X$ can be expressed as

$$\int_V div_x X \, dV = \int_V |g|^{-1/2} \partial_i \{|g|^{1/2} X^i\} |g|^{1/2} dx^1 \dots dx^N$$

$$= \int_V \partial_i \{|g|^{1/2} X^i\} dx^1 \dots dx^N ,$$

which can be transformed into a flux integral through ∂V.

The answers to the questions (g) and (h) can be found in any textbook of theoretical or mathematical physics.

Problem 1.10. *Geodesic and harmonic coordinate systems*

The solution of several important problems is greatly simplified by the use of a special coordinate system. We now present two such systems.

(a) **Geodesic coordinate system:** since Γ^i_{jk} are not tensor fields, their value 0 in a given coordinate system does not imply that they vanish in all frames. We therefore choose a point ξ of our manifold and show that locally at ξ the Γ's can be chosen to be 0.

Consider formula (5.1) from Level 1,

$$\Gamma'^s_{ki}(x')\frac{\partial x^l}{\partial x'^s} = \frac{\partial^2 x^l}{\partial x'^i \partial x'^k} + \frac{\partial x^p}{\partial x'^i}\frac{\partial x^q}{\partial x'^k}\Gamma^l_{pq}(x) \;, \tag{I}$$

and choose now a point transformation $x \rightarrow x'$:

$$x'^l = x^l - \xi^l + \frac{1}{2}A^l_{mn}(x^m - \xi^m)(x^n - \xi^n) \;, \tag{II}$$

where the coefficient A^k_{lm} are constants. We notice also that the right-hand side of (2) vanishes for $x = \xi$.

Show, using (II), that

$$\frac{\partial x^i}{\partial x'^k}\Big|_{x=\xi} = \delta^i_k \;, \qquad \frac{\partial^2 x^i}{\partial x'^l \partial x'^k}\Big|_{x=\xi} = -A^i_{lk} \;,$$

and insert these results in (I) to obtain for the limit $x = \xi$

$$A^l_{ik} = \Gamma^l_{ik}(\xi) \;.$$

We have thus proved that given a point $\xi \in \mathcal{M}_N$, we can always determine a frame in which the Γ-symbols vanish.

Show that at the previous chosen point $\varepsilon \in \mathcal{M}_N$, all tensors are invariant and the geodesic lines equations are reduced to

$$\frac{d^2 x^i}{ds^2}(\xi) = 0 \;.$$

They are thus locally straight lines. Such a frame is also known under the name of locally geodesic frame. It plays an important part in general relativity since the gravitational field that acts on a point can be eliminated by a convenient choice of a new frame which is precisely the geodesic frame. The above considerations constitute the equivalence principle of general relativity.

Show finally that the metric tensor $\mathbf{G}(x)$ is stationary at the point $\xi \in \mathcal{M}_N$.

Determine the point transformation (II) for a sphere S_2 in spherical coordinates.

(b) **The harmonic frame:** for a Riemannian manifold $\mathcal{M}_N$, define the N-uplet of fields

$$\Gamma^i(x) = g^{jk}(x)\Gamma^i{}_{jk}(x) \; ,$$

which are not the components of a contravariant field (except in the case where only linear coordinate transformations are to be considered).

Show that

$$\Gamma^i = g^{kl}\Gamma^i{}_{kl}$$
$$= \frac{1}{2}g^{kl}g^{im}(\partial_k g_{ml} + \partial_l g_{mk} - \partial_m g_{kl})$$

and

$$\Gamma^i = -|g|^{-1/2}\partial_k\{|g|^{1/2}g^{ki}\} \; .$$

Then, following considerations similar to the ones developed in (a), show that there exists a point transformation which defines a system of coordinates where all the Γ^i are zero: such a system constitutes a harmonic frame.

The term "harmonic frame" can be justified by showing that for any function $\varphi(x^1 \ldots x^N)$ one has in such a frame

$$\Delta_2\varphi = g^{ij}\partial_i\partial_j\varphi \; ,$$

where the Beltrami's coefficient Δ_2 has been defined in the preceding exercise.

Hints: (a) As shown by formula (9.4) of Level 1,

$$\nabla_l g_{ik} = 0 = \partial_l g_{ik} - \Gamma_{kil} - \Gamma_{ikl} \; ,$$

which means that in a geodesic frame $\partial_l g_{ik} = 0$.

(b) Use formula (9.7) Level 1 and then the second formula of Part (a) of the preceding exercise. Take finally the derivative of $g_{is}g^{sk} = \delta_i^k$ to obtain

$$g^{is}\partial_l g_{sk} = -(\partial_l g^{is})g_{sk}$$

and also the second formula of (b).

Transform the definition of Δ_2 into the following:

$$\Delta_2\varphi = g^{ij}\partial_i\partial_j\varphi - \Gamma^i\partial_i\varphi ,$$

from which one obtains the last formula of (b).

$\boxed{Problem\ 1.11.}$ *The Bianchi identities*

(a) Show that for a Riemannian manifold, one has

$$\nabla_m R^i{}_{jkl} + \nabla_l R^i{}_{jmk} + \nabla_k R^i{}_{jlm} = 0 , \tag{1}$$

$$\nabla_i \left\{ R^i{}_k - \frac{1}{2}g^i_k R \right\} = 0 , \tag{2}$$

$$\nabla_i R^i_k = \frac{1}{2}\partial_k R . \tag{3}$$

(b) Show that the tensor $S_{ij} = R_{ij} - (1/4)g_{ij}R$ is traceless.

Hints: Define at the point $x \in \mathcal{M}_N$ a geodesic system of coordinates and apply to the tensor field R^i_{jk} the covariant derivative rule as given just after the formulae (5.20) of Level 1. Since all $\Gamma : .(x) = 0$, one gets

$$\nabla_m R^i{}_{jkl} = \partial_m R^i{}_{jkl} + \partial_m\partial_k\Gamma^i{}_{jl} - \partial_m\partial_l\Gamma^i{}_{jk} ;$$

permute then the indices k, l to obtain formula (1). Contract the indices i, k and l and use the property $\nabla_i g_{ik} = 0$ and the symmetry properties of $R : .$ to show that

$$\nabla_m R_{jl} - \nabla_l R_{jm} + \nabla_i R^i{}_{jlm} = 0 .$$

Contract also j and l to get (notice $\nabla_k g_{ij} = 0$)

$$\nabla_m g^{lj} R_{jl} - \nabla_l g^{lj} R_{jm} + \nabla_i g^{lj} R^i{}_{jlm} = \nabla_m R - 2\nabla_i R^i{}_m = 0 \, ,$$

which can be written as (2). Formula (3) is an immediate consequence of (2).

| Problem 1.12. | *Commutator of covariant derivative*

Prove formulae (10.4a), (10.4b), (10.4c) of Level 1.

Solution:

Consider formula (10.4b) for instance:

$$[\nabla_k, \nabla_l] A_i(x) = -R^m{}_{ikl}(x) A_m(x) \, ,$$

and put $\nabla_l A_i = T_{li}$, then by (5.20) (Level 1),

$$\nabla_k T_{li} = \partial_k T_{li} - \Gamma^m{}_{lk} T_{ml} - \Gamma^m{}_{ik} T_{lm} \, .$$

Using formula (5.5) (Level 1), one obtains

$$\begin{aligned}
\nabla_k T_{li} = \nabla_k \nabla_l A_i &= \partial_k \{\partial_l A_i - \Gamma^m{}_{li} A_m) - \Gamma^m{}_{lk} \\
&\times (\partial_m A_i - \Gamma^n{}_{mi} A_n) - \Gamma^m{}_{ik}(\partial_l A_m - \Gamma^n{}_{lm} A_n) \\
&= \partial_k \partial_l A_i - (\partial_k \Gamma^m{}_{il}) A_m - \Gamma^m{}_{li} \partial_k A_m - \Gamma^m{}_{lk} \partial_m A_i \\
&+ \Gamma^m{}_{lk} \Gamma^n{}_{mi} A_n - \Gamma^m{}_{ik} \partial_l A_m + \Gamma^m{}_{ik} \Gamma^n{}_{lm} A_n \, ,
\end{aligned}$$

where only the second and the last term are not symmetric or do not enter the summation in a symmetric way with respect to k and l since in a Riemannian manifold $\Gamma^i{}_{jk} = \Gamma^i{}_{kj}$. Then only these terms and their symmetric will remain in the expression of $[\nabla_k, \nabla_l] A_i$. Apply finally the expression of $R^n{}_{ikl}$ given by formula (7.4) to obtain the required result. There is another way for the proof of (10.4b): one lowers the index i in (10.3) and use $\nabla_k g_{ij} = 0$. (For the proof, see question (a) of Problem 1.16.)

The proof of (10.3c) runs along the same lines: we put as before $T_{lij} = \nabla_l A_{ij}$ and notice that in the expression of $[\nabla_k, \nabla_l]A_{ij}$ only the following (not symmetric) terms remain:

$$[\nabla_k, \nabla_l]A_{ij} = - (\partial_k \Gamma^n{}_{jl} - \partial_l \Gamma^n{}_{jk} + \Gamma^m{}_{jl}\Gamma^n{}_{mk} - \Gamma^m{}_{jk}\Gamma^n{}_{ml})A_{in}$$

$$- (\partial_l \Gamma^n{}_{ik} - \partial_k \Gamma^n{}_{il} + \Gamma^m{}_{ik}\Gamma^n{}_{ml} - \Gamma^m{}_{il}\Gamma^n{}_{mk})A_{nj} \, ,$$

a formula which leads to the required result.

| Problem 1.13. | *Parallel displacement of a vector field*

(a) Consider the manifold defined in the affine vector space $\mathcal{E}_3$ in spherical coordinates:

$$x = r\sin\theta\cos\varphi \, , \quad y = r\sin\theta\sin\varphi \, , \quad z = r\cos\theta \, ,$$
$$ds^2 = dx^2 + dy^2 + dz^2 = dr^2 + r^2 d\theta^2 + r^2 \sin^2\theta d\varphi^2 \, .$$

To each point $M \in \mathcal{E}_3$ we associate the triplet $(x^1 = r, x^2 = \theta, x^3 = \varphi)$. Show that the only Christoffel symbols of such a manifold that are different from 0 are the following:

$$\Gamma^1{}_{22} = -r \, , \qquad \Gamma^1{}_{33} = r\sin^2\theta \, , \quad \Gamma^2{}_{12} = r \, ,$$
$$\Gamma^2{}_{33} = -\sin\theta\cos\theta \, , \quad \Gamma^3{}_{13} = \frac{1}{r} \, , \qquad \Gamma^3{}_{23} = \cotg\theta \, .$$

(b) Let S^2 be the sphere of radius 1, consider its equator as a curve of $\mathcal{E}_3$ of equations $x^1 = 1, x^2 = \pi/2, x^3 = \varphi$ and any of its meridians $x^1 = 1, x^2 = \theta, x^3 = \alpha(\text{cst})$.

Show that a contravariant vector field of components $(X^1(x^1, x^2, x^3), X^2(x^1, x^2, x^3), X^3(x^1, x^2, x^3))$, where $x^1 = r, x^2 = \theta, x^3 = \varphi$, when parallel-displaced along the equator satisfies the equations

$$\frac{dX^1}{d\varphi} - X^3 = 0 \, , \quad \frac{dX^2}{d\varphi} = 0 \, , \quad \frac{dX^3}{d\varphi} + X^1 = 0 \, , \tag{I}$$

and the same field parallel-displaced along any meridian satisfies the equations

$$\frac{dX^1}{d\theta} - X^2 = 0 \ , \quad \frac{dX^2}{d\theta} + X^1 = 0 \ , \quad \frac{dX^3}{d\theta} = 0 \ . \tag{II}$$

Show that one has the two corresponding systems of solutions

$$X^1 = A_1 \cos\varphi + B_1 \sin\varphi \ , \quad X^2 = k = \text{cst.} \ , X^3 = -A_1 \sin\varphi + B_1 \cos\varphi \ , \tag{I'}$$

$$X^1 = A_2 \cos\theta + B_2 \sin\theta \ , \quad X^2 = -A_2 \sin\theta + B_2 \cos\theta \ , \quad X^3 = k' = \text{cst.} \tag{II'}$$

(c) In order to display the significance of the two systems of differential equations (I) and (II), consider a meridian Γ characterized by its azimuthal angle φ_0 and two of its points $M_0(1, \theta_0, \varphi_0)$ and $M(1, \theta, \varphi_0)$, and let $(X^1(M_0), X^2(M_0), X^3(M_0))$ be an arbitrary vector given at M_0. Show that a parallel displacement of this vector from M_0 to M along Γ amounts to a rotation $\theta - \theta_0$ around the Oz axis and give the corresponding rotation matrix.

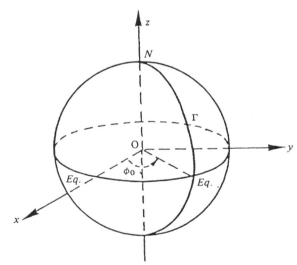

(d) Solve the same problem for the equator $E_q : (1, \pi/2, \varphi)$ and show that a similar displacement amounts to a rotation of angle $\varphi - \varphi_0$ around the Oy axis.

(e) Consider now a spherical triangle ABN built up by two meridians and the equator: starting from the vertex $A(1, \pi/2, 0)$ with an arbitrary vector $(X^1(A), X^2(A), X^3(A))$, show that its parallel displacement along AB, where $B = (1, \pi/2, 0)$, followed by displacement along BN, where $N = (1, 0, 0)$, and finally back to A amounts to a rotation around the Oy axis.

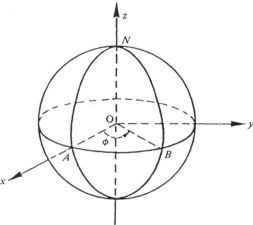

Hints: (a) See formula (3.28) Level 0.

(b) The equator is a curve on S^2 of parametric equations $x^1 = r = 1, x^2 = \theta = \pi/2, x^3 = \varphi$ while any meridian is of equations $x^1 = 1, x^2 = \theta, x^3 = \phi = cst$. A straightforward application of Eq. (6.8b) of Level 1 gives the differential systems (I) and (II).

(c) The initial conditions $X^1(1, \theta_0, \varphi_0) = A_2 \cos \theta_0 + B_2 \sin \theta_0$, $X^2(1, \theta_0, \varphi_0)$ $= -A_2 \sin \theta_0 + B_2 \cos \theta_0, X^3(1, \theta_0, \varphi_0) = k'$ can be solved to give $A_2 = X^1(M_0) \cos \theta_0 - X^2(M_0) \sin \theta_0$, $B_2 = X^1(M_0) \sin \theta_0 + X^2(M_0) \cos \theta_0$ and finally

$$X^1(M) = X^1(M_0) \cos(\theta - \theta_0) + X^2(M_0) \sin(\theta - \theta_0) ,$$
$$X^2(M) = -X^1(M_0) \sin(\theta - \theta_0) + X^2(M_0) \cos(\theta - \theta_0) ,$$

which can be written in matrix form as

$$\begin{pmatrix} X^1(M) \\ X^2(M) \\ X^3(M) \end{pmatrix} = \begin{pmatrix} \cos(\theta - \theta_0) & \sin(\theta - \theta_0) & 0 \\ -\sin(\theta - \theta_0) & \cos(\theta - \theta_0) & 0 \\ 0 & 0 & 1 \end{pmatrix} , \begin{pmatrix} X^1(M_0) \\ X^2(M_0) \\ X^3(M_0) \end{pmatrix} ,$$

a matrix which represents a rotation of angle $\theta - \theta_0$ around Oz.

Similar calculations apply to questions (d) and (e).

Problem 1.14. *Weyl tensor — Number of independent components of $R_{\alpha\beta\gamma\delta}, R_{\alpha\beta}, S_{\alpha\beta}, C_{\alpha\beta\gamma\delta}$*

Starting from the Riemannian tensor $R_{\alpha\beta\gamma\delta}$, we have constructed a traceless tensor $S_{\alpha\beta}$ and a scalar R: both of these tensor fields are irreducible (see Ex. 1.4). We investigate now the properties of a third irreducible tensor field $C_{\alpha\beta\gamma\delta}$, called the **Weyl tensor** or **conformal tensor**. We shall be working in $\mathcal{M}_4$ and count the number of independent components of these tensors.

(a) Define $C_{\alpha\beta\gamma\delta}$ as follows:

$$C_{\alpha\beta\gamma\delta} = R_{\alpha\beta\gamma\delta} + \frac{1}{2}(g_{\alpha\delta}R_{\beta\gamma} - g_{\alpha\gamma}R_{\beta\delta} + g_{\beta\gamma}R_{\alpha\delta}$$
$$- g_{\beta\delta}R_{\alpha\gamma}) + \frac{1}{6}(g_{\alpha\gamma}g_{\beta\delta} - g_{\alpha\delta}g_{\beta\gamma})R ;$$

show that

$$C_{\alpha\beta\gamma\delta} = R_{\alpha\beta\gamma\delta} + \frac{1}{2}(g_{\alpha\delta}S_{\beta\gamma} - g_{\alpha\gamma}S_{\beta\delta} + g_{\beta\gamma}S_{\alpha\delta}$$
$$- g_{\beta\delta}S_{\alpha\gamma}) + \frac{1}{12}(g_{\alpha\delta}g_{\beta\gamma} - g_{\alpha\gamma}g_{\beta\delta})R ,$$

where $S_{\alpha\beta} = R_{\alpha\beta} - (1/4)g_{\alpha\beta}R$ (Ex. 1.4) is traceless. Show also that $C_{\alpha\beta\gamma\delta}$ has the same symmetry properties as $R_{\alpha\beta\gamma\delta}$ and that

$$C^{\alpha}{}_{\beta\alpha\delta} = C^{\beta}{}_{\alpha\gamma\beta} = 0 ,$$

a statement which proves that $C_{\alpha\beta\gamma\delta}$ is an irreducible tensor too.

(b) Count the number of independent components of the following tensor fields $R_{\alpha\beta\gamma\delta}, R_{\alpha\beta}, S_{\alpha\beta}$ and $C_{\alpha\beta\gamma\delta}$ defined on a Riemannian $\mathcal{M}_4$ manifold.

(c) Follow the same method as in (b) to show that in a 3-dimensional manifold $\mathcal{M}_3$ the Riemannian and Ricci tensors have 6 independent components and that the Weyl tensor vanishes.

Hint: (a) Simple algebra transformations.
(b) As noticed in Sec. 7, Level 1 (in fine), the tensor $R^\alpha_{\beta\gamma\delta}$ has 96 independent components. This number will be reduced for a $R_{\alpha\beta\gamma\delta}$ defined in a Riemannian space since it has additional symmetries (see Sec. 10, Level 1 just after formula (10.5)). Because of its antisymmetry with respect to its first two indices (or to its last two indices) the first and the last couple of indices can take only 6 values: 01, 02, 03, 12, 13, 23. Because at its symmetry with respect to two couples of indices (α, β) and (γ, β), $R_{\alpha\beta\gamma\delta}$ behaves like a symmetric 6×6 matrix. It has therefore 15 off-diagonal terms and 6-diagonal, in all 21 elements. But there is another relation given by formula (10.4) of Level 1, namely,

$$R_{0123} + R_{0231} + R_{0312} = 0 \; ,$$

which limits its number of independent components to 20. $R_{\alpha\beta}$ is a 4×4 symmetric matrix; it has therefore 10 independent components. $S_{\alpha\beta}$ is also a 4 by 4 symmetric and trace-free matrix, has only 9 independent components. We now come to $C_{\alpha\beta\gamma\delta}$; since it has the same symmetries as $R_{\alpha\beta\gamma\delta}$, it has at most 20 independent components. But it satisfies the trace-free conditions. The first kind of conditions are of the type $C^\rho_{\beta\rho\delta} = 0, \beta$ and δ varying from 0 to 3, giving at most 16 conditions. But there is a second kind of condition: from $C_{\alpha\beta\gamma\delta} = C_{\gamma\delta\alpha\beta}$, we also have $C^\rho_{\beta\rho\delta} = 0$. The 4×4 matrix $C^\rho_{\beta\rho\delta}$ is then symmetric in β and δ: the 16 conditions reduce to 10 and $C_{\alpha\beta\gamma\delta}$ has finally 10 independent components.

Compare now the number of independent components of $R_{\alpha\beta\gamma\delta}$ (20 in number) with those of $C_{\alpha\beta\gamma\delta}$ (10), $S_{\alpha\beta}$ (9), and R (1 independent component). The equality thus exhibited leads to the notation

$$R_{\alpha\beta\gamma\delta} = C_{\alpha\beta\gamma\delta} \oplus S_{\alpha\beta} \oplus R \; ,$$

which reminds one of the notation of direct sum in the theory of vector spaces.

All the formulae given above can be generalized for $\mathcal{M}_N$, in which case the coefficients 1/2 and 1/6 in the definition of $C_{\alpha\beta\gamma\delta}$ are to be replaced by $1/(N-2)$ and $1/(N-1)(N-2)$ respectively.

Problem 1.15. *Conformal mapping and conformal (Weyl) tensor*

Let A and B be two vector fields in a Riemannian space, consider the bilinear form (written below for $\mathcal{M}_4$ although the considerations remain valid for any dimension)

$$(A, B) = \frac{A_\mu B^\mu}{\sqrt{|A_\mu A^\mu||B_\mu B^\mu|}} \ .$$

(a) Show that if the metric is positive, then

$$|(A, B)| \leq 1 \ ,$$

and it may be taken as a definition of $\cos \alpha$, α being the "angle" between the vectors A and B.

(b) In a given system of coordinates, consider two distinct Riemannian spaces with metric tensors $\mathbf{G}(x)$ and $\overline{\mathbf{G}}(x)$ such that

$$\overline{g}_{\mu\nu}(x) = e^{2\sigma} g_{\mu\nu}(x) \ ,$$

where $\sigma(x)$ is a real function of the point $(x^1 \ldots x^N)$: such spaces are called **conformal spaces**. Show that if (A, B) and (A', B') are two couples of vectors belonging respectively to each of the spaces, then

$$(A', B') = (A, B) \ .$$

It is clear that the conformal transformations build up a group and that, in an Euclidean space, they leave invariant the angle of two vectors A and B. We want to show that the Weyl tensor is one of the invariants of this group.

(c) Show that under the conformal transformation

$$\overline{g}^{\alpha\beta} = e^{-2\sigma} g^{\alpha\beta}(x) \ ,$$

one has

$$\overline{\Gamma}_{\alpha\beta\gamma} = e^{2\sigma}(\Gamma_{\alpha\beta\gamma} + g_{\alpha\beta}\partial_\gamma\sigma + g_{\alpha\gamma}\partial_\beta\sigma - g_{\beta\gamma}\partial_\alpha\sigma) \ .$$

(d) Show that

$$\begin{aligned}
\overline{R}_{\alpha\beta\gamma\delta} = e^{2\sigma}[R_{\alpha\beta\gamma\delta} &+ (g_{\alpha\delta}\sigma_{\beta\gamma} + g_{\beta\gamma}\sigma_{\alpha\delta} - g_{\alpha\gamma}\sigma_{\beta\delta} \\
&- g_{\beta\delta}\sigma_{\alpha\gamma} + (g_{\alpha\delta}g_{\beta\gamma} - g_{\alpha\gamma}g_{\beta\delta})\Delta_1\sigma] \ ,
\end{aligned}$$

where $\Delta_1 \sigma$ is the first Beltrami's parameter (Ex. 1.9)

$$\Delta_\mu \sigma = g^{\beta\nu}(\nabla_\mu \sigma)(\Delta_\nu \sigma) = g^{\mu\nu}(\partial_\mu \sigma)(\partial_\nu \sigma)$$

and

$$\sigma_{\alpha\beta} = \nabla_\alpha \nabla_\beta \sigma - (\nabla_\alpha \sigma)(\nabla_\beta \sigma) : \sigma_{\alpha\beta} = \sigma_{\beta\alpha} .$$

(e) Derive the expression of the Ricci tensor to show that

$$\overline{R}_{\alpha\beta} = \overline{g}^{\rho\sigma}\overline{R}_{\rho\alpha\sigma\beta} = R_{\alpha\beta} + 2\sigma_{\alpha\beta} - (\Delta_2\sigma + 2\Delta_1\sigma)g_{\alpha\beta} ,$$

where Δ_2 is the second Beltrami's parameter or generalized d'Alembertian:

$$\Delta_2 \sigma = \Box\sigma = g^{\mu\nu}\nabla_\mu\nabla_\nu\sigma .$$

(f) Derive the expression of the scalar curvature to show

$$\overline{R} = \overline{g}^{\alpha\beta}\overline{R}_{\alpha\beta} = e^{-2\sigma}(R + 6\Delta_2\sigma + 6\Delta_1\sigma) .$$

(g) Show that the Weyl tensor is invariant under any conformal transformation.

Hint: (a) We use the notations $\langle A, B\rangle = A_\mu B^\mu, \|A\|^2 = \langle A, A\rangle$ *and as the metric is positive definite then for any* $\lambda \in \mathbb{R}$

$$0 \le \|A + \lambda B\|^2 = \|A\|^2 + 2\lambda\langle A, B\rangle + \lambda^2\|B\|^2 ,$$

which implies $\langle A, B\rangle^2 - \|A\|^2\|B\|^2 \le 0$, *the required inequality.*
(c), (d), (e) and (f) require somewhat lengthy calculations but with little difficulty.
(g) We form

$$\overline{g}_{\alpha\beta}\overline{R} = g_{\alpha\beta}(R + 6\Delta_2\sigma + 6\Delta_1\sigma)$$

and eliminate $\Delta_2\sigma$ *between this expression and* $\overline{R}_{\alpha\beta}$ *to get*

$$\sigma_{\alpha\beta} = \frac{1}{2}(\overline{R}_{\alpha\beta} - R_{\alpha\beta}) - \frac{1}{12}(\overline{g}_{\alpha\beta}\overline{R} - g_{\alpha\beta}R) - \frac{1}{2}g_{\alpha\beta}\Delta_1\sigma .$$

We then bring into $\overline{R}_{\alpha\beta\gamma\delta}$ *the expression of* $\sigma_{\mu\nu}$ *and arrive at* $\overline{C}_{\alpha\beta\gamma\delta} = C_{\alpha\beta\gamma\delta}$. *Since the calculations are nevertheless rather involved, it is better to proceed as follows: raise the index* α *in* $R_{\alpha\beta\gamma\delta}$ *and note that* $g^\mu_\nu = \delta^\mu_\nu$. *One gets then*

$$\overline{R}^\alpha_{\ \beta\gamma\delta} = R^\alpha_{\ \beta\gamma\delta} - \delta^\alpha_\delta \sigma_{\beta\gamma} - \delta^\alpha_\gamma \sigma_{\beta\delta} + g^{\alpha\rho}(g_{\beta\gamma}\sigma_{\rho\delta}$$
$$- g_{\beta\delta}\sigma_{\rho\alpha}) + (\delta^\alpha_\delta g_{\beta\gamma} - \delta^\alpha_\gamma g_{\beta\delta})\Delta_1\sigma .$$

Substitute $\sigma_{\mu\nu}$ *as given before to arrive at* $\overline{C}^\alpha_{\beta\gamma\delta} = C^\alpha_{\beta\gamma\delta}$. *The results can be extended to* $\mathcal{M}_N$, *see [Bibl. 5], p. 90.*

Problem 1.16.   *Symmetric spaces**

(a) Show that the formula (10.3) of Level 1,

$$[\nabla_k, \nabla_l]A^i(x) = R^i{}_{nkl}A^n(x) \,,$$

takes for a covariant field vector the following form:

$$[\nabla_k, \nabla_l]A_i(x) = -R^n{}_{ikl}A_n(x) \,. \tag{I}$$

(b) Show that the Killing equation for a given infinitesimal transformation

$$\nabla_i\xi_j + \nabla_j\xi_i = 0 \tag{II}$$

can be written as

$$\nabla_i\nabla_j\xi_k = -R^n{}_{ijk}\xi_n \,. \tag{II'}$$

(c) Show that by choosing for a point $X = (X^1 \ldots X^N)$ the values of the $N(N-1)/2$ derivatives of the type $\partial_i\xi_j$ and the N components $(\xi_1 \ldots \xi_N)$ of any Killing vector, one can define an analytic solution on the real axis of the Killing equation depending on $N(N+1)/2$ arbitrary constants. However, $N(N+1)/2$ represents the maximum number of constants which can be arbitrarily chosen since it may happen that the structure of the metric tensor is such that this number is less than $N(N+1)/2$, for instance one may not be able to choose $N(N-1)/2$ derivatives $\partial_i\xi_j$ at the given point y.

(d) Show that the Killing equations admit at most $N(N+1)/2$ linearly independent solutions which correspond to $N(N+1)/2$ Killing vectors $\xi^{(n)}(x)$.

(e) A Riemannian manifold is said to be a **homogeneous** space if there exist infinitesimal isometries of the type (11.1) of Level 1 that carry any given point X into any other point of the neighborhood of X. A space is said to be **isotropic** at a given point X if there exist infinitesimal isometries that leave the point X invariant and for which $\nabla_l\xi_p(x)|_{x=X}$ take arbitrary values and are antisymmetric in the indices l and p (Killing equations). Show that for any homogeneous Riemannian space $\mathcal{M}_N$ there exists a set of N linearly independent Killing vectors; if $\mathcal{M}_N$ is also isotropic at X then

---

*Since this problem is more difficult than the preceding ones, we give a detailed solution. See also Problem 1.1.

the total number of the corresponding Killing vectors is $N(N+1)/2$. Such a space is called a **maximally symmetric space**.

(c) Show that for a totally symmetric space $\mathcal{M}_N$, the Riemann tensor takes the following form

$$R_{plij} = \frac{R}{N(N-1)}(g_{pi}g_{lj} - g_{pj}g_{li}) \,,$$

where $R(x)$ is the scalar curvature. Furthermore, R is constant if $N > 2$.[*]

The case $N = 2$ has been dealt with at some length at Level 0, Sec. 3.

Solution:

(a) One can write

$$[\nabla_k, \nabla_l]g^{im}A_m = R^i{}_{nkl}g^{nm}A_m \,,$$

and since $\nabla_k g_{il} = 0$, one has also

$$g^{im}[\nabla_k, \nabla_l]A_m = R^i{}_{nkl}g^{nm}A_m \,.$$

Multiplying both sides by g_{pi}, one obtains the required result.

(b) Applying formula (1) successively to ξ_i, ξ_k, ξ_j one has

$$\begin{aligned}
\nabla_k\nabla_j\xi_i - \nabla_j\nabla_k\xi_i &= -R^n{}_{ijk}\xi_n \,, \\
\nabla_j\nabla_i\xi_k - \nabla_i\nabla_j\xi_k &= -R^n{}_{kij}\xi_n \,, \qquad\qquad \text{(III)} \\
\nabla_i\nabla_k\xi_j - \nabla_k\nabla_i\xi_j &= -R^n{}_{jki}\xi_n \,.
\end{aligned}$$

Adding and using formula (10.4), $R^n{}_{ijk} + R^n{}_{kij} + R^n{}_{jki} = 0$, one obtains

$$0 = \nabla_k\{\nabla_j\xi_i - \nabla_i\xi_j\} + \nabla_j\{\nabla_i\xi_k - \nabla_k\xi_i\} + \nabla_i\{\nabla_k\xi_j - \nabla_j\xi_k) \,.$$

Taking, furthermore, into account the Killing equation, the preceding relation can be brought into the form

$$\nabla_k\nabla_j\xi_i - \nabla_j\nabla_k\xi_i = \nabla_i\nabla_j\xi_k \,,$$

which when substituted in (III) gives the expected result.

---

[*]The study of maximally symmetric spaces is an important one, since the metric of such spaces can be defined as a solution of Einstein's equations. For more details on the subject, see [Bibl. 7], Chap. 13. The generalization of the constant K to any coordinate system (K becomes then a function of x) can also be given, see [Bibl. 5], p. 79.

(c) Consider the Killing Eqs. (II) as a set of relations between ordinary partial derivatives. Using the technics of Ex. 1.12, one sees that $\nabla_i \nabla_j \xi_k(x)$ is of the form

$$\nabla_i \nabla_j \xi_k(x) = \partial_i \partial_j \xi_k(x) + f_{ijk}^m(\mathbf{G}(x))\xi_m(x) + h_{ijk}^{ml}(\mathbf{G}(x))\partial_l \xi_m(x) ,$$

where f and h are functions of the matrix elements $g_{pq}(x)$ (symbolized by $\mathbf{G}(x)$). Putting that expression in (II'), one gets an ordinary partial derivative equation of second order:

$$\partial_i \partial_j \xi_k(x) = F_{ijk}^m(\mathbf{G}(x))\xi_m(x) + H_{ijk}^{ml}(\mathbf{G}(x))\partial_l \xi_m(x) .$$

By taking successive derivatives of that relation, one gets all the higher derivatives of $\xi_k(x)$ as a linear combination of $\xi_m(x)$ and $\partial_l \xi_m(x)$. Let X be any given point of $\mathcal{M}_N$ around which $\xi_i(x)$ can be expanded in a Taylor series on the real axis. $\xi_i(x)$ can then be expressed as

$$\xi_i(x) = \sum_p (x^{s_1} - X^{s_1})\ldots(x^{s_p} - X^{s_p})F_{s_1\ldots s_p}^m(\mathbf{G})\xi_m(x)$$

$$+ \sum_p (x^{s_1} - X^{s_1})\ldots(x^{s_p} - X^{s_p})H_{s_1\ldots s_p}^{ml}(\mathbf{G})\frac{\partial \xi_m(X)}{\partial X^l} .$$

We note that we may replace $\partial_1 \xi_m(X)$ by $\nabla_l \xi_m(X)$ if we modify the first term of the expansion of $\xi_i(x)$. Collecting, finally, terms together, we may write the expression of $\xi_i(x)$ in the neighbourhood of the point X as follows:

$$\xi_i(x, X) = \varphi_i^m(x, X)\xi_m(X) + \psi_i^{nl}(x, X)\nabla_l \xi_n(X) . \tag{IV}$$

Consider now the Killing equation at the same fixed point X:

$$\frac{\partial \xi_j(X)}{\partial X^i} + \frac{\partial \xi_i(X)}{\partial X^j} = -2\Gamma^k{}_{ij}(X)\xi_k(X) . \tag{V}$$

This is a set of linear equations (with $\partial_i \xi_j(X)$ as unknowns) symmetric in i and j; their number is then $N(N+1)/2$ for a N-dimensional manifold. But the unknowns are not only the N^2 terms $\partial_i \xi_j(X)$ but also the N terms $\xi_k(X)$, in all $N(N+1)$ unknown quantities in a set of $N(N+1)/2$ equations. Then $N(N+1)/2$ of these known quantities can be arbitrarily chosen and their values fixed at the point $X \in \mathcal{M}_N$.

This remark shows that in the expression of $\xi_i(x)$ given above the sum on m runs from 1 to N and the sum on n runs from 1 to $(N(N+1)/2) - N = N(N-1)/2$: then any Killing vector $\xi(x)$ depends effectively at most on $N(N+1)/2$ arbitrary constants.

The proof given here depends on the analyticity of the functions which were considered: this is too strong an assumption but it can be released by using other special technics.

(d) Put $M = N(N+1)/2$ and call $C^1 \ldots C^M$ the N quantities $\xi_m(X)$ and the $N(N-1)/2$ quantities $\partial_i \xi_j(X)$. Since they should be arbitrarily chosen, they are linearly independent and may be considered as defining a basis of an M-dimensional vector space. Consider then the set of Killing vectors as introduced in (c):

$$\xi^p(x, X) = \varphi^{(p)m}(x, X)\xi_m(X) + \psi^{(p)nl}(x, X)\nabla_l \xi_n(X) . \qquad (VI)$$

This is a formula that can be abbreviated into the form

$$\xi^{(p)}(x, X) = C^q f_q(x, X) .$$

$\xi^p(x, X)$ is therefore a vector of the vector space with the basis $C^1 \ldots C^M$ and one cannot define more than M linearly independent ξ^p. The adverb "more" in the last sentence is justified by the last remark in (c).

(e) Consider a homogeneous manifold at X. From its very definition it should admit in X any arbitrarily chosen Killing vectors $\xi^{(n)}(X)$; we show now that n runs from 1 to N. Indeed, it suffices to choose the component of $\xi^{(n)}(X)$ equal to δ_i^n. Such a set of N vectors is clearly linearly independent and, using the method of question (c), one can build up N linear independent Killing vectors. Consider now an isotropic space $\mathcal{M}_N$ at the point X: by its very definition, there should exist a null Killing vector $\xi(X) = 0$ (which leaves X invariant). On the other hand, consider Eq. (IV). Since the space is both homogeneous and isotropic, the $\xi_m(X)$ and the $(\nabla_l \xi_m(x))_{x=X}$ can take any arbitrary value, say a_m and A_{nl} (which depend on X which is otherwise fixed). We then write linear combinations

$$\xi_{(a,A)}(x, X) = \varphi^m(a, X)a_m + \psi_m^{nl}(x, X)A_{nl}$$

and it turns out that $\xi_{(a,A)}(x, X)$ may take at $x = X$ any value we choose: they are, thus, linear independent at X. But then a simple consideration shows that one can build up linearly independent combinations of $\xi(a, A)$

for any point x in a neighbourhood of X (use the same method as the well-known one for linear differential equations). But since we have proved that the maximum number of Killing vectors is $N(N + 1)/2$, we may conclude that a homogeneous and isotropic Riemannian space admits $N(N + 1)/2$ Killing vectors.

(f) Apply formula (10.3c) to the tensor field $\nabla_i \xi_j(x)$,

$$[\nabla_k, \nabla_l]\nabla_i \xi_j = R^n{}_{jkl}\nabla_i \xi_n - R^n{}_{ilk}\nabla_n \xi_j \ , \tag{VII}$$

and formula (II) of this problem,

$$\nabla_i \nabla_j \xi_k = -R^n{}_{ijk}\xi_n \ , \tag{VIII}$$

and consider a maximally symmetric space (see below) such that n runs from 1 to $N(N + 1)/2$ while the other indices run from 1 to N. The right-hand side of (VIII), taking (VII) into account, can be written as

$$\nabla_k \nabla_l \nabla_i \xi_j - \nabla_l \nabla_k \nabla_i \xi_j = -\nabla_k\{R^n{}_{lij}\xi_n\} + \nabla_l\{R^n{}_{kij}\xi_n\} \ .$$

Developing the derivatives and collecting the terms, one gets

$$(-\nabla_k R^n{}_{lij} + \nabla_l R^n{}_{kij})\xi_{in} = R^n{}_{lij}\nabla_k \xi_n - R^n{}_{kij}\nabla_l \xi_n$$
$$+ R^n{}_{jkl}\nabla_l \xi_n - R^n{}_{ilk}\nabla_n \xi_j \ .$$

Because of Killing equations we may replace the factor $\nabla_n \xi_j$ in the last term by $-\nabla_j \xi_n$. Then the above relation takes the form

$$(-\nabla_k R^n{}_{lij} + \nabla_l R^n{}_{kij})\xi_n = (R^n{}_{lij}\delta^p_k - R^n{}_{kij}\delta^p_l + R^n{}_{jkl}\delta^p_i + R^n{}_{ilk}\delta^p_j)\nabla_p \xi_n \ . \tag{IX}$$

At this point we take into account the property of $\mathcal{M}_N$ of being totally symmetric at x and choose for $\xi(x)$ the value 0. On the other hand, $\nabla_p \xi_n$ is antisymmetric, then only the antisymmetric part of the bracket in $p(= 1 \ldots N)$ and $n(= 1 \ldots N(N + 1)/2)$ contribute and one has

$$R^n{}_{lij}\delta^p_k - R^n{}_{kij}\delta^p_l + R^n{}_{jkl}\delta^p_i + R^n{}_{ilk}\delta^p_j$$
$$= R^p{}_{lij}\delta^n_k - R^p{}_{kij}\delta^n_l + R^p{}_{jkl}\delta^n_i + R^p{}_{ilk}\delta^n_j \ ,$$

since the left-hand side of (IX) vanishes.

Contract then the indices p and k to get

$$N R_{ilj}^n - R^n{}_{lij} - R^n{}_{jil} + R^n{}_{ilj} = R^n{}_{lij} + R_{jl}\delta_i^n - R_{il}\delta_j^m$$

and, because of the cyclicity, $R^n{}_{lji} + R^n{}_{jil} + R^i{}_{ilj} = 0$, one finally has

$$(N-1)R^n{}_{lij} = R_{jl}\delta_n^i - R_{il}\delta_j^n \ .$$

Multiplying then both sides by g_{pn}, one gets

$$(N-1)R_{plij} = g_{pi}R_{jl} - g_{pj}R_{il} \ . \tag{X}$$

Now since R_{plij} is antisymmetric in p and l, one may obtain a further relation for the right-hand side:

$$g_{pi}R_{jl} - g_{pj}R_{il} = -g_{li}R_{jp} + g_{lj}R_{ip} \ .$$

Contract further p and i to obtain

$$N R_{jl} - R_{jl} = -g_{li}R_j^i + g_{lj}R \ ,$$

i.e.

$$N R_{jl} = g_{lj}R \ . \tag{XI}$$

Finally Eq. (X) takes the form

$$R_{plij} = \frac{R}{N(N-1)}(g_{pi}g_{lj} - g_{pj}g_{li}) \ . \tag{XII}$$

We may then notice that for a totally symmetric $\mathcal{M}_N$ with $N > 2$, R is independent of x. Indeed the Bianchi identity (3) of Ex. 1.11,

$$\nabla_i\left\{R_k^i - \frac{1}{2}g_k^i R\right\} = 0 \ ,$$

can be transformed into

$$0 = \nabla_i\left\{g^i{}_k\frac{R}{N} - \frac{1}{2}g^i{}_k R\right\} = \left(\frac{1}{N} - \frac{1}{2}\right)g_k^i\nabla_i R$$
$$= \left(\frac{1}{N} - \frac{1}{2}\right)\partial_k R \ ,$$

taking Eq. (X) into account. This result shows that R is constant if $N > 2$. We define a Gaussian curvature K as in Level 0, Sec. 3:

$$R(x) = N(N-1)K(x) ,$$

then

$$R_{jl} = (N-1)g_{jl}K , \quad R_{plij} = K(g_{pi}g_{lj} - g_{pj}g_{li}) .$$

| Problem 1.17. | *Geodesic deviation and the Riemann tensor*

Let C be a curve and $x = x(s)$ its intrinsic equation. We have defined in (5.19) (Level 1) the derivative of a vector field $A(x)$ with respect to another vector field $v(x)$. Let us take $v(x) = dx/ds$, then

$$\nabla_{dx/ds} A^k(x) = \frac{dA^k(x(s))}{ds} + \Gamma^k_{ij}(x(s))\frac{dx^j}{ds}A^i(x(s)) .$$

Such a derivative is called the derivative of the field $A(x)$ along the curve C and is denoted by $DA^k(x)/Ds$.

Consider two curves C and $C + dC$ with intrinsic equations

$$x = x(s) , \quad x + \delta x = x(s) + \delta x(s) .$$

If C and C' are geodesics, their respective equations are

$$\frac{d^2 x^i}{ds^2} + \Gamma^i_{jk}(x(s))\frac{dx^j}{ds}\frac{dx^k}{ds} = 0$$

and

$$\frac{d^2}{ds^2}\{x^i(s) + \delta x^i(s)\} + \Gamma^i_{jk}(x(s) + \delta x(s))\frac{d}{ds}\{x^j(s) + \delta x^j(s)\}$$
$$\times \frac{d}{ds}\{x^k(s) + \delta x^k(s)\} = 0 .$$

Develop the terms of the second equation up to the first order in δx and take the first equation into account to show that

$$\frac{d^2}{ds^2}\delta x^i + 2\Gamma^i_{jk}\frac{dx^j}{ds}\delta x^k + (\partial_l\Gamma^i_{jk})\frac{dx^j}{ds}\frac{dx^k}{ds}\delta x^l = 0 . \tag{I}$$

Using finally the previous definition of D/Ds, show that

$$\frac{D^2}{Ds^2}\delta x^i = R^i{}_{jkl}\frac{dx^j}{ds}\frac{dx^l}{ds}\delta x^k \ . \tag{II}$$

This is a relation bringing out a new aspect of the Riemann tensor, namely its connection with the geodesic deviation δx.

Hints: Equation (I) results from a straightforward calculation. In order to prove Eq. (II), calculate $D^2\delta x^i/Ds^2$ and use the expression of $d^2\delta x^i/ds^2$ obtained from (I).

Applications to Physics

| Problem 1.18. | *Lorentz transformations and Lorentz group*[*]

The manifold $\mathcal{M}_N$ of special relativity is Riemannian with a diagonal metric η whose covariant components $\eta_{\mu\nu}$ are $(+1,-1,-1,-1)$. Alternatively its points can be parametrized by the frame (e_0, e_1, e_2, e_3) such that $M(x) \in \mathcal{M}_4$:

$$x = x^\mu e_\mu \ ,$$

$$\langle e_0, e_0 \rangle = 1 \ , \quad \langle e_0, e_i \rangle = 0 \quad \begin{matrix} \mu = 0,1,2,3 \\ i,j = 1,2,3 \end{matrix}$$

$$\langle e_i, e_j \rangle = -\delta_{ij} \ .$$

A **Lorentz transformation** is a linear transformation

$$x'^\alpha = L^\alpha{}_\mu x^\mu \ ,$$

which is an isometry of the space of special relativity.

(a) Show that the contravariant components of η are also diagonal and with $(+1,-1,-1,-1)$ as values and that the conditions on L are of the form

$$\eta_{\mu\nu} = L^\rho_\mu L^\sigma_\nu \eta_{\rho\sigma} \ .$$

(b) Relations between $L^0{}_i$ and $L_0{}^i$, $L^i{}_j$ and $L_i{}^j$. Show that $(L_0^0, L_0^1, L_0{}^2, L_0{}^3)$ are the contravariant components of a vector u and denote these components as follows:

$$L^0{}_0 = u^0 = \gamma \ , \quad L^i{}_0 = u^i = \gamma\beta^i \ ,$$

---

[*]The solution is complete.

where the (β^i) form a triplet of real numbers. Calculate the covariant components of u as functions of β^r and $\beta_r = -\beta^r$. Show that

$$\gamma^2 = \frac{1}{1 + \beta_r \beta^r} = \frac{1}{1 + \beta^2}$$

with $\beta^2 = \beta_r \beta^r$.

(c) Using the values of $\eta_{\alpha\beta}$, derive several relations between γ, β_r and $L^\mu{}_\nu$.

(d) Consider the case of **pure Lorentz** transformations or **boost** such that*

$$L^i{}_j = \delta^i_j + f \beta^i \beta_j \ ,$$

where f is a function of u. Determine $f(u)$ and $L^i{}_j$. Considering the special case where for $i \neq k, \beta^i = 0$, i.e. only $\beta^k \neq 0$, show that $\beta^k = v^k/C$, where $v^i = 0$ for $i \neq k$.

(e) Under which convention can one replace the condition in (a) by a matrix relation

$$L^T \eta L = \eta \ ?$$

Show that the Lorentz transformations form a group and use this last formula to classify the set of all Lorentz transformations.

Solution:

(a) *Since L describes an isometry:* $\eta_{\mu\nu} = \eta'_{\mu\nu}$ *(Sec 12, Level 1) and*

$$\eta'_{\mu\nu} = \eta_{\mu\nu} = L^\rho{}_\mu L^\sigma{}_\nu \eta_{\rho\sigma} \ .$$

(b) *One has*

$$L^0{}_i = \eta^{0\rho} \eta_{i\sigma} L^\sigma_\rho = -L_0{}^i \ ,$$
$$L^i{}_j = \eta^{i\rho} \eta_{j\sigma} L_\rho{}^j = L^i{}_j \ .$$

Consider the condition (a); it can be also expressed as

$$\eta_{\mu\nu} = L^\rho{}_\mu L^\sigma{}_\nu \eta_{\rho\sigma} = L^\rho{}_\mu(\eta_{\rho\sigma} L^\sigma{}_\nu) = L^0{}_\mu L^0_\nu - \sum_r L^r{}_\mu L^r{}_\nu \ .$$

Consider its (0, 0) component

$$\eta_{00} = 1 = (L^0{}_0)^2 - \sum_r (L^r{}_0)^2 = u_0^2 - \sum_r (u^r)^2 \ .$$

---

*Note $L^i{}_j = L_j{}^i$, and also $L^i{}_0 = L_0{}^i$.

This relation shows that u is a vector. Its covariant component u_i can be expressed as

$$U_i = \eta_{i\mu} u^\mu = -u^i = -\gamma\beta^i = \gamma\beta_i \ ,$$

and the expression of η_{00} shows that

$$1 = (u^0)^2 + u_r u^r = \gamma^2 + \gamma^2 \beta_r \beta^r = \gamma^2(1 + \beta^2) \ .$$

(c)　$\eta_{0i} = 0 = L^0{}_0 L^0{}_i - \sum_r L_0{}^r L_r{}^i = -\gamma^2 \beta^i - \gamma\beta^r L_r{}^i \ .$

In the same way

$$\eta_{ii} = -1 = (L^0{}_i)^2 - L_i{}^r L^r{}_i = \gamma^2(\beta_i)^2 - L_i{}^r L_r{}^i \ ,$$
$$i \neq j : \eta_{ij} = 0 = L^0{}_i L^0{}_j - L_i{}^r L_r{}^j = \gamma^2 \beta_i \beta_j - L_i{}^r L_i{}^j \ .$$

These two last formulae can be summarised into the single formula (notice that $L^r{}_i = L_i{}^r$ which will be simply denoted as $L^i{}_j$):

$$\gamma^2 \beta_i \beta^j + L_i^r L_r^j = \delta_i^j \ ,$$

to which we add the condition on η_{0i}:

$$\gamma\beta^i + \beta^r L^i{}_r = 0 \ .$$

(d) *The first of these two last equations gives the relation*

$$\gamma^2 \beta_i \beta^j + 2f\beta_i \beta^j + f^2 \beta^2 \beta_i \beta^j = 0 \ ,$$

which implies

$$\beta^2 f^2 + 2f + \gamma^2 = 0$$

with solutions

$$f = \frac{-1 \pm \sqrt{1 - \gamma^2 \beta^2}}{\beta^2} = \frac{-1 \pm \gamma}{\beta^2} \ .$$

Only the solution $-(\gamma+1)/\beta^2$ satisfies the next to last equation in (c). One may thus write

$$L^i{}_j = \delta^i_j - \frac{\gamma + 1}{\beta^2} \beta^i \beta_j \ ,$$

and also

$$x'^0 = L^0{}_0 x^0 + L^0{}_r x^r = \gamma(x^0 + \beta_r x^r) = \gamma(x^0 - \sum \beta^r x^r) \ ,$$

$$x'^i = L^i{}_0 x^0 + L^i{}_r x^r = \gamma \beta^i x^0 + \left(\delta^i_r - \frac{\gamma+1}{\beta^2} \beta^i \beta_r \right) x^r$$

$$= x^i + \gamma \beta^i x^0 - \frac{\gamma+1}{\beta^2} \beta^i \beta_r x^r \ .$$

We now examine the case where $\beta^i = 0$ *for* $i \neq k$, *and* $\beta^k \neq 0$. *Then* $L^i_0 = 0$ *if* $i \neq k$, $L^k{}_0 = \gamma \beta^k$, *and*

$$x'^0 = \gamma(x^0 - \beta^k x^k) \ , \quad x'^i = x^i : i \neq k \ ,$$

$$x'^k = x^k + \gamma \beta x^0 - \gamma x^k - x^k$$

$$= \gamma(-x^k + \beta^k x^0) \ .$$

Another expression of the preceding formulae is obtained by replacing the frame we have been using by its symmetric with respect to the origin O, i.e. replacing x^k *and* β^k *by* $-x^k$ *and* $-\beta^k$, *then the last formulae become*

$$x'^0 = \gamma(x^0 - \beta^k x^k) \ , \quad x'^k = \gamma(x^k - \beta^k x^0) \ .$$

Both sets of formulae are well-known to all physicists (who use, by the way, a quicker and simpler line of reasoning).

(e) Given a second order tensor A, one defines the corresponding matrix (A) as the matrix having $A^\alpha{}_\beta$ *as elements, the index* α *indicating the row and* β *the column. In condition (a) we have to interpret the factor* $\eta_{\mu\nu}$ *as a matrix: let the index* μ *indicate the row and* ν *the column, then the product* $L^\mu_\alpha \eta_{\mu\nu} L^\nu_\beta$ *cannot be interpreted as a matrix product. However, since* $\Sigma_\mu (L)^T{}^\alpha{}_\mu \eta_{\mu\nu} L^\nu{}_\beta$ *is the* (α, β) *element of the product* $(L)^T (\eta)(L)$, *condition (a) can be brought in a matrix form*

$$(L)^T (\eta)(L) = (\eta)$$

(recall that $(L)^{T\alpha}{}_\beta = L^\beta_\alpha$). *To show that the Lorentz transformations form a group, we consider*

$$x \xrightarrow{\ L\ } x' \xrightarrow{\ L\ } x''$$
$$\underrightarrow{\qquad L''\qquad}$$

$$x''^\alpha = L'^\alpha{}_\nu x'^\nu = L'^\alpha{}_\nu (L^\nu_\mu x^\mu)$$

$$= L''^\alpha{}_\mu x^\mu$$

and the product of two such transformations given by

$$L''^{\alpha}_{\ \mu} = L'^{\alpha}_{\ \nu} L^{\nu}_{\ \mu} \ .$$

Consider the condition (a) in its matrix form: taking the det of both sides, one gets

$$(\det L)^2 = 1 : \ \det L = \pm 1 \ ,$$

then calculate η_{00}:

$$\eta_{00} = (L^0_0)^2 + L^k_{\ 0} L^0_{\ k} = (L^0_{\ 0})^2 + \sum (L^k_{\ 0})^2 \ ,$$

i.e.

$$(L^0_{\ 0})^2 < 1 \ .$$

Clearly these two last relations divide the set of all Lorentz transformations into four distinct sets, since there is no way to go from the class $\det L = +1$ to the class $\det L = -1$ by a continuous transformation (with respect to L^{α}_{β}). We may distinguish between the **orthochronous** *Lorentz subgroup $L^{\uparrow}$ of all Lorentz transformations corresponding to $L^0_0 > 1$ and the* **antichronous** *Lorentz transformations $L^{\downarrow}$ corresponding to $L^0_0 < 1$. The group $L^{\uparrow}$ is made up by the subgroup $L^{\uparrow}_+$ of proper Lorentz transformations characterized by $\det L = +1$; the class of transformations corresponding to $\det L = -1$ contains all transformations which are products of a proper Lorentz transformation by a spatial symmetry. The class $L^{\downarrow}$ contains all transformations with $\det L = +1$ (class $L^{\downarrow}_+$) (all total symmetries $x_{\mu} \rightarrow -x_{\mu}$) and the transformations $L^{\downarrow}_-$ corresponding to $\det L = -1$ (time symmetries $x^0 \rightarrow -x^0$).*

Note that physicists often use the metric $(-1, +1, +1,)$ with the advantage that the contravariant spatial coordinates are equal to the covariant components of x.

Problem 1.19. *Killing vectors and angular momenta**

(a) Prove formulae (11.9), (11), (10) of Level 1.

(b) Consider the manifold $\mathcal{M}_3 = \mathcal{E}_3$ with $(+1, +1, +1)$ as metric. In order to find its isometries, look for the Killing equations and show that

---

*Except for some of the questions, the hints have been developed as full solutions.

they can be divided into two systems with simple solutions. Give geometric interpretation.

(c) The isometries determined in (b) are translations and rotations in $\mathcal{E}_3$. Give for pure rotations the Lie derivatives, first of a scalar $F(\mathbf{x})$ field and then of a vector field, as expressions of the differential angular momenta and the matrix angular momenta.

(d) Extend the previous considerations to special relativity.

(e) Instead of the space $\mathcal{E}_3$ as in (b) and (c), consider the plane $\mathcal{E}_2$ with $(+1, +1)$ as a metric.

Solution:

(b) Since $\mathcal{E}_3$ is Euclidean, all Γ's vanish and the Killing equations become partial differential equations

$$\partial_i \xi_j(x) + \partial_j \xi_i(x) = 0 .$$

They can be divided into the two systems

$$\partial_1 \xi_1(x) = 0 , \quad \partial_2 \xi_2(x) = 0 , \quad \partial_3 \xi_3(x) = 0 : i = j$$

and

$$\partial_i \xi_j(x) + \partial_j \xi_i(x) = 0 : i \neq j ,$$

of $N(N+1)/2 = 6$ equations. Their solutions are simple: those of the first system are

$$\xi_1 = \xi_1(x^2, x^3) , \quad \xi_2 = \xi_2(x^1, x^3) , \quad \xi_3 = \xi_3(x^1, x^2) ;$$

a solution of the second system can be defined by

$$\xi_k = \varepsilon_k + \varepsilon_{kp} x^p ; \quad \varepsilon_{kp} - \varepsilon_{pk} ,$$

where ε_k and ε_{kp} are independent of x. This is also a solution of the first system and represents the product of infinitesimal translations and rotations in $\mathcal{E}_3$.

(c) Begin with a scalar $F(\mathbf{x})$; from Eq. (11.6) (Level 1) it turns out

$$L_{\xi} F(\mathbf{x}) = \xi^k(\mathbf{x}) \partial_k F(\mathbf{x}) = \varepsilon^{kj} x_j \partial_k F(\mathbf{x})$$
$$= \frac{1}{2} \varepsilon^{kj} \{ x_j \partial_k - x_k \partial_j \} F(\mathbf{x})$$

and we notice that with the metric $(+1, +1, +1)$ *we can raise or lower the indices without changing the value of the tensor. One can also write*

$$L_\xi F(\mathbf{x}) = \sum_{k<j} \varepsilon^{kj} \{x_j \partial_k - x_k \partial_j\} F(\mathbf{x}) \ .$$

It is then customary to introduce as in formulae (26.42), (26.46) of Level 2

$$L_1 = -i\{x_2 \partial_3 - x_3 \partial_2\} \ , \quad L_2 = -i\{x_3 \partial_1 - x_1 \partial_3\} \ , \quad L_3 = -i\{x_1 \partial_2 - x_2 \partial_1\} \ ,$$

to form a vector differential operator $\mathbf{L}(\partial)$, *and a vector (pseudo)* $\delta\alpha$ *with components*

$$\delta\alpha_1 = \varepsilon_{23} \ , \quad \delta\alpha_2 = \varepsilon_{31} \ , \quad \delta\alpha_3 = \varepsilon_{12} \ ,$$

to obtain

$$L_\xi F(\mathbf{x}) = -i\delta\alpha, \mathbf{L}(\partial) F(\mathbf{x}) \ ,$$

where $\delta\alpha_1, \delta\alpha_2, \delta\alpha_3$ *are infinitesimal rotation angles around the* $\mathbf{e}_1, \mathbf{e}_2, \mathbf{e}_3$ *axes respectively.*

We now turn our attention to a vector field $\mathbf{X}(\mathbf{x})$: *from Eq. (1.7) it turns out that*

$$\begin{aligned} L_\xi X^i(\mathbf{x}) &= \xi^k \partial_k X^i - X^k \partial_k \xi^i \\ &= \frac{1}{2}\varepsilon^{kj}\{x_j\partial_k - x_k\partial_j\}X^1(\mathbf{x}) - \varepsilon^i_k X^k(\mathbf{x}) \ , \end{aligned}$$

or

$$L_\xi X^i(\mathbf{x}) = \sum_{k<j} \varepsilon^{kj}\{x_j\partial_k - x_k\partial_j\}X^i(\mathbf{x}) - \varepsilon^i_k X^k(\mathbf{x}) \ ,$$

which can be written (cum grano salis) in the compact form

$$L_\xi \mathbf{X}(\mathbf{x}) = -i(\delta\alpha \cdot \{\mathbf{L}(\partial) - \mathbf{J}\})\mathbf{X}(\mathbf{x}) \ ,$$

where J *is a matrix with* ε^{ij} *as elements. It is customary to express the 3 by 3 antisymmetric* J *matrix as follows:*

$$J = \begin{pmatrix} 0 & \varepsilon^{12} & \varepsilon^{13} \\ -\varepsilon^{12} & 0 & \varepsilon^{23} \\ -\varepsilon^{13} & -\varepsilon^{23} & 0 \end{pmatrix} = \varepsilon^{12}\begin{pmatrix} 0 & 1 & 0 \\ -1 & 0 & 0 \\ 0 & 0 & 0 \end{pmatrix} + \varepsilon^{13}\begin{pmatrix} 0 & 0 & 0 \\ -1 & 0 & 0 \\ 0 & 0 & 0 \end{pmatrix}$$

$$+ \varepsilon^{23}\begin{pmatrix} 0 & 0 & 0 \\ 0 & 0 & 1 \\ 0 & -1 & 0 \end{pmatrix} = \varepsilon^{12}J_3 + \varepsilon^{13}J_2 + \varepsilon^{23}J_1 \ .$$

In physics, one wants to deal with hermitian matrices, and thus defines the spin one (1) matrices:

$$S_3 = -iJ_3 , \quad S_2 = iJ_2 , \quad S_3 = -iJ_1 ;$$

taking into account the previous definition of $\delta\boldsymbol{\alpha}$, one may write

$$L_{\boldsymbol{\xi}}\mathbf{X}(\mathbf{x}) = -i(\delta\alpha^i \{L_i(\partial) \otimes I + S_i\})\mathbf{X}(\mathbf{x}) ,$$

where $\mathbf{L}(\partial) = -i\mathbf{x} \wedge \boldsymbol{\nabla}$. The differential operators and the matrices in the brackets are now hermitian* and the commutation relations of the matrices S_i are well-known:

$$[S_i, S_j] = i\varepsilon_{ij}{}^k S_k .$$

We are now on ground familiar to all physicists: we do not want to go into the properties of the rotation group $SO(3)$ (see however Problems 2.10 and 2.11), its spinorial representation and the theory of scalar and vector harmonics.

(e) The metric is given by $ds^2 = \eta_{\mu\nu}dx^\mu dx^\nu$ and the matrices with elements $\eta_{\mu\nu}$ and $\eta^{\mu\nu}$ are equal, diagonal and have as nonvanishing elements the diagonal elements $+1, -1, -1, -1$.

The Killing equations remain the same as in (d) with Latin indices $i = 1,2,3$ replaced by Greek indices running from 0 to 3. We have as before

$$\xi_\alpha(x) = \varepsilon_\alpha + \varepsilon_{\alpha\lambda}x^\lambda : \varepsilon_{\alpha\beta} = -\varepsilon_{\beta\alpha}$$

and the matrix $(\varepsilon_{\alpha\beta})$:

$$(\varepsilon_{\alpha\beta}) = \begin{pmatrix} 0 & \varepsilon_{01} & \varepsilon_{02} & \varepsilon_{03} \\ -\varepsilon_{01} & 0 & \varepsilon_{12} & \varepsilon_{13} \\ -\varepsilon_{02} & -\varepsilon_{12} & 0 & \varepsilon_{23} \\ -\varepsilon_{03} & -\varepsilon_{13} & -\varepsilon_{23} & 0 \end{pmatrix} .$$

The contravariant components are then given by

$$\xi^\alpha(x) = \varepsilon^\alpha + \varepsilon^\alpha_\lambda x^\lambda$$

---

*It should be reminded that, compared to the definition of the hermiticity as given by mathematicians, the one admitted by physicists appears coarse and particular.

and

$$\varepsilon^\alpha = \eta^{\alpha\lambda}\varepsilon_\lambda$$

$$\varepsilon^\alpha{}_\beta = \eta^{\alpha\lambda}\varepsilon_{\lambda\beta} : \quad \varepsilon^i{}_k = \eta^{i\lambda}\varepsilon_{\lambda k} = -\varepsilon_{ik}$$

$$\varepsilon_\beta{}^\alpha = \eta^{\alpha\lambda}\varepsilon_{\beta\lambda} : \quad \varepsilon_k{}^i = \eta^{i\lambda}\varepsilon_{ki} = -\varepsilon_{ki}$$

$$\varepsilon^0{}_\beta = \eta^{0\lambda}\varepsilon_{\lambda\beta} = \varepsilon_{0\beta} = \varepsilon^\beta{}_0 .$$

Then,

$$(\varepsilon^\alpha{}_\beta) = \begin{pmatrix} 0 & \varepsilon^0{}_1 & \varepsilon^0{}_2 & \varepsilon^0{}_3 \\ \varepsilon^1{}_0 & 0 & \varepsilon^1{}_2 & \varepsilon^1{}_3 \\ \varepsilon^2{}_0 & \varepsilon^2{}_1 & 0 & \varepsilon^2{}_3 \\ \varepsilon^3{}_0 & \varepsilon^3{}_1 & \varepsilon^3{}_2 & 0 \end{pmatrix} = \begin{pmatrix} 0 & \varepsilon_{01} & \varepsilon_{02} & \varepsilon_{03} \\ \varepsilon_{01} & 0 & -\varepsilon_{12} & -\varepsilon_{13} \\ \varepsilon_{02} & \varepsilon_{12} & 0 & -\varepsilon_{23} \\ \varepsilon_{03} & \varepsilon_{13} & \varepsilon_{23} & 0 \end{pmatrix} ,$$

which can also be written as

$$(\varepsilon^\alpha{}_\beta) = \varepsilon_{01} \begin{pmatrix} 0 & 1 & 0 & 0 \\ 1 & 0 & 0 & 0 \\ 0 & 0 & 0 & 0 \\ 0 & 0 & 0 & 0 \end{pmatrix} + \varepsilon_{02} \begin{pmatrix} 0 & 0 & 1 & 0 \\ 0 & 0 & 0 & 0 \\ 1 & 0 & 0 & 0 \\ 0 & 0 & 0 & 0 \end{pmatrix}$$

$$+ \varepsilon_{03} \begin{pmatrix} 0 & 0 & 0 & 1 \\ 0 & 0 & 0 & 0 \\ 0 & 0 & 0 & 0 \\ 1 & 0 & 0 & 0 \end{pmatrix} + \varepsilon_{12} \begin{pmatrix} 0 & 0 & 0 & 0 \\ 0 & 0 & -1 & 0 \\ 0 & 1 & 0 & 0 \\ 0 & 0 & 0 & 0 \end{pmatrix}$$

$$+ \varepsilon_{13} \begin{pmatrix} 0 & 0 & 0 & 0 \\ 0 & 0 & 0 & -1 \\ 0 & 0 & 0 & 0 \\ 0 & 1 & 0 & 0 \end{pmatrix} + \varepsilon_{23} \begin{pmatrix} 0 & 0 & 0 & 0 \\ 0 & 0 & 0 & 0 \\ 0 & 0 & 0 & -1 \\ 0 & 0 & 1 & 0 \end{pmatrix}$$

$$= \varepsilon_{01}J_{01} + \varepsilon_{02}J_{02} + \varepsilon_{03}J_{03} + \varepsilon_{12}J_{12} + \varepsilon_{13}J_{13} + \varepsilon_{23}J_{23} .$$

Putting

$$K_i = iJ_{0i} , \quad J_1 = iJ_{23} , \quad J_2 = iJ_{31} , \quad J_3 = iJ_{12} ,$$

we have the known commutation relations

$$[J_i, J_j] = i\varepsilon_{ijk}J^k , \quad [K_i, K_j] = i\varepsilon_{ijk}K^k$$

$$[J_i, K_j] = -i\varepsilon_{ijk}K^k ,$$

valid for Lorentz or Poincaré transformations. Their complete study is left to the reader.

Problem 1.20. *Maxwell's equations on a curved space (MKSA and natural units)*

Although the reader is supposed to be familiar with the theory of Maxwell's equations, we give here a resumé of some results.

A. The space of special relativity

We shall first take as a curved space the manifold of special relativity with the metric

$$ds^2 = \eta_{\mu\nu}dx^\mu dx^\nu = (dx^0)^2 - (dx^1)^2 - (dx^2)^2 - (dx^3)^2 = (dx^0)^2 - d\mathbf{x}^2 \ ,$$

$$\langle x, y \rangle = \eta_{\mu\nu}x^\mu y^\nu = x^0 y^0 - x^1 y^1 - x^2 y^2 - x^3 y^3 = x^0 y^0 - \mathbf{x} \cdot \mathbf{y} \ ,$$

where $(\eta_{\mu\nu})$ and $(\eta^{\mu\nu})$ are diagonal matrices $(+1, -1, -1, -1)$ and

$$\eta_{\mu\nu} = \langle e_\mu, e_\nu \rangle \ , \quad x = \begin{pmatrix} x^0 \\ \mathbf{x} \end{pmatrix} \ , \quad \mathbf{x} = \begin{pmatrix} x^1 \\ x^2 \\ x^3 \end{pmatrix} = \begin{pmatrix} x \\ y \\ z \end{pmatrix} \ ,$$

and let the differential operator ∂ denote the quadruplet of differential operators:

$$\partial = \{\partial_0, \partial_1, \partial_2, \partial_3\} \left\{ \frac{\partial}{\partial x^0} \ , \quad -\frac{\partial}{\partial x^1} \ , \quad -\frac{\partial}{\partial x^2} \ , \quad -\frac{\partial}{\partial x^3} \right\} = \{\partial_0, \boldsymbol{\nabla}\} \ .$$

As is well-known, the description of the electromagnetic field requires the following fields: $\mathbf{E}(x), \mathbf{B}(x)$, the electric and magnetic fields $\mathbf{D}(x), \mathbf{H}(x)$, the electric induction and magnetic excitation and $\rho(x)$, the density of electric charge and $\mathbf{J}(x)$, the current. With ρ and $\mathbf{J}$ we form the following column and row matrices:

$$j(x) = \begin{pmatrix} j^0(x) \\ j^1(x) \\ j^2(x) \\ j^3(x) \end{pmatrix} = \begin{pmatrix} \rho(x) \\ \frac{1}{c}\mathbf{J}(x) \end{pmatrix} \ ,$$

$$\underline{j}(x) = (j_0(x), j_1(x), j_2(x), j_3(x)) = \left(\rho(x), -\frac{1}{c}\mathbf{J}(x) \right) \ ,$$

and with the electric potential $V(x)$ and the magnetic vector potential $\mathbf{A}(x)$:

$$\mathbf{B} = \boldsymbol{\nabla} \wedge \mathbf{A} = \operatorname{curl} \mathbf{A} \ , \quad \mathbf{E} = -\frac{\partial \mathbf{A}}{\partial t} - \operatorname{grad} V \ ,$$

we form

$$\phi(x) = \begin{pmatrix} V(x) \\ c\mathbf{A}(x) \end{pmatrix} , \quad \underline{\phi}(x) = (V(x) , \quad -c\mathbf{A}(x)) .$$

We further introduce

$$F_{\alpha\beta} = \partial_\alpha \phi_\beta - \partial_\beta \phi_\alpha$$

and the matrices

$$(F_{\alpha\beta}) = \begin{pmatrix} 0 & E_x & E_y & E_z \\ -E_x & 0 & -cB_z & cB_y \\ -E_y & cB_z & 0 & -cB_x \\ -E_z & -cB_y & cB_x & 0 \end{pmatrix} ,$$

$$(F^{\alpha\beta}(x)) = \begin{pmatrix} 0 & -E_x & -E_y & -E_z \\ E_x & 0 & -cB_z & cB_y \\ E_y & cB_z & 0 & -cB_x \\ E_z & -cB_y & cB_x & 0 \end{pmatrix} ,$$

$E_x, E_y, E_z ; B_x, B_y, B_z$ being the orthogonal (ordinary) components of the fields $\mathbf{E}$ and $\mathbf{B}$.

With all these notations Maxwell's equations read

$$\nabla \wedge \mathbf{H} = \frac{\partial \mathbf{D}}{\partial t} + J , \quad \nabla \cdot \mathbf{D} = \rho$$

$$\nabla \wedge \mathbf{E} = \frac{\partial \mathbf{B}}{\partial t} , \qquad \nabla \cdot \mathbf{B} = 0 .$$

They can also be brought into an apparently covariant form (special relativity):*

$$\partial_\mu j^\mu(x) = 0 , \qquad \text{(I)} \qquad\qquad \partial_\mu \phi^\mu(x) = 0 , \qquad \text{(II)}$$

$$\partial_\nu F^{\nu\mu}(x) = \frac{1}{\varepsilon_0} j^\mu(x) , \qquad \text{(III)} \qquad\qquad \varepsilon^{\alpha\lambda\mu\nu} \partial_\lambda F_{\mu\nu} = 0 . \qquad \text{(IV)}$$

---

*And a propagation equation

$$\Box \phi^\alpha(x) = \frac{1}{\varepsilon_0} j^\alpha(x) .$$

(IV) can also be written as

$$\partial_\alpha F_{\beta\gamma} + \partial_\beta F_{\gamma\alpha} + \partial_\gamma F_{\alpha\beta} = 0 \ . \tag{V}$$

In what follows, we shall use natural units characterized by $c = 1$, $\varepsilon_0 = 1$ and consider the case of the vacuum as the natural medium (then $\mathbf{B} = \mathbf{H}$).

(a) Show that

$$F_{\alpha\beta} F^{\alpha\beta} = \mathbf{E}^2 - \mathbf{B}^2 \ ,$$
$$\det(F^{\alpha\beta}) = -(\mathbf{E}, \mathbf{B})^2 = -\det(F_{\alpha\beta}) \ .$$

(b) Define

$$F^*_{\ \alpha\beta} = \frac{1}{2}\varepsilon_{\alpha\beta\mu\nu} F^{\mu\nu}$$

and show that

$$(F^*_{\alpha\beta} = \begin{pmatrix} 0 & -B_x & -B_y & -B_z \\ B_x & 0 & -E_z & -E_y \\ B_y & E_z & 0 & -E_x \\ B_z & -E_y & E_x & 0 \end{pmatrix}$$

and

$$F^*_{\mu\nu} F^{\mu\nu} = \mathbf{E} \cdot \mathbf{B} \ .$$

(b) Curved space

We now go over to a curved space, i.e., a Riemannian manifold $\mathcal{M}_4$.

(c) Show that the definition of $F_{\mu\nu}(x)$ can still be written as

$$F_{\mu\nu}(x) = \nabla_\mu \phi_\nu(x) - \nabla_\nu \phi_\mu(x) \ ,$$

and that Eq. (IV) takes the form

$$\nabla_\alpha F_{\beta\gamma} + \nabla_\beta F_{\gamma\alpha} + \nabla_\gamma F_{\alpha\beta} = 0 \ .$$

(d) As it is well-known Eq. (IV) is equivalent to the second pair of Maxwell's equations given at the end of the introduction to the problem (just before formulae (I)–(V)), the second pair is conveniently summarized by (III). The generalization of (III) in vacuum leads to an equation where the ∂_ν operator is replaced by ∇_ρ, i.e.,

$$\nabla_\rho F^{\rho\mu} = j^\mu \ .$$

The raising and lowering of the indices now follow the law

$$F^{\alpha\beta} = g^{\alpha\mu} g^{\beta\nu} F_{\mu\nu} \ .$$

Show that this last equation can also be written as

$$\frac{1}{\sqrt{|g|}} \partial_\rho \{\sqrt{|g|} F^{\rho\mu}(x)\} = j^\mu(x) \ .$$

(e) In classical electrodynamics, the motion of a charged particle in an electromagnetic field in vacuum is given by the equation

$$m \frac{du^\alpha}{ds} = e F^\alpha{}_\mu u^\mu \ ,$$

where $u^\alpha(x)$ is the velocity of the particle along its trajectory, and

$$ds = \sqrt{(dx^0)^2 - \mathbf{dx}^2} = dx^0 \sqrt{1 - \beta^2}, \ \beta = \|\mathbf{u}\| \ ,$$

we generalize the equation for a curved space by replacing du^α/ds by the covariant derivative of u^α along the trajectory of the particle, i.e., $\nabla_u u^\alpha$ (see formula (5.19), Level 1). For a Lagrangian formulation the reader may be referred to [Bibl. 6], p. 254 or [Bibl. 8], p. 105. He will also find there that the generalized conserved current law is $\nabla_\mu J^\mu(x) = 0$.

Hints: (a) and (b) result from very simple calculations. (c) Use the symmetry of the Γ-symbols. (d) We use formula (5.20), Level 1:

$$\nabla_\nu F^{\nu\mu} = \partial_\nu F^{\nu\mu} + \Gamma^\nu{}_{\rho\mu} F^{\rho\mu} + \Gamma^\mu{}_{\rho\nu} F^{\nu\rho} \ .$$

The last term vanishes since its two factors are respectively symmetric and antisymmetric in ρ, ν. Taking into account the second formulae of question (c), Problem 1.9, one gets

$$\nabla_\nu F^{\nu\mu} = \partial_\nu F^{\nu\mu} + (\partial_\rho \{\log |g|^{1/2}\}) F^{\rho\mu} = \partial_\nu F^{\nu\mu} + \frac{1}{2|g|} \partial_\rho |g| \ ,$$

but one has also

$$\frac{1}{\sqrt{|g|}} \partial_\nu \{|g|^{1/2} F^{\nu\mu}\} = \partial_\nu F^{\nu\mu} + \frac{1}{2|g|} \partial_\nu |g| \ ,$$

which proves the last formula of question (d).

(e) The equation of motion becomes

$$m\nabla_u u^\alpha = eF^\alpha{}_\mu u^\mu \ ,$$

where

$$\nabla_u u^\alpha = u^\mu(\partial_\mu u^\alpha + \Gamma^\alpha{}_{\mu\sigma} u^\sigma) = \frac{dx^\mu}{ds}\frac{du^\alpha}{dx^\mu} + \Gamma^\alpha{}_{\mu\sigma} u^\mu \mu^\sigma \ ,$$

which is the required equation.

$\boxed{\textit{Problem 1.21.}}$ *Variational methods and applications to general relativity*[*]

Before studying general relativity, we first consider the variational formulation of Maxwell's equations on a curved space.

(a) Let $F^{\alpha\beta}(x)$ be the contravariant Maxwell field (Ex. 1.20), show that

$$\nabla_\mu F^{\alpha\mu} = \frac{1}{\sqrt{|g|}}\partial_\mu\{\sqrt{|g|}F^{\alpha\mu}\} \ .$$

(b) Given the Lagrangian density

$$\mathcal{L} = -\frac{\varepsilon_0}{4}\sqrt{|g|}F_{\mu\nu}F^{\mu\nu} + \sqrt{|g|}j^\mu\phi_\mu \ ,$$

derive the generalized Maxwell's equations and the equation of continuity for $j(x)$.

We now turn our attention to general relativity and begin with a few preliminaries.

(c) Calculate from (7.4) Level 1 the variation $\delta R^j{}_{ikl}$ when each of the Γ is submitted to a variation $\delta\Gamma$ and show by an explicit calculation that

$$\delta R^j{}_{ikl} = \nabla_k\delta\Gamma^j{}_{kl} - \nabla_l\delta\Gamma^j{}_{ik} + (\Gamma^m{}_{lk} - \Gamma^m{}_{kl})\delta\Gamma^j{}_{im} \ ,$$

after proving that $\delta\Gamma$ is a tensor field indeed (contrary to the Γ's).

These relations are known as **Palatini's identities**.

---

[*]We shall give a full solution of this problem.

(d) We restrict ourselves to point transformations $x \to x'$ which preserve the orientation of the considered manifold (Level 2, Sec. 13 at the beginning), i.e. such that $\partial(x')/\partial(x) > 0$; show that if $\mathcal{T}$ is a tensor field density of weight W then $|g|^{-\frac{W}{2}}\mathcal{T}$ is a tensor field.

(e) Starting from the preceding remark, find the covariant derivatives of a vector field density and a double contravariant tensor field density.

(f) Show that for the purpose of partial integration the covariant derivative can be treated as an ordinary derivation, provided only tensor densities of weight 1 figure in the integrand.

Although the first of these questions concern any $\mathcal{M}_N$, we shall restrict ourselves to considering the hyperbolic manifold $\mathcal{M}_4$ with (x^0, x^1, x^2, x^3) as generalized coordinates of $M \in \mathcal{M}_4$; in this special case $|g| = -g$. This is the manifold of general relativity.

We shall consider in the following questions a scalar field density $\mathcal{L}(g_{\alpha\beta}, \partial g_{\alpha\beta})$, where $\partial g_{\alpha\beta}$ symbolizes partial derivatives of $g_{\alpha\beta}$ and the integral*

$$\mathfrak{I}[\mathbf{G}] = \int_{\mathcal{M}_4} \mathcal{L}(g_{\alpha\beta}, \partial g_{\alpha\beta}, \dots) d_4 x \ ,$$

where the integrand is supposed to be invariant under the point transformations we are going to consider.

We state furthermore that $\mathcal{L}$ is such that a permissible variation $\delta g_{\mu\nu}(x)$ of $\mathbf{G}$ induces a variation given by the scalar $\delta\mathfrak{I}[\mathbf{G}]$ which can be brought into the form

$$\delta\mathfrak{I}[\mathbf{G}] = \int_{\mathcal{M}_4'} \frac{\partial\mathfrak{I}[\mathbf{G}]}{\delta g_{\mu\nu}(x)} dg_{\mu\nu}(x) d_4 x \ .$$

The first factor in the integrand gives the definition of a **functional derivative** and is supposed to be a double contravariant field density such that the preceding integrand is again an invariant.

(g) Let $\delta g_{\mu\beta}$ be induced by an infinitesimal transformation $x \to x + \delta x$, show that if furthermore $\mathfrak{I}[\mathbf{G}]$ is also an invariant one can prove

$$\nabla_\mu \frac{\delta\mathfrak{I}[\mathbf{G}]}{\delta g_{\alpha\mu}(x)} = 0 \ ,$$

which is a generalized conservation law.

---

*The symbol $\mathbf{G}$ denotes the matrix with elements $g_{\alpha\beta}$.

(h) Choose $\mathcal{L}^G = R\sqrt{-g}$ $(-g = |g|$ for a hyperbolic manifold in general relativity) where R is the total curvature. Consider the action integral

$$\mathcal{A}^G[\mathbf{G}] = \int R\sqrt{-g}\,d_4x \ ,$$

where the domain of integration is the whole space-time of general relativity. Show that the stationarity of $\mathcal{A}^G[g]$ for arbitrary $\delta g_{\alpha\beta}$, vanishing on the boundaries of the considered domain, leads to Einstein's equations for a space-time domain free of matter.

(i) Choose $\mathcal{L}^M = \sqrt{-g}T(g_{\alpha\beta}, \delta g_{\alpha\beta})$, where T is any scalar field, and consider the action**

$$\mathcal{A}^M[\mathbf{G}] = \int \sqrt{-g}T(g_{\alpha\beta}, \partial g_{\alpha\beta})d_4x \ .$$

Show that there exists a twice-contravariant symmetric tensor field $T^{\alpha\beta}$, the **energy-momentum** or **matter tensor**, such that the variation $\delta \mathcal{A}^M[g]$ can be brought into the form

$$\delta\mathcal{A}^M[\mathbf{G}] = \int T^{\alpha\beta}\delta g_{\alpha\beta}\sqrt{-g}\,d_4x \ .$$

(j) Consider the total action

$$\mathcal{A}[\mathbf{G}] = A^G[\mathbf{G}] - \chi A^M[\mathbf{G}] \ ,$$

where χ is the **coupling constant** between gravitation and matter, and derive the general **Einstein's equations**. Show that the Bianchi identities imply a general conservation law for the matter tensor $T^{\alpha\beta}$.

Several important problems such as initial values problems, choice of $T^{\alpha\beta}, \ldots$ are physically important in general relativity. The reader should consult any textbook on the physics of this chapter: only very few have been quoted in the bibliography. See also Problem 1.23.

Solution:
 (a) One has

$$\nabla_\mu F^{\alpha\mu} = \partial_\mu F^{\alpha\mu} + \Gamma^\alpha_{\ \rho\sigma}F^{\rho\sigma} + \Gamma^\rho_{\ \rho\sigma}F^{\alpha\sigma} \ .$$

**The upper indices G and M refer respectively to gravitation and matter field.

We notice that $F^{\alpha\beta}$ (as well as $F_{\alpha\beta}$) is antisymmetric; indeed,

$$F^{\alpha\beta} = g^{\alpha\mu}g^{\beta\nu}(\partial_\mu\phi_\nu - \partial_\nu\phi_\mu) = g^{\beta\rho}\partial^\alpha\phi_\rho - g^{\alpha\rho}\partial^\beta\phi_\rho$$
$$= \partial^\alpha\phi^\beta - \partial^\beta\phi^\alpha - (\partial^\alpha g^{\beta\rho})\phi_\rho + (\partial^\beta g^{\alpha\rho})\phi_\rho \ ,$$

a formula which manifests clearly its antisymmetry. It also implies that the second term of the expression of $\nabla_\mu F^{\alpha\mu}$ vanishes. The first and last terms remain, but as shown in question (c) of Ex. 1.9, one has

$$\Gamma^\mu{}_{\mu\sigma} = \partial_\sigma\{\log|g|^{1/2}\} = \frac{1}{2\sqrt{|g|}}\partial_\alpha\sqrt{|g|}$$

and

$$\nabla_\mu F^{\alpha\mu} = \partial_\mu F^{\alpha\mu} + \frac{1}{2\sqrt{|g|}}(\partial_\sigma\sqrt{|g|})F^{\alpha\sigma} \ ,$$

which is equal to

$$\frac{1}{\sqrt{|g|}}\partial_\mu\{\sqrt{|g|}F^{\alpha\mu}\} \ .$$

(b) The Euler-Lagrange equations are obtained, remarking that the derivatives should be taken with respect to ϕ_α and $\phi_{\alpha,\beta}$:

$$\frac{\partial\mathcal{L}}{\partial\phi_\alpha} = \sqrt{|g|}j^\alpha \ , \quad \frac{\partial\mathcal{L}}{\partial\phi_{\alpha,\beta}} = \varepsilon_0\sqrt{|g|}F^{\alpha\beta} \ .$$

The field equations are

$$\frac{1}{\sqrt{|g|}}\partial_\mu\{\sqrt{|g|}F^{\alpha\mu}\} = \varepsilon_0 j^\alpha \ .$$

One also has, using question (a),

$$\nabla_\mu F^{\alpha\mu} = \varepsilon_0 j^\alpha \ .$$

The equation of continuity of j can be obtained from the second equation of this paragraph by taking its derivative ∂_α. Then because of the symmetry properties, one gets

$$\partial_\alpha\{\sqrt{|g|}j^\alpha\} = 0 \ ,$$

or taking into account question (c) of Ex. 1.9,

$$\nabla_\mu j^\mu = 0 \ .$$

The cyclic relation (a) of the previous exercise and the equation of motion have to be postulated.

(c) The variation $\delta R^j{}_{ikl}$ as calculated from (7.4) Level 1 reads

$$\delta R^j{}_{ikl} = \partial_k \delta \Gamma^j{}_{il} - \partial_l \delta \Gamma^j{}_{ik} + \delta \Gamma^j{}_{mk} \Gamma^m{}_{il}$$
$$+ \Gamma^j{}_{mk} \delta \Gamma^m{}_{il} - \delta \Gamma^j{}_{ml} \Gamma^m{}_{ik} - \Gamma^j{}_{ml} \delta \Gamma^m{}_{ik} \; .$$

Let us prove then that $\delta \Gamma$ is a tensor field: consider indeed formula (5.12) Level 1 defining the variance of Γ^j_{ki} and perform a variation $\delta \Gamma$ while retaining the same coordinate systems (the primed and nonprimed). It turns out that the variance of $\delta \Gamma$ is given by

$$\delta \Gamma'^j{}_{ki}(x') = \frac{\partial x'^j}{\partial x^l} \frac{\partial x^p}{\partial x'^i} \frac{\partial x^q}{\partial x'^k} \delta \Gamma^l{}_{pq}(x) \; ,$$

which is precisely a criterion of tensoriality. Also the covariant character of Palatini's identity is fully justified. It suffices now to perform these covariant derivations to obtain the previous expression of δR^j_{ikl}.

(d) Let us denote $|g|^{-\frac{W}{2}} \mathcal{T}$ by t. If t is to be a tensor field, then under a point transformation $x \to x'$ one should have

$$t'^{\alpha \cdots}_{\beta \cdots}(x') = \frac{\partial x'^\alpha}{\partial x^{\mu_1}} \cdots \frac{\partial x^{\nu_1}}{\partial x'^\beta} \cdots t^{\mu_1 \cdots}_{\nu_1 \cdots}(x) \; .$$

Using formulae (8.13) and (4.5) of Level 1, the right-hand side of this relation takes the form

$$\left| \left(\frac{\partial(x)}{\partial(x')} \right)^2 g(x) \right|^{-\frac{W}{2}} \frac{\partial x'^\alpha}{\partial x^{\mu_1}} \cdots \frac{\partial x^{\nu_1}}{\partial x'^\beta} \cdots \left(\frac{\partial(x)}{\partial(x')} \right)^W \mathcal{T}^{\mu_1 \cdots}_{\nu_1 \cdots}(x)$$

$$= |g(x)|^{-\frac{W}{2}} \frac{\partial x'^\alpha}{\partial x^{\mu_1}} \cdots \frac{\partial x^{\nu_1}}{\partial x'^\beta} \cdots \mathcal{T}^{\mu_1 \cdots}_{\nu_1 \cdots}(x) \; ,$$

since $(\partial(x')/\partial(x)) > 0$.

(e) Let $a^\alpha(x)$ be a vector field density of weight W, then

$$\nabla_\beta a^\alpha = \nabla_\beta \{ |g|^{\frac{W}{2}} |g|^{-\frac{W}{2}} a^\alpha \} = |g|^{\frac{W}{2}} \nabla_\beta \{ |g|^{-\frac{W}{2}} a^\alpha \} \; ,$$

since ∇_β and $|g|$ commute. As $|g|^{-\frac{W}{2}} a^\alpha$ is a vector field, we have

$$\nabla_\beta a^\alpha = |g|^{\frac{W}{2}} \left[\partial_\beta \{ |g|^{-\frac{W}{2}} a^\alpha \} + \Gamma^\alpha{}_{\beta\mu} |g|^{-\frac{W}{2}} a^\mu \right]$$

$$= |g|^{\frac{W}{2}} \left[-\frac{W}{2} |g|^{-\frac{W}{2}-1} (\partial_\beta |g|) a^\alpha + \Gamma^\alpha{}_{\beta\mu} |g|^{-\frac{W}{2}} a^\mu \right]$$

$$= \partial_\beta a^\alpha + \Gamma^\alpha{}_{\beta\mu} a^\mu - \frac{W}{2} \frac{\partial_\beta |g|}{|g|} a^\alpha \; ,$$

where the last term can be written as $-W\partial_\beta\{\log|g|^{1/2}\}a^\alpha$ or, taking into account question (b) of Ex. 1.9, as $-W\Gamma^\mu{}_{\mu\beta}a^\alpha$.

(f) Following the remark after the present question (f), we shall limit ourselves, as in (d) and (e), to $\mathcal{M}_4$. We notice first that if $\mathcal{F}$ is any integrable scalar density then $\int\mathcal{F}d_4x$ is invariant under any point transformation preserving $\mathcal{M}_4$ and its orientation, since such a transformation introduces the factor $(\partial(x')/\partial(x)) > 0$ in the new integral as required by elementary integration theory. We notice also that it results from question (e) that for a vector field density a^α of weight 1, one has

$$\nabla_\mu a^\mu = \partial_\mu a^\mu \ .$$

Suppose now that one considers an integral

$$\mathfrak{I} = \int_{\mathcal{M}_4} a^{\cdots}\nabla_\alpha\mathcal{B}^{\cdots}d_4x \ ,$$

where the integrand is supposed to be a scalar field density (α and $\mathcal{B}$ being of weight 1), it is then clear that $a^{\cdots}\mathcal{B}^{\cdots}$ is a contravariant vector field density $C^\alpha(x)$ and one has

$$\nabla_\alpha C^\alpha = (\nabla_\alpha a^{\cdots})b^{\cdots} - a^{\cdots}\nabla_\alpha\mathcal{B}^{\cdots} \ .$$

If furthermore there is a summation on α, one obtains

$$\mathfrak{I} = \int_{\mathcal{M}_4} \partial_\alpha C^\alpha d_4x - \int_{\mathcal{M}_4} (\nabla_\alpha a^{\cdots})\mathcal{B}^{\cdots}d_4x \ ,$$

the first term being the integral of an ordinary divergence which can be integrated and give 0 for important applications (see below).

(g) Using formula (12.5), Level 1, one can write $\delta\mathfrak{I}[\mathbf{G}]$ as

$$\delta\mathfrak{I}[\mathbf{G}] = -\int \frac{\delta\mathfrak{I}[\mathbf{G}]}{\delta g_{\mu\nu}(x)}(\nabla_\mu\delta x_\nu + \nabla_\nu\delta x_\mu)d_4x \ ,$$

and, because of the symmetry of $\delta g_{\alpha\beta}$, one can also bring $\delta\mathfrak{I}[\mathbf{G}]$ into the form

$$\delta\mathfrak{I}[\mathbf{G}] = -2\int_{\mathcal{M}_4'} \frac{\delta\mathfrak{I}[\mathbf{G}]}{\delta g_{\mu\nu}(x)}(\nabla_\mu\delta x^\nu)d_4x \ .$$

We apply now the results of question (f) and obtain

$$\delta \mathfrak{I}[\mathbf{G}] = -2 \int_{\mathcal{M}_4'} \partial_\mu \left\{ \frac{\partial \mathfrak{I}[\mathbf{G}]}{\delta g_{\mu\nu}(x)} \delta x^\nu \right\} d_4 x$$

$$+ \int_{\mathcal{M}_4'} \left(\nabla_\mu \frac{\delta \mathfrak{I}[\mathbf{G}]}{\delta g_{\mu\nu}(x)} \right) \delta x^\nu d_4 x \ .$$

We suppose also that $\mathcal{M}_4 = \mathcal{M}_4'$ *(in general relativity* $\mathcal{M}_4$ *is the whole space-time) and* δx_ν *vanishes on its boundaries, which is a common supposition to all variational problems, then because of the divergence the first integral on the right-hand side of* $\delta \mathfrak{I}$ *vanishes. The second integral remains; if* $\delta \mathfrak{I}$ *is to be an invariant with respect to the choice of* δx, *then our proof is complete.*

(h) Under the variations $\delta g_{\alpha\beta}$, *the action* $\mathcal{A}^G$ *becomes* $\mathcal{A}^G + \delta \mathcal{A}^G$, *where*

$$\delta \mathcal{A}^G[\mathbf{G}] = \int \delta\{\sqrt{-g} g^{\alpha\beta}\} R_{\alpha\beta} d_4 x + \int \sqrt{-g} g^{\alpha\beta} \delta R_{\alpha\beta} d_4 x \ .$$

Take the second integral and apply the Palatini's identities of question (c), remarking that the last term of the right-hand side vanishes since the manifold is Riemann. Then, applying the method of partial integration as given in question (f), one verifies easily that this integral vanishes. We are now left with

$$\delta \mathcal{A}^G[\mathbf{G}] = \int R_{\alpha\beta} \delta\{\sqrt{-g} g^{\alpha\beta}\} d_4 x \ .$$

But

$$\delta\{\sqrt{-g} g^{\alpha\beta}\} = (\delta\sqrt{-g}) g^{\alpha\beta} + \sqrt{-g} \delta g^{\alpha\beta} \ ,$$

and from Ex. 1.9 (a), it turns out that

$$\delta\{\sqrt{-g} g^{\alpha\beta}\} = \frac{1}{2} g^{\alpha\beta} \delta\{\sqrt{-g}\} + \sqrt{-g} \delta g^{\alpha\beta}$$

$$= -\frac{1}{2} g^{\alpha\beta} \sqrt{-g} g_{\mu\lambda} \delta g^{\mu\alpha} + \sqrt{-g} \delta g^{\alpha\beta} \ .$$

One gets finally the relation

$$\delta \mathcal{A}^G[\mathbf{G}] = \int \sqrt{-g} \left(R_{\alpha\beta} \delta g^{\alpha\beta} - \frac{1}{2} R_{\alpha\beta} g^{\alpha\beta} g_{\mu\lambda} \delta g^{\mu\lambda} \right) d_4 x$$

$$= \int \sqrt{-g} \left(R_{\alpha\beta} - \frac{1}{2} R g_{\alpha\beta} \right) \delta g^{\alpha\beta} d_4 x \ ,$$

from which one deduces Einstein's equations in free space

$$R_{\alpha\beta} - \frac{1}{2}Rg_{\alpha\beta} = 0 \ .$$

(i) The variation $\delta\mathcal{A}^M[g]$ can be written as

$$\delta\mathcal{A}^M[\mathbf{G}] = \int \left(\frac{\partial}{\partial g_{\alpha\beta}}\{\sqrt{-g}T\}\delta g_{\alpha\beta} + \frac{\partial}{\partial g_{\alpha\beta,\lambda}}\{\sqrt{-g}T\}\delta g_{\alpha\beta,\lambda} \right) d_4 x \ ,$$

with $g_{\alpha\beta,\lambda} = \partial_\lambda g_{\alpha\beta}$ and $\delta g_{\alpha\beta,\lambda} = \partial_\lambda \delta g_{\alpha\beta}$. Now, we assume that the $\delta g_{\alpha\beta}$ vanish on the boundary of the space-time domain; performing then the usual partial integration on the second term of the right-hand side of $\delta\mathcal{A}^M$, one obtains

$$\delta\mathcal{A}^M[\mathbf{G}] = \int \left[\frac{\partial}{\partial g_{\alpha\beta}}\{\sqrt{-g}T\} - \partial_\lambda \frac{\partial}{\partial g_{\alpha\beta,\lambda}}\{\sqrt{-g}T\} \right] \delta g_{\alpha\beta} d_4 x \ .$$

Defining

$$T^{\alpha\beta} = \frac{1}{\sqrt{-g}} \left(\frac{\partial}{\partial g_{\alpha\beta}}\{\sqrt{-g}T\} - \partial_\lambda \frac{\partial}{\partial g_{\alpha\beta,\lambda}}\{\sqrt{-g}T\} \right) \ ,$$

one arrives at the required result.

(j) We deduce Einstein's equations by requiring $\delta\mathcal{A}[g] = 0$; taking furthermore questions (h) and (i) into account, one obtains

$$\delta\mathcal{A}[\mathbf{G}] = \int \sqrt{-g} \left(R^{\alpha\beta} - \frac{1}{2}g^{\alpha\beta}R - \chi T^{\alpha\beta} \right) \delta g_{\alpha\beta} d_4 x$$

and Einstein's equations

$$R^{\alpha\beta} - \frac{1}{2}g^{\alpha\beta}R = \chi T^{\alpha\beta} \ .$$

The Bianchi identity (2) of Ex. 1.11 proves that the matter tensor $T^{\alpha\beta}$ is subject to a supplementary condition

$$\nabla_\mu T^{\mu\beta} = 0 \ .$$

It should be noticed that this is not a conservation law since we have a covariant derivative instead of an ordinary one. One calls it a generalized conservation law.

| Problem 1.22. | *The energy momentum tensor in electrodynamics and fluid dynamics**

(a) As far as electrodynamics is concerned, i.e. interaction of the electromagnetic field with electric current (or more basically, electron-photon interaction), we saw in the preceding problem that the variables which have to be varied are the generalized potentials ϕ.

We suppose now that the right-hand side of Einstein's equations is due to electromagnetic interaction and follow the precise prescription given in question (i) of Problem 1.21. More precisely, the interaction of electromagnetic field will be described by the tensor density of question (b) of Problem 1.21 with $j^\mu = 0$. Derive the energy momentum tensor $T_{\alpha\beta}$ following the prescription of question (i) of the same problem.

(b) We now consider fluid dynamics: this is an important case, since the universe at large is described in cosmological theories as a perfect fluid (see next problem). We consider then a fluid at rest with respect to a certain orthornormal Euclidean frame $(e_1 e_2 e_3)$ and an elementary surface dS at M with $\mathbf{N}$ as unit normal. We suppose the fluid to be **perfect**, i.e., the internal force exerted by the fluid on dS is of the form

$$\mathbf{f}(M) = p(M)\mathbf{N}(M)dS(M) = \mathbf{t}(M)dS(M) ,$$

where $p(M)$ is the pressure of the fluid at M. The length of $\mathbf{t}(M)$ is $\|\mathbf{t}(M)\| = p(M)$, same for any direction $\mathbf{N}$: the fluid is said to be **isotropic** at M. If $\mathbf{N}$ is along the e_j axis, then the vector $\mathbf{t}(M)$ has for components

$$t_i^j(M) = -\delta_i^j p(M) : \; i = 1, 2, 3 ,$$
$$\mathbf{t}(M) = -t_i^j(M)N^i(M)e_j = p(M)N^j(M)e_j = p(M)\mathbf{N}(M) ,$$

and in the given frame t_i^j defines a tensor field.

If the fluid is in motion, we assume that the velocity of the fluid particle M is $\mathbf{v}(M)$, and the velocity field is isotropic. In a reference frame located at M moving also with the velocity $\mathbf{v}(M)$ (**comoving frame**), we assume the same relation between $t^{ji}(M)$ and the pressure $p(M)$ as above. The extension to a relativistic fluid is simple: instead of having a diagonal 3×3 matrix with elements t^{ji}, we have a diagonal 4×4 matrix with the additional term

$$t^0{}_0(M) = \mu(M)c^2 = \varepsilon(M) ,$$

---

*A complete solution is given.

$\mu(M)$ being the mass of the unit volume and c the velocity of light. Starting from $t^i{}_j$, calculate the corresponding tensor $T^i j$ with respect to a frame at rest in the laboratory and finally T_{ij} using "ordinary" coordinates.

Generalizations:

(c) Show that the symmetric mixed tensor field

$$P^\alpha_\beta = g^\alpha_\beta - U^\alpha U_\beta \ , \quad U_\mu = \frac{dx^\mu}{ds} \ , \quad U_\mu U^\mu = 1 \ ,$$

is a projection operator P_U (with respect to the vector U).

(d) If the energy momentum tensor $T_{\alpha\beta}$ for a fluid is used as the right-hand side of Einstein's equations, then it is to be submitted to the compatibility condition $\nabla_\mu T^{\mu\beta} = 0$. Show that for an incoherent fluid ($p = 0$), such conditions imply that any particle of the fluid describes a geodesic.

(e) See question (d) of the next problem.

(f) The condition $\nabla_\mu T^{\mu\beta} = 0$ does not lead to a conserved energy momentum vector P^α. (Recall that in special relativity, the condition $\partial_\mu T^{\mu\beta} = 0$ leads to a conserved energy momentum vector P^α, independent of t.) However, show that for a manifold $\mathcal{M}_4$ which admits a Killing vector $\xi_\alpha(x)$, one can always define a conserved vector. Show that the vector field $W^\alpha(x) = T^{\alpha\rho}(x)\xi_\rho(x)$ is conserved. Consider the case of special relativity. Comment.

Solution:

(a) Following the procedure given in this same question, we choose, according to question (b) of Problem 1.21, the density

$$\sqrt{|g|}T = -\frac{\varepsilon_0}{4}\sqrt{-g}g^{\mu\rho}g^{\nu\sigma}F_{\mu\nu}F_{\rho\sigma} \ .$$

Then $T_{\alpha\beta}$ can be deduced, following the results of question (i) of Problem 1.21:

$$T_{\alpha\beta} = \frac{1}{\sqrt{-g}}\left(\frac{\partial}{\partial g^{\alpha\beta}}\{\sqrt{-g}T\} - \partial_\lambda\frac{\partial}{\partial g^{\alpha\beta},\lambda}\{\sqrt{-g}T\}\right) \ .$$

The second term of the right-hand side does not contribute since it is a divergence.

We use a formula of Ex. 1.9(a): $\partial g/\partial^{ik} = -gg_{ik}$ to obtain

$$T_{\alpha\beta} = \frac{-1}{\sqrt{-g}}\frac{\varepsilon_0}{4}\frac{\partial}{\partial g^{\alpha\beta}}\{\sqrt{-g}g^{\mu\rho}g^{\nu\sigma}\}F_{\mu\nu}F_{\rho\sigma}$$

$$= \frac{-1}{\sqrt{-g}}\frac{\varepsilon_0}{4}[-\sqrt{-g}g_{\alpha\beta}F^\mu{}_\sigma F^\sigma{}_\mu + \sqrt{-g}F_{\alpha\nu}g^{\nu\sigma}F_{\beta\sigma}$$

$$+ \sqrt{-g}F_{\mu\alpha}g^{\mu\rho}F_{\rho\beta}] \ ,$$

since

$$\frac{\partial}{\partial g^{\alpha\beta}} g^{\mu\rho} = \delta_\alpha^\mu \delta_\beta^\rho \ .$$

Consider the second term of the bracket: it can be written as

$$F_{\alpha\nu} F^\nu{}_\beta \ ,$$

and the contribution of the third term is

$$-F_{\alpha\mu} F^\mu{}_\beta = F_{\alpha\mu} F_\beta{}^\mu \ .$$

Both results are consequences of formula (13.8) Level 1 and the antisymmetry of $F_{\alpha\beta}$ gives

$$F^\mu{}_\beta = g^{\mu\sigma} F_{\sigma\beta} = -g^{\mu\sigma} F_{\beta\sigma} = -F_\beta{}^\mu \ .$$

Collecting all the previous results, one obtains

$$\begin{aligned}
T_{\alpha\beta} &= \frac{\varepsilon_0}{4} \left(\frac{1}{2} g_{\alpha\beta} F_{\mu\nu} F^{\mu\nu} - 2 F_{\alpha\mu} F_\beta{}^\mu \right) \\
&= \frac{\varepsilon_0}{2} \left(\frac{1}{4} g_{\alpha\beta} F_{\mu\nu} F^{\mu\nu} - F_{\alpha\mu} F_\beta{}^\mu \right) \ .
\end{aligned}$$

$T_{\alpha\beta}$ is thus symmetric and furthermore

$$T^\mu{}_\mu = T_\mu{}^\mu = 0 \ ,$$

which amounts to Tr $\{T_{\alpha\beta}\} = 0$.

(b) The tensor field $T^\alpha{}_\beta$ will be obtained by a pure Lorentz transformation (boost). Indeed, we have

$$T^\alpha{}_\beta = L^\alpha{}_\mu L^\nu{}_\beta t^\mu{}_\nu$$

and $t_\nu{}^\mu = t^\mu{}_\nu = t_\nu^\mu$. We shall use the notations of Problem 1.18 and recall that

$$\begin{aligned}
& L^0{}_0 = \gamma \ , && L^i{}_0 = \gamma\beta^i \\
& L^0{}_i = \gamma\beta_i = -\gamma\beta^i : && \beta_i = \eta_{i\mu}\beta^\mu = -\beta^i \\
& \beta^2 = \beta_r\beta^r = -\boldsymbol{\beta}^2 \ , && \gamma = \frac{1}{1+\beta^2} \ , && L^i{}_j = \delta^i_j - \frac{\gamma+1}{\beta^2} \delta^i_j \ .
\end{aligned}$$

We first have

$$T^0{}_0 = L^0{}_\mu L^\nu{}_0 t^\mu{}_\nu = (L^0{}_0)^2 t^0{}_0 + L^0{}_i L^j{}_0 t^i{}_j$$
$$= \gamma^2 \varepsilon - \gamma^2 \beta_i \beta^j \delta^i_{jp} = \gamma^2(\varepsilon - \beta_i \beta^i p)$$
$$T^0{}_i = L^0{}_\mu L^\nu{}_i t^\mu{}_\nu = L^0{}_0 L^0{}_i t^0{}_0 + L^0{}_k L^l{}_i t^k{}_l$$
$$= \gamma^2 \beta_i \varepsilon - \gamma \beta_k \left(\delta^l_i - \frac{\gamma+1}{\beta^2} \beta^l \beta_i \right) \delta^k{}_l p$$
$$= \gamma^2 \beta_i \varepsilon - \gamma \beta_i p + \gamma \frac{\gamma+1}{\beta^2} \beta_k \beta^k \beta_i p$$
$$= \gamma^2 \beta_i \varepsilon + \gamma^2 \beta_i p = \gamma^2(\varepsilon + p)\beta_i \ .$$

Finally, since $t^i_j = -\delta^i_j p$*, we have*

$$T^i{}_j = L^i{}_\mu L^\nu{}_j t^\mu{}_\nu = L^i{}_0 L^0{}_j t^0{}_0 + L^i{}_k L^l{}_j t^k{}_l$$
$$= \gamma^2 \beta^i \beta_j \varepsilon - \left(\delta^i_k - \frac{\gamma+1}{\beta^2} \beta^i \beta_k \right) \left(\delta^l_j - \frac{\gamma+1}{\beta^2} \beta^l \beta_j \right) \times \delta^k_l p$$
$$= \gamma^2 \beta^i \beta_j \varepsilon - \delta^i_j p + 2 \frac{\gamma+1}{\beta^2} \beta_k \beta^k \beta^i \beta^j p - \frac{(\gamma+1)^2}{\beta^4} \beta_k \beta^k \beta^i \beta_j p$$
$$= \delta^i_j p + (\varepsilon + p) \gamma^2 \beta^i \beta_j$$

and all components of $T^i_j = T_j{}^i$*. We now go over to the components of* $T_{\alpha\beta}$*:*

$$T_{00} = \gamma^2(\varepsilon - \beta^2 p) \ , \quad T_{0i} = \gamma^2(\varepsilon + p)\beta_i$$
$$T_{ij} = -\eta_{ij} p + (\varepsilon + p)\gamma^2 \beta_i \beta_j \ .$$

We now use the same notations as in Problem 1.18, question (d) and define the vector β *with components* $\{\beta^1 \beta^2 \beta^3\}$*:*

$$\beta^r = -\beta_r = -\frac{v_r}{c} \ , \quad \beta^2 = \boldsymbol{\beta}^2 \frac{\mathbf{v}^2}{c^2} \ , \quad \gamma^2 = \frac{1}{1 - \beta^2} \ .$$

Finally we have

$$T_{00} = \gamma^2 \left(\varepsilon + \frac{\mathbf{v}^2}{c^2} p \right) \ , \quad T_{0i} = \gamma^2(\varepsilon + p)\frac{v_i}{c}$$
$$T_{ij} = \delta_{ij} p + (\varepsilon + p)\gamma^2 \frac{v_i v_j}{c^2} \ ,$$

which is a tensor in an Euclidean space with metric $(+, +, +)$. *Summarize the preceding formulae as a single one:*

$$T_{\alpha\beta}(x) = (\varepsilon + p)U_\alpha U_\beta - p\eta_{\alpha\beta} \ ,$$

where the generalized fluid velocity is

$$U^\alpha = \frac{dx^\alpha}{ds} = \gamma \frac{dx^\alpha}{dx^0} \ , \quad U_\alpha = \frac{dx_\alpha}{ds} = \gamma \frac{dx_\alpha}{dx^0} \ ,$$

with

$$\gamma^2 = 1 - \left(\frac{d\mathbf{x}}{dx^0} \right)^2 = 1 - \frac{\mathbf{v}^2}{c^2} \ .$$

Finally, for a fluid in general relativity an obvious generalization of the previous formula reads

$$T_{\alpha\beta}(x) = (\varepsilon(x) + p(x))U_\alpha U_\beta - pg_{\alpha\beta} \ ,$$

where ε *and* p *are, as before, the energy density and pressure measured in the comoving frame.*

(c) Since from its definition such an operator is symmetric, let us call P^α_β *the operator* $P^\alpha{}_\beta = P_\beta{}^\alpha$ *and show that* P^α_β *is an idempotent operator,*

$$P^\alpha_\mu P^\mu_\beta = (g^\alpha_\mu - U^\alpha U_\beta)(g^\mu_\beta - U^\mu U_\beta) = P^\alpha_\beta \ .$$

It is also nilpotent, since

$$(P_U)^2 = P^\rho_\lambda P^\lambda_\rho = (g^\rho_\lambda - U^\rho U_\lambda)(g^\lambda_\rho - U^\lambda U_\rho) = 0 \ .$$

It is furthermore easy to show that $P^\alpha_\lambda U^\lambda = 0$: *the vector* U *belongs to the kernel of* P_U.

(d) The condition $\nabla_\mu T^{\mu\beta} = 0$ *can be written as follows:*

$$\nabla_\mu T^{\mu\beta} = \nabla_\mu \{(\varepsilon + p)U^\mu U^\beta - pg^{\mu\beta}\} = 0 \ ,$$

which implies

$$U_\rho \nabla_\mu T^{\mu\rho} = U_\rho \nabla_\mu \{(\varepsilon + p)U^\mu\}U^\rho + U_\rho(\varepsilon + p)U^\mu \nabla U_\rho - U_\rho g^{\mu\rho} \nabla_{\mu\rho} \ ,$$

since $\nabla_\mu g^{\mu\rho} = 0$. *But one also has*

$$U_\rho \nabla_\mu U^\rho = \frac{1}{2} \nabla_\mu \{U_\rho U^\rho\} \ ,$$

and, finally,

$$\nabla_\mu\{(\varepsilon + p)U^\mu\} - U^\mu\nabla_\mu p = 0 \ . \tag{1}$$

We form, on the other hand, the covariant vector field

$$P^\alpha_\lambda\nabla_\mu T^\mu{}_\lambda = P^\alpha_\lambda\nabla_\mu\{(\varepsilon + p)U^\mu U_\lambda - pg^\mu_\lambda\}$$
$$= P^\alpha_\lambda\nabla_\mu\{(\varepsilon + p)U^\mu\}U_\lambda + P^\alpha_\lambda(\varepsilon + p)U^\mu\nabla_\mu U_\lambda - g^\mu_\alpha\nabla_\mu p = 0 \ ;$$

the first term in the second line vanishes since $P^\lambda_\alpha U^\alpha = 0$ and we are left with two terms

$$(\varepsilon + p)U^\mu P^\lambda_\alpha U_\lambda - g^\mu{}_\alpha\nabla_\mu p = 0 \ .$$

Introducing in the first term the expression of P^λ_α:

$$(\varepsilon + p)U^\mu(g^\lambda_\alpha - U^\lambda U_\alpha)\nabla_\mu U_\lambda = (\varepsilon + p)U^\mu\nabla_\mu U_\alpha + (\varepsilon + p)U^\mu U_\alpha U^\lambda\nabla_\mu U_\lambda \ ,$$

where again the second term vanishes $(U^\lambda U_\lambda = 1)$, we therefore have another relation

$$(\varepsilon + p)U^\mu\nabla_\mu U_\alpha - g^\mu_\alpha\nabla_\mu p = 0 \ . \tag{2}$$

Consider now an incoherent fluid $(p = 0)$. Relation (1) leads to a generalized conservation law:

$$\nabla_\mu\{\varepsilon U^\mu\} = 0 \ ,$$

while relation (2) can be written as

$$U^\alpha\nabla_\mu U_\alpha = U^\mu(\partial_\mu U^\alpha + \Gamma^\alpha{}_{\mu\rho}U^\rho)$$
$$= \frac{dx^\mu}{ds}\partial_\mu U^\alpha + \Gamma^\alpha{}_{\mu\rho}U^\mu U^\rho = 0 \ .$$

The last relation can be written as

$$\frac{d^2x^\alpha}{ds^2} + \Gamma^\alpha{}_{\mu\rho}\frac{dx^\mu}{ds}\frac{dx^\rho}{ds} = 0$$

and is clearly the equation of a geodesic.

   (f) We note first that

$$\nabla_\mu W^\mu = \nabla_\mu\{T^{\mu\rho}\xi_\rho\} = T^{\mu\rho}\nabla_\mu\xi_\rho$$

and, because of the symmetry of $T^{\mu\rho}$, we also have

$$\nabla_\mu W^\mu = \frac{1}{2}T^{\mu\rho}(\nabla_\mu\xi_\rho + \nabla_\rho\xi_\mu) = 0 \ .$$

Then for any volume V of $\mathcal{M}_4$,

$$\int_V (\nabla^\mu W^\mu) \sqrt{|g|} d_4 x = \int_V \nabla_\mu \{ \sqrt{|g|} W^\mu \} d_4 x = 0 \ .$$

But $\sqrt{|g|} W^\mu$ *is a vector field density, then from questions (e) and (f) and the preceding problem, it turns out that*

$$\nabla_\mu \{ \sqrt{|g|} W^\mu \} = \partial_\mu \{ \sqrt{|g|} W^\mu \} = 0 \ .$$

The integrand of the last integral over V is then an ordinary four-dimensional divergence of the vector field W. It is then well-known in classical field theory that one can define a conserved energy-momentum four vector.

Consider the case of special relativity. We have shown in Problem 1.19 that a solution of the Killing equations is

$$\xi_\mu(x) = \varepsilon_\mu + \varepsilon_{\mu\rho} x^\rho \ .$$

Accordingly, we define the vector field:

$$W^\alpha = T^{\alpha\mu}(\varepsilon_\mu + \varepsilon_{\mu\rho} x^\rho) = T^{\alpha\mu}\varepsilon_\mu + \frac{1}{2}\varepsilon_{\mu\rho}(T^{\alpha\mu} x^\rho - T^{\alpha\rho} x_\mu)$$

and since in this case $|g| = 1$, *then*

$$\varepsilon_\mu \oint_S T^{\mu\lambda} dS_\lambda + \frac{1}{2}\varepsilon_{\mu\rho} \oint_S (T^{\lambda\mu} x^\rho - T^{\lambda\rho} x^\mu) dS_\lambda = 0 \ ,$$

a relation which leads to the definitions of the energy-momentum vector P^α *and to angular momenta tensor field* $M^{\mu\rho}$.

$\boxed{\textit{Problem 1.23.}}$ *The hyperbolic manifold of general relativity*

One assumes in general relativity that the Riemannian metric we deal with can always be reduced by a convenient point transformation to the quadratic form $\eta_{\mu\nu} dx^\mu dx^\nu$. Any such manifold is called **hyperbolic** and its properties have been rapidly surveyed at the end of Sec. 8, Level 1.

Consider a frame built up by four vectors: e_0 is a time-like vector while (e_1, e_2, e_3) are space-like. One says that such a frame is well adapted to the hyperbolic structure concerned and several of its properties will be proved below.

(a) Show that in such a frame $g_{00} > 0$ and $g_{ij} dx^i dx^j < 0$.

(b) Define

$$\gamma_{ij} = \frac{g_{0i} g_{0j}}{g_{00}} - g_{ij}$$

and show that

$$\gamma_{ij} dx^i dx^j > 0 .$$

(c) Show that the ds^2 can be brought in a well-adapted frame to the form

$$ds^2 = \left(\frac{g_{0\mu}}{\sqrt{g_{00}}} dx^\mu \right)^2 - \gamma_{ij} dx^i dx^j .$$

(d) Show that under conditions (a), one has $T_{00} > 0$.

No such statement can be made for T_0^0.

Solution:

(a) e_0 time-like, $g_{00} = \langle e_0, e_0 \rangle > 0$; e_i being space-like, $g_{ii} = \langle e_i, e_i \rangle < 0$. Consider the spatial part $g_{ij} dx^i dx^j$ of the ds^2; it may be reduced to a negative quadratic form $\overset{3}{\sum} \langle e_i, e_i \rangle (dx^i)^2$ and so the second condition (a) is proved.

(b) Indeed,

$$\gamma_{ij} dx^i dx^j = \frac{(g_{0i} dx^i)^2}{g_{00}} - g_{ij} dx^i dx^j .$$

Both terms of the right-hand side are positive.

(c) We first remark that

$$\left(\frac{g_{0\mu} dx^\mu}{\sqrt{g_{00}}} \right)^2 = \left(\sqrt{g_{00}} dx^0 + \frac{g_{0i} dx^i}{\sqrt{g_{00}}} \right)^2 .$$

One then verifies that the expression

$$\left(\sqrt{g_{00}} dx^0 + \frac{g_{0i}}{\sqrt{g_{00}}} dx^i \right)^2 - \gamma_{ij} dx^i dx^j$$

gives ds^2 as a result.

(d) One has

$$T_{00} = (\varepsilon + p)(U_0)^2 - p g_{00} = \varepsilon (U_0)^2 + p((U_0)^2 - g_{00}) ,$$

the first term of the right-hand side being positive. For the second term, one has

$$(U_0)^2 - g_{00} = \left(\frac{g_{0\mu} dx^\mu}{ds} \right)^2 - g_{00} = g_{00} \gamma_{ij} dx^i dx^j \ ,$$

where we have used conditions (c). This last relation proves that $((U_0)^2 - g_{00})p$ is also positive.

| Problem 1.24. | *An exact solution of Einstein's system of equations: the Robertson-Walker solution**

Usually the Robertson-Walker metric is presented on the grounds of the homogeneity and isotropy of the universe viewed at large. A deeper study leads to the conclusion that the manifold representing such a universe is a maximally symmetric space. Such an introduction is used in several text books, see for instance [Bibl. 7], Chaps. 14 and 15. We cannot go into these considerations but wish to introduce the problem we are dealing with in the shortest possible way.

We may begin with the 3-dimensional metric given in Problem 1.1, question (h) which describes a manifold with a constant curvature. Such a metric reads

$$\widehat{ds}^2 = \frac{1}{|K|} \left(\frac{dr^2}{1 - kr^2} + r^2 d\sigma^2 \right) : \ d\sigma^2 = d\Theta^2 + r^2 \sin^2 \Theta d\Theta^2 \ ,$$

the variable r, Θ, φ being defined as follows:

$$x^1 = r \sin \Theta \cos \varphi \ , \quad x^2 = r \sin \Theta \sin \varphi \ , \quad x^3 r \cos \Theta \ .$$

(r, Θ, φ) would have been spherical coordinates if (x^1, x^2, x^3) were the rectangular coordinates of a point. It has also been shown in Problem 1.1, question (g), that the correponding Ricci tensor is

$$\hat{R}_{ij} = 2k \hat{g}_{ij} \ .$$

We will describe the manifold $\mathcal{M}_4$ of general relativity by using a normal metric (no terms in g_{0i}!) and we replace the number $1/|K|$ by a function

---

*With a complete solution.

$R^2(x^0)$ of the time coordinate only (not to be confused with the scalar curvature R of the manifold $\mathcal{M}_4$!). We thus assume the metric to be

$$ds^2 = (dx^0)^2 - R^2(x^0)\left(\frac{dr^2}{1 - kr^2} + r^2 d\sigma^2\right) .$$

This is the Robertson-Walker metric and our main purpose is to show that $R(x^0)$ can be chosen in such a way that the corresponding $g_{\mu\nu}(x)$ are gravitational potentials, i.e., solutions of Einstein's equations. We have then

$$g_{00} = 1 , \quad g_{11} = -\frac{R^2}{1 - kr^2} , \quad g_{22} = -R^2 r^2 , \quad g_{33} = -R^2 r^2 \sin^2 \Theta .$$

We also recall that, for a diagonal metric $g_{\alpha\beta}$, it was shown in formulae (3.28) and (3.29) of Level 0 that the Γ symbols can be expressed as follows:

$$\Gamma^\alpha{}_{\beta\gamma} = 0 : \alpha \neq \beta \neq \gamma , \quad \Gamma^\alpha{}_{\beta\beta} = -\frac{1}{2g_{\alpha\alpha}}\partial_\alpha g_{\beta\beta} : \quad \alpha \neq \beta ,$$

$$\Gamma^\alpha{}_{\beta\alpha} = \frac{1}{2g_{\alpha\alpha}}\partial_\beta g_{\alpha\alpha} : \beta \neq \alpha , \beta = \alpha .$$

(a) Let $\hat{g}_{ij}$ be defined as $g_{ij} = -R^2 \hat{g}_{ij}$ and calculate g^{ij} as functions of $\hat{g}^{ij}$. Apply the above formulae to the calculation of $\Gamma^\alpha{}_{\beta\gamma}$.

(b) Apply formula (7.13) Level 1 to find R_{ik}, R_{00} and R_{0k}.

(c) Consider Einstein's equations and express the second term of the left-hand side by means of Tr T.

(d) Taking for $T_{\alpha\beta}$ the energy-momentum tensor of a perfect fluid, show that $R(x^0)$ is a solution of a given system of differential equations and conclude that the Robertson-Walker metric is compatible with Einstein's equations.

(e) Find geodesics of the Robertson-Walker metric. Give physical comments.

Solution:

(a) One has

$$\delta^i_j = g^{i\mu}g_{\mu j} = g^{ip}g_{pj} = -R^2 g^{ip}\hat{g}_{pj} ,$$

then

$$g^{ii} = -\frac{1}{R^2}\frac{1}{\hat{g}_{ii}} .$$

If we define

$$\hat{g}_{ij} = 0 : i \neq j , \quad \hat{g}^{ii} = -\frac{1}{R^2}\frac{1}{\hat{g}_{ii}} ,$$

we may write

$$g^{ij} = -\frac{1}{R^2}\hat{g}^{ij} .$$

On the other hand, the nonvanishing Γ symbols are

$$\Gamma^0{}_{ii} = -\frac{1}{2g_{00}}\partial_0\{-R^2\hat{g}_{ii}\} = R\mathring{R} ,$$

$$\Gamma^i{}_{0i} = \frac{1}{2g_{ii}}\partial_0 g_{ii} = -\frac{1}{2R^2\hat{g}_{ii}}\partial_0\{-R^2\hat{g}_{ii}\} = \frac{\mathring{R}}{R} ,$$

where $\mathring{R}$ denotes the derivative with respect to x^0. We can summarize these formulae as follows:

$$\Gamma^0{}_{ij} = R\mathring{R}\hat{g}_{ij} , \quad \Gamma^i{}_{0j} = \frac{\mathring{R}}{R}\delta^i_j . \tag{I}$$

Finally from formula (9.7), Level 1, one gets

$$\begin{aligned}
\Gamma^i{}_{kl} &= g^{i\mu}\Gamma_{\mu kl} = g^{ip}\Gamma_{pkl} = -\frac{1}{2R^2}\hat{g}^{ip}(\partial_l g_{kp} + \partial_k g_{lp} - \partial_p g_{kl}) \\
&= \frac{1}{2}\hat{g}^{ip}(\partial_l\hat{g}_{kp} + \partial_k\hat{g}_{lp} - \partial_p\hat{g}_{kl}) = \hat{\Gamma}^i kl .
\end{aligned} \tag{II}$$

(b) Formula (7.13) of Level 1 reads

$$R_{ik} = \partial_\mu\Gamma^\mu{}_{ik} - \partial_k\Gamma^\mu{}_{i\mu} + \Gamma^\mu{}_{ik}\Gamma^\nu{}_{\mu\nu} - \Gamma^\nu{}_{i\mu}\Gamma^\mu{}_{k\nu} .$$

Retaining only the nonvanishing terms, we can also write

$$\begin{aligned}
R_{ik} &= \partial_0\Gamma^0{}_{ik} + \partial_p\Gamma^p{}_{ik} - \partial_k\Gamma^p{}_{ip} + \Gamma^0{}_{ik}\Gamma^0{}_{pp} + \Gamma^l{}_{ik}\Gamma^l{}_{pp} \\
&\quad - \Gamma^0{}_{i\mu}\Gamma^\mu{}_{k0} - \Gamma^p{}_{i\mu}\Gamma^\mu{}_{kp} .
\end{aligned}$$

Finally the two last terms can be expressed as follows:

$$-\Gamma^0{}_{ip}\Gamma^k{}_{k0} - \Gamma^p{}_{i0}\Gamma^0{}_{kp} - \Gamma^p{}_{iq}\Gamma^q{}_{kp} .$$

Let us collect terms which depend on x^0 and terms which do not:

$$R_{ik} = [\partial_0\{R\mathring{R}\} + \mathring{R}^2\delta_p^p - 2\mathring{R}^2]\hat{g}_{ik}$$
$$+ \partial_p\Gamma^p{}_{ik} - \partial_k\Gamma^p{}_{ip} + \Gamma^l{}_{ik}\Gamma^p{}_{lp} - \Gamma^p{}_{iq}\Gamma^q{}_{ip} .$$

The first line of this relation can be reduced to

$$(R\mathring{\mathring{R}} + 2\mathring{R}^2)\hat{g}_{ik} .$$

We now use relation (II) of question (a) to bring the second line of the same relation into the form

$$\partial_p\hat{\Gamma}^p{}_{ik} - \partial_k\hat{\Gamma}^p{}_{ip} + \hat{\Gamma}^l{}_{ik}\hat{\Gamma}^p{}_{lp} - \hat{\Gamma}^p{}_{iq}\hat{\Gamma}^q{}_{kp} .$$

Clearly this is the Ricci tensor R_{ik} of a tridimensional manifold with

$$\widehat{ds}^2 = \hat{g}_{11}(dx^1)^2 + \hat{g}_{22}(dx^2)^2 + g_{33}(dx^3)^2 ,$$

i.e.,

$$\widehat{ds}^2 = \frac{dr^2}{1 - kr^2} + r^2 d\sigma^2 .$$

Manifolds with this kind of metric have been considered in Problem 1.1, question (h), and the corresponding Ricci tensor was given in question (g) as $R_{ik} = K(N-1)g_{ik}$. We also introduced the constant $k : K = |K|k = \pm 1$, so the corresponding Ricci tensor is

$$\hat{R}_{ik} = 2k\hat{g}_{ik} ,$$

since $|K| = 1$. We collect the two terms we have been calculating to get, finally,

$$R_{ij} = (R\mathring{R} + 2\mathring{R}^2 + 2k)\hat{g}_{ij} .$$

Similar calculations lead to

$$R_{00} = -\frac{3\mathring{\mathring{R}}}{R} , \quad R_{0j} = 0 .$$

(c) Einstein's equations read

$$R_{\alpha\beta} - \frac{1}{2}g_{\alpha\beta}R = \chi T_{\alpha\beta} : \quad \chi = \frac{8\pi}{c^4}G ,$$

where the value χ has been obtained from the Newtonian limit and G is the gravitational constant given by

$$f = G\frac{mm'}{d^2} \ .$$

One also has

$$R_\beta^\alpha - \frac{1}{2}\delta_\beta^\alpha R = \chi T_\beta^\alpha \ ,$$

taking $\alpha = \beta$; one gets

$$R = -\chi T_\lambda^\lambda = -\chi Tr \cdot T \ .$$

Finally, Einstein's equations can be written as

$$R_{\alpha\beta} = \chi\left(T_{\alpha\beta} - \frac{1}{2}g_{\alpha\beta}T_\lambda^\lambda\right) \ .$$

(d) Consider

$$T_{\alpha\beta} = (\varepsilon + p)U_\alpha U_\beta - pg_{\alpha\beta} \ ,$$

from which one deduces $(U_\lambda U^\lambda = 1)$

$$T_\lambda^\lambda = \varepsilon - 3p \ .$$

Then the right-hand side of the last equation in (c) can be brought into the form

$$T_{\alpha\beta} = \frac{1}{2}g_{\alpha\beta}T_\lambda^\lambda = (\varepsilon + p)U_\alpha U_\beta - \frac{1}{2}(\varepsilon - p)g_{\alpha\beta} \ .$$

Finally Einstein's equations read

$$R_{\alpha\beta} = \chi\left((\varepsilon + p)U_\alpha U_\beta - \frac{1}{2}(\varepsilon - p)g_{\alpha\beta}\right) \ .$$

We first consider its R_{0j} components as given in (b):

$$R_{0j} = 0 = \chi(\varepsilon + p)U_0 U_j \ ,$$

which implies $U_j = 0$ and $U_0 = 1$.

Next, consider R_{00}:

$$R_{00} = -3\frac{\overset{\circ\circ}{R}}{R} = \chi\left(\varepsilon + p - \frac{1}{2}(\varepsilon - p)\right) = \frac{\chi}{2}(\varepsilon + 3p) \ .$$

Finally, R_{ij} leads to the equation

$$R_{ij} = (R\overset{\circ\circ}{R} + 2\overset{\circ}{R}^2 + 2k)\hat{g}_{ij} = -\frac{\chi}{2}(\varepsilon - p)g_{ij}$$
$$= \frac{\chi}{2}(\varepsilon - p)R^2 \hat{g}_{ij} .$$

We are thus left with the two equations

$$3\overset{\circ\circ}{R} = -\chi(\varepsilon + 3p)R , \quad R\overset{\circ\circ}{R} + 2\overset{\circ}{R}^2 + 2k = \frac{\chi}{2}(\varepsilon - p)R^2 .$$

The elimination of R between these two last equations gives a first order-differential equation for R

$$\overset{\circ}{R}^2 + k = \frac{\chi}{3}\varepsilon R^2 .$$

This shows that the Robertson-Walker metric is compatible with Einstein's equations.

(e) We gave in (13.3), Level 1 the equations of a geodesic

$$\frac{d^2 x^\alpha}{ds^2} + \Gamma^\alpha{}_{\mu\nu}\frac{dx^\mu}{ds}\frac{dx^\nu}{ds} = 0 .$$

Consider now a curve $C : x^k = a^k$, where the components a^k define a constant vector. For such a curve the left-hand side of these equations is reduced to

$$\frac{d^2 x^0}{ds^2} + \Gamma^0{}_{00}\frac{dx^0}{ds}\frac{dx^0}{ds} = 0 .$$

For the Robertson-Walker metric, $\Gamma^\alpha_{00} = 0$ and the ds^2 along the curve C is $ds^2 = dx_0^2$; then both the terms of the previous formula vanish and it turns out that C is a geodesic for such a metric.

The physical implications of such an assertion are important. The galaxies are supposed to move along geodesics. Then, if we use a geodesic system of coordinates, their spatial generalized coordinates x^k are independent of x^0, the cosmic time. Galaxies are spatially motionless in a geodesic coordinates system.

The Robertson-Walker metric plays a central role in cosmology, we can only refer the reader to textbooks in gravitation theory, for instance, the one quoted at the beginning of this problem. We shall find another study of the Robertson-Walker metric in Problem 2.7.

CHAPTER III — LEVEL 2
The Intrinsic Approach

The division of the present text into three levels is meant to correspond to the degree of familiarity of the average physicist with differential geometry; Level 0 is the closest to the intuition that originates in everyday experience.

This division is actually quite close to the main stages of the historical development of differential geometry: Level 0 corresponds in the main to the work of C.F. Gauss (Disquisitiones generales circa superficies curvas, 1827) and of his immediate predecessors and successors who left their names on several of the basic theorems of infinitesimal geometry. This set of ideas is developed in the majority of treatises on classical analysis that were published before 1950 [Bibl. 1,2,3]. Work in this domain was continued throughout the beginning of the 20th century.

Level 1 is dominated by the contributions of B. Riemann (Ueber die Hypothesen welche der Geometrie zugrunde liegen 1854) who introduced the "N-fold extended" quantities and the "multiplicities of N dimensions" which have since become our N-dimensional manifolds. Riemann asks also the question of fundamental importance for the physicists, namely, whether our ambient space can be modeled by the Euclidean affine space, $\mathcal{E}_3$, or whether it has a non-zero curvature? This was a daring question at a time when Euclidean geometry reigned in the minds of mathematicians as well as of physicists, and also philosophers. Riemann's question could not be answered at that time since the technical means which were to make possible the detection of phenomena testing this hypothesis were not yet available. Riemann was followed by E. B. Christoffel whose name has remained attached to the coefficients of affine connection (1869), and also by J.F. Ricci (to whom we owe the name of tensors) who introduced the concept of absolute or covariant, derivative. Afterwards T. Levi-Cività gave its own geometric interpretation of a covariant derivative with the help of the concept of parallel displacement along a curve (1917). Finally, the appearance in 1916 of Einstein's general relativity opened the doors of tensor analysis to physicists.

Level 2 starts with E. Cartan (from 1918 on); his work marks the beginning of modern differential geometry. H. Weyl, at the same time, spurred by the relativists working on the unified theory of gravitation and of electromagnetism, distinguished clearly between the metrical aspects of differential geometry and the parts depending on the concept of connection.

We should also mention another important aspect of the intrinsic approach which sets it apart from works of the preceding periods. We have already remarked that most of the results obtained by using the taxonomic approach are local in character, i.e. limited to a point and its immediate neighborhood. Straightforward extensions of local results to global situations are not always correct; we have seen examples in the study of global properties of geodesics and in the discussion of the curvature of a manifold. In order to state and prove global results one has to change the methods of attack of the taxonomic approach. The intrinsic method offers ways and means for an efficient attack on global problems.

We arrive finally at the present time where one puts much weight on the refinement of concepts and on the precision of the language. Unfortunately, since the vocabulary of differential geometry is not infinitely extensible (as we have already noticed) the mathematicians have given to traditional names new meanings, different from the ones that the physicists were used to. This way of proceeding has created, and still is creating, obstacles to the understanding and assimilation of differential geometric ideas by the physicists. This remark is the motivation of the text that follows.

A. Differentiable Manifolds
1. *Differentiable Manifold*

The concept of manifold has been introduced at Level 1 where some elementary notions of topology were recalled and in Part (a) of Sec. 2

where, without showing undue concern for mathematical rigour, we tried to make clear the intuitive content of the concept of manifold.

Let us consider, for instance, any part of a sphere in $\mathcal{E}_3$: it is defined as a set of **points**. For the physicist, that term refers to an object without dimensions, a central concept of Euclid's geometry. For the mathematician, the points of a given manifold are abstract objects: they can represent elements of an affine space, vector or topological space, as well as maps between such spaces, endomorphisms,...: the mathematician uses that concept in its utmost generality. This is admittedly a very trivial remark, which may be kept in mind. A **manifold** considered as a set of points without any other specification is fundamentally without any organization. We intend, in the present paragraph, to endow such a set of points with more and more properties, transforming it into an essential tool of differential geometry. First of all, the manifolds $\mathcal{M}$, which we will consider, are **topological spaces** (i.e. such that each of their points admits a neighborhood); they are also **separate spaces** such that given two of their points each point admits a neighborhood and both neighborhoods can be disjoint. We shall also state that there exists an **homeomorphism** φ (a bijective, bicontinuous map) between neighborhoods of $\mathcal{M}$ and neighborhoods of $\mathbb{R}^N$: to each point $M \in \mathcal{M}$, there will correspond an N-uplet of real numbers

$$x = \left\{ x^1 \dots x^N \right\} ,$$

an element of $\mathbb{R}^N$ such that

$$\varphi(M) = x , \tag{1.1}$$

the numbers x^j being the **local coordinates** (similar to the generalized coordinates of points of a mechanical system). Such a manifold endowed with such an homeomorphism will be denoted by $\mathcal{M}_N$ and N will be its **dimension**. Let us then consider two homeomorphisms φ and φ' and two open sets U and W of $\mathcal{M}_N$ and we suppose also that

$$U \cap W \neq \phi .$$

Let M be a point of that intersection. Its local coordinates with respect to the homeomorphisms φ and φ',

$$\varphi(M) = x , \quad \varphi'(M) = x' \tag{1.2}$$

obey clearly the relation

$$x' = \varphi' \circ \varphi^{-1}(x) = \phi(x) : \ x = \phi^{-1}(x') \ . \tag{1.3}$$

The map

$$\phi = \varphi' \circ \varphi^{-1}$$

transforms the N-uplet x into the N-uplet x'; it is therefore a map $\mathbb{R}^N \to \mathbb{R}^N$ and (1.3) reads

$$x'^i = \phi^i(x^1 \dots x^N) \ , \quad x^i = (\phi^{-1})^i(x'^1 \dots x'^N) \ . \tag{1.4}$$

The sets $\{\phi^i\}$ and $\{(\phi^{-1})^i\}$ are N real functions of N real coordinates of the point $M \in \mathcal{M}_N$. Therefore, one may state that there exists a homeomorphism ϕ between open sets $U_{\mathbb{R}^N}(x)$ and $U_{\mathbb{R}^N}$ of $\mathbb{R}^N$. We suppose, furthermore, that $\phi(x)$ and $\phi^{-1}(x)$ are C^∞, i.e., both these functions admit an infinite number of partial derivatives. Such homeomorphisms are called **diffeomorphisms** and this concept will be extended in Sec. 6 to maps $\mathcal{M}_N \to \mathcal{M}'_N$.

Another map is also of interest: the one which maps the N-uplet x into one of the coordinates x^i. Let X^i be such a map: $\mathbb{R}^N \to \mathbb{R}$, then

$$x^i = X^i(x) = X^i(x^1 \dots x^N) : \ i = 1 \dots N \ . \tag{1.5}$$

Using X^i and the homeomorphism φ, one may also define a map $\hat{\varphi}^i :$ $\mathcal{M}_N \to \mathbb{R}$ (called sometimes **coordinate map**) such that

$$x^i = X^i(x) = X^i \circ \varphi^{-1}(M) = \hat{\varphi}^i(M) \ . \tag{1.6}$$

Let $U_\mathcal{M}(M)$ be a neighborhood of M; the pair $\{U_\mathcal{M}(M), \varphi\}$ defines a **chart** of $U_\mathcal{M}(M)$ on $\mathbb{R}^N$. We now consider the set of all charts corresponding to all points of $\mathcal{M}_N$; if they cover $\mathcal{M}_N$, they build up an **atlas** of $\mathcal{M}_N$. Both terms chart and atlas are particularly well chosen since they correspond exactly to their common meanings. As a conclusion to the present paragraph we define the concept of a **C^∞ atlas** and of a **C^∞ manifold** called also **differentiable manifold**. But first of all, let us point out that most of our considerations about C^∞ differentiability are also valid for a C^p differentiability (p partial derivatives).

We begin by the definition of C^∞ atlas: we consider an atlas built up by a set of charts $\{(U, \varphi)\}$, by definition the union of the domains U is supposed to cover $\mathcal{M}_N$. Let us pick up two of the charts $(U_{\mathcal{M}}(M), \varphi)$ and $(U_{\mathcal{M}}(P), \psi)$ with a nonempty intersection: if the coordinates of any point of that intersection, considered as belonging to the first chart, are C^∞ functions of the coordinates of the same point considered as belonging to the second one, then such an atlas is said to be C^∞ (or C^p) differentiable and we complete such a definition by stating that two $C^\infty(C^p)$ atlases are equivalent if their union is a $C^\infty(C^p)$ atlas and requiring that equivalent atlases on $\mathcal{M}_N$ define the same $C^\infty(C^p)$ manifold.

Remark: It is appropriate at this point to remind the reader of the definition of an **equivalence relation**. Let x, y be elements of a set X, an equivalence relation is a subset $\mathcal{R} \subset X \times X$ such that

$$\{x, x\} \in \mathcal{R} \ ,$$
$$\text{if } \{x, y\} \in \mathcal{R} \text{ and } \{y, z\} \in R, \text{ then } \{x, y\} \in \mathcal{R} \ ,$$
$$\text{if } \{x, y\} \in \mathcal{R}, \text{ and } \{y, x\} \in \mathcal{R} \ .$$

One often replaces the notation $\{x, y\} \in \mathcal{R}$ by $x\mathcal{R}y$ or $x \sim y$, mod $\mathcal{R}$. An **equivalence class** of representative $x \in X$ is the subset of all elements $y \in X$ such that $\{y : x\mathcal{R}y\}$. Then X is the disjoint union of equivalence classes and $X/\mathcal{R}$ has for elements all the equivalence classes of X. The set $X/\mathcal{R}$ is the **quotient set** of X by the relation $\mathcal{R}$. A similar notion, the one of **quotient group** will be studied in the remarks ending the paragraph 2 of Chap. 4.

2. *Functions and Curves Defined on a Manifold*

The scope we are pursuing in the present Level 2 is a presentation of only intrinsic concepts and by that we mean concepts defined independently of any chart of $\mathcal{M}_N$, or, if they are defined in a given chart, we want to show that they are independent of the homeomorphism φ defining the chart.

In what follows, we shall find examples of such a procedure. Let $U_{\mathcal{M}}(P)$ be a neighborhood of $P \in \mathcal{M}_N$. We consider the map $\hat{F}$:

$$U_{\mathcal{M}_N}(P) \overset{\hat{F}}{\to} \mathcal{I} \subset \mathbb{R} \ , \tag{2.1a}$$

then for any $M \in U_{\mathcal{M}}(P)$, one has

$$\hat{F}(M) \in \mathcal{I} \ . \tag{2.1b}$$

Let us consider two charts defined by the homeomorphisms φ and φ', since

$$\hat{F}(M) = \hat{F} \circ \varphi^{-1}(x) = \hat{F} \circ \varphi'^{-1}(x') , \qquad (2.2a)$$

we may define the two maps F and F': $\mathbb{R}^N \to \mathbb{R}$ such that

$$F = \hat{F} \circ \varphi^{-1} , \quad F' = \hat{F} \circ \varphi'^{-1} , \qquad (2.2b)$$

and obtain the relation

$$\hat{F}(M) = F(x) = F'(x') , \qquad (2.3)$$

which is the transformation law of $\hat{F}$ (its variance). In physics, the function $F(x)$ is called **a scalar field**. We now state that the function F at M is C^∞ if $F(X)$ is C^∞ at $x = \varphi(M)$, and, because of the chain rule for derivatives, the differentiability of $\hat{F}$ does not depend on the chart, i.e. it is intrinsic. Indeed, one may write

$$\hat{F} \circ \varphi'^{-1} = (\hat{F} \circ \varphi^{-1}) \circ (\varphi \circ \varphi'^{-1}) \qquad (2.4)$$

and remark that the left-hand side is differentiable at M since the first bracket on the right-hand side is differentiable by definition and so is the second one as $\mathcal{M}_N$ is a differentiable manifold.

We are now ready to define a **curve** Γ on $\mathcal{M}_N$. Let $\mathcal{I} \subset \mathbb{R}$ and $\lambda \in I$, we may consider the following map

$$\mathcal{I} \xrightarrow{\hat{\mathcal{F}}} U_M(P) , \qquad (2.5)$$

i.e. for any $M \in U_{\mathcal{M}}(P)$,

$$\hat{\mathcal{F}}(\lambda) = M \qquad (2.6)$$

and when λ describes the interval $\mathcal{I}$, M is said to describe the curve $\Gamma \subset U_{\mathcal{M}}(P)$. Instead of $\hat{\mathcal{F}}(\lambda)$, the shorter notation $\tilde{\mathcal{F}}_\lambda$ is also used.

The local description of Γ is given by N function $\mathcal{F}^1(\lambda) \ldots \mathcal{F}^N(\lambda)$ by the following procedure. One may indeed write

$$\varphi(M) = x = \varphi \circ \hat{\mathcal{F}}(\lambda) = \mathcal{F}(\lambda) , \qquad (2.7)$$

or in matrix form

$$x = \begin{pmatrix} x^1 \\ \vdots \\ x^N \end{pmatrix} = \begin{pmatrix} \mathcal{F}^1(\lambda) \\ \vdots \\ \mathcal{F}^N(\lambda) \end{pmatrix} . \qquad (2.8)$$

This is a representation of the curve $C \subset \mathbb{R}^N$ which is the transform of $\Gamma \in U_{\mathcal{M}}$ by the homeomorphism φ. One often adopts the notation

$$C : \quad x = x(\lambda) \tag{2.9}$$

as we proceeded above when we introduced the concept of the differentiability of a function; we may also consider the differentiability of a curve: the extension is straightforward.

B. Tangent Vector Space

3. *Tangent Vector Space*

Because of their long familiarity with differential and partial derivative equations, physicists are expert users of differential operators. Such operators map the set of C^∞ functions on $\mathbb{R}^N$ (such a set is denoted by C^∞ ($\mathbb{R}^N$)) into itself. In particular, the set of first-order differential operators may be endowed with a vector space structure, a structure of noncommutative algebra, where a new product of elements defined by the commutator of such operators may be added to the first one. These properties are generally well-known to physicists. Appropriate adaptation and generalization of such first-order differential operators on a manifold $\mathcal{M}_N$ will be the cornerstone of the concept of **tangent vector space at M $\in \mathcal{M}_N$**. We may summarize our method as follows.

First, we shall give to the space $C^\infty(\mathcal{M}_N)$ of functions (see Sec. 2) a double structure: a vector space structure and a commutative algebra structure.

Second, we shall define a local map of $C^\infty(\mathcal{M}_N)$ into itself denoted by v_M, which will retain all the fundamental properties of the differential operators on $\mathbb{R}^N$, and will show in the next paragraph that the local expression of v_M is indeed a first-order differential operator on $\mathbb{R}^N$. We now proceed along these lines.

(a) *Algebra of C^∞ functions on $\mathcal{M}_N$.* It is clear that C^∞ (or C^p) functions $\hat{F}$ build up a vector space. If α and β are real numbers, we define $\alpha\hat{F} + \beta\hat{G}$ by its value on $\mathcal{M}_N$:

$$(\alpha\hat{F} + \beta\hat{G})(M) = \alpha\hat{F}(M) + \beta\hat{G}(M) = \alpha F(x) + \beta G(x) . \tag{3.2a}$$

Such a vector space can be transformed into an algebra by defining an internal operation, the **product** FG, by

$$\hat{F}\hat{G}(M) = \hat{F}(M)\hat{G}(M) = F(x)G(x) . \tag{3.2b}$$

(b) The tangent vector v_M to $\mathcal{M}_N$ as a linear map. Let $C^\infty(U_\mathcal{M}(M))$ be the class of C^∞ functions defined in the neighborhood $U_\mathcal{M}(M)$ (**germs of differentiable functions at M**) and let v_P be the linear map

$$C^\infty(U_\mathcal{M}(M)) \overset{v_p}{\to} \mathbb{R}\ ,\quad P \in U_\mathcal{M}(M)\ , \tag{3.3}$$

such that

$$v_P(\hat{F}) = \hat{F}'(P) = F'(x)\ , \tag{3.4}$$

where $\hat{F}$ and $\hat{F}'$ are functions on $U_\mathcal{M}(M)$. We postulate that such maps obey the Leibniz rule

$$\begin{aligned}
v_P(\hat{G}\hat{F}) &= (v_P\hat{F})\hat{G}(P) + \hat{F}(P)v_P\hat{G} \\
&= F'(y)G(y) + F(y)G'(y)\ ,
\end{aligned} \tag{3.5}$$

where $y = \{y^1 \ldots y^N\}$ are the coordinates of P. In addition to (3.5), we also ask that for any constant k independent of P

$$v_P k = 0\ . \tag{3.6}$$

Consider now the set $\{v_M\}$: when endowed with a vector space structure, it is called the **tangent vector space** $T_\mathcal{M}(P)$ at $P \in \mathcal{M}_N$ and its elements, the v_M's, are called **tangent vectors** or simply **vectors** of $\mathcal{M}_N$.

We also will define a vector field as the set of maps $T_\mathcal{M}$:

$$U_\mathcal{M}(p) \overset{T_\mathcal{M}}{\to} U_\mathcal{M}(P) \times T_\mathcal{M}(P)\ .$$

In other words, a vector field is the space of all couples $\{M, v_N\}$, $M \in \mathcal{M}_N$. We shall see in Chap. 4, that the set $\{T_\mathcal{M}\}$ defines a **fibre bundle** called the **tangent bundle** at $\mathcal{M}_N$. As already noticed elsewhere, physicists do use such concepts under different names: as pointed out in Sec. 2, Level 1, they consider the configuration space of a dynamical system as the space whose points are parametrized $\{q^1(t) \ldots q^N(t)\}$ and the phase space (which corresponds to the tangent bundle) of dimension $2N$ parametrized by the q_k's and p_k's; see also Problem 2.5.

The concept of a *p*-**contravariant tensor** can now be readily defined, using the method of tensor product of vector spaces $T_\mathcal{M} \otimes T_\mathcal{M} \otimes \ldots (p$ factors): any of its elements is now a *p*-contravariant tensor.

4. *Local Expression of a Tangent Vector* v_M

We consider again (3.4) and change slightly our notation:

$$v_p \hat{F} = \hat{F}'(p) = \hat{F}' \circ \varphi^{-1}(y) = F'(y) = d_y^v F(y) \tag{4.1}$$

with $y = \varphi(P)$. The definition of d_y^v is given by (4.1); indeed d_y^v will turn out to be a differential operator. The linearity of v_p and the Leibniz rule lead to

$$d_y^v \{\alpha F(y) + \beta G(y)\} = \alpha d_y^v F(y) + \beta d_y^v G(y) , \tag{4.2}$$

$$d_y^v \{F(y)G(y)\} = (d_y^v F(y))G(y) + F(y)d_y^v G(y) . \tag{4.3}$$

Formula (3.6) leads also to

$$d_y^v k = 0 , \tag{4.4}$$

where k is a constant, independent of y.

In the neighborhood $U_\mathcal{M}(M)$, the mean value theorem reads

$$F(y) = F(x) + (y^i - x^i)F_i(x, \theta) , \tag{4.5a}$$

where $y = \varphi(P), x = \varphi(M); \Theta$ is the N-uplet of real numbers $\{\Theta^1 \ldots \Theta^N\}$ and

$$F_i(x, \Theta) = \left. \frac{\partial F(\xi)}{\partial \xi} \right|_{\xi^i = x^i + \Theta^i} . \tag{4.5b}$$

Now, since d_y^v is linear and obeys the Leibniz rule,

$$\begin{aligned} d_y^v(F(y) &= d_y^v F(x) + d_y^v \{(y^i - x^i)F_i(x, \Theta)\} \\ &= (d_y^v y^i)F_i(x, \Theta) + (y^i - x^i)d_y^v F_i(x, \Theta) . \end{aligned} \tag{4.6}$$

For $y = x$, the second term of the right-hand side vanishes and the preceding relation becomes

$$d_y^v F(x) = v_M \hat{F} + (d_x^v x^i) \left. \frac{\partial F(\xi)}{\partial \xi^i} \right|_{\xi^i = x^i} . \tag{4.7}$$

We now denote $d_x^v(x^i)$ by $v^i(x)$ which defines the i-th **component of the vector** $\mathbf{v_M}$ at the point $M = \varphi^{-1}(x)$. One usually introduces also the set of N vectors

$$\{\varepsilon_1^M \ldots \varepsilon_N^M\} ,$$

such that

$$\varepsilon_i^M \hat{F} = \frac{\partial F(x)}{\partial x^i} = \partial_i F(x) \ , \tag{4.8}$$

and writes

$$v_M = v^i(x)\varepsilon_i^M : \ x = \varphi(M) \ . \tag{4.9}$$

The N vectors ε_i^M are linearly independent since the x^k's can be independently chosen and their set is the **natural basis** of the tangent vector space $T_{\mathcal{M}}(M)$. One also says that $\{\varepsilon_i^M \ldots \varepsilon_N^M\}$ build up a **moving frame**, an expression which means that two elements are required in the definition of such a concept: a point $M \in \mathcal{M}_N$ and a basis (frame) attached to M. The notion of a moving frame is by no means alien to the classical theory of space curves and surface. For a curve $C \subset \mathcal{E}_3$, we met the Frenet-Serret frame built up by the unit tangent vector, the unit principal normal and the unit binormal (see Problem 0.2). The definition of the tangent vector τ_M to a curve $\Gamma \subset \mathcal{M}_N$ can be given along similar lines:

$$\tau_{M_\lambda} \hat{F}(M_\lambda) = \frac{dF(x(\lambda))}{d\lambda} = \frac{dx^i}{d\lambda} \partial_i F(x) \ , \tag{4.10a}$$

a formula from which one derives the definition

$$\tau_{M_\lambda} = \frac{dx^i}{d\lambda} \varepsilon_i^M \lambda \ . \tag{4.10b}$$

The **natural components** of τ_M are

$$v^i(x) = \frac{dx^i}{d\lambda} \ ,$$

being components of the tangent to a curve $C \in C^\infty(\mathbb{R}^N)$, image of Γ.

This last remark provides an alternative definition of the vector v_M : $C \subset \mathbb{R}^N$ being in some chart the image of $\Gamma \subset \mathcal{M}_N$ with $dx/d\lambda$ as a tangent at the point $x = \varphi(M)$, let us consider all curves going through M and having this same tangent; the set of such curves is an equivalence class. Then a vector v_M at $M \in \mathcal{M}_N$ is one of the equivalence classes and v_M is given by (4.10).

The concept of **pfaffian derivative*** is relative to a regular matrix $S(x)$ whose elements are the functions $S_j^i(x)$ of class $C^\infty(U_{\mathcal{M}}(P))$:

$$\mathbf{D}_i = S_i^j(x)\partial_j = S_i^j(x)\varepsilon_j^M \ . \tag{4.11}$$

---

*See also Problem 2.12 for the definition of a covariant derivative in gauge theories.

This is just a generalization of (4.8) since for $S^i_j = \delta^i_j$ one goes back to (4.9). Such a notation leads to the following problem familiar to physicists: since

$$[\partial_i, \partial_j] = 0 , \tag{4.12}$$

where

$$[\partial_i, \partial_j] = \partial_i \partial_j - \partial_j \partial_i ,$$

what is $[\mathbf{D}_i, \mathbf{D}_j]$? A straightforward calculation shows that

$$[\mathbf{D}_i, \mathbf{D}_j] F = \left(S^k_i \frac{\partial S^l_j}{\partial x^k} = S^k_j \frac{\partial S^l_i}{\partial x^k} \right) \frac{\partial F(x)}{\partial x^l} , \tag{4.13}$$

a formula which may be brought into the form

$$[\mathbf{D}_i, \mathbf{D}_j] = c^m_{ij} \mathbf{D}_m , \tag{4.14}$$

where the **structure coefficients** c^m_{ij} are

$$c^m_{ij} = \left(S^k_i \frac{\partial S^l_j}{\partial x^k} - S^k_j \frac{\partial S^l_i}{\partial x^k} \right) (S^{-1})^m_l . \tag{4.15}$$

As is easily verified, these coefficients are antisymmetric with respect to i, j and they verify **Jacobi's identity**

$$c^i_{jk} c^m_{il} + c^i_{kl} c^m_{ij} + c^i_{jl} c^m_{ik} = 0 . \tag{4.16}$$

Although the product $\mathbf{D}_i \mathbf{D}_j$ cannot be expressed as a linear combination of $\mathbf{D}_k$, their commutator belongs to the algebra of pfaffian derivatives. By definition, this is a **Lie algebra** and the algebra's product is given by the commutator (4.15). The tangent vector space $T_{\mathcal{M}}(M)$ has been parametrized above by the basis $\{\varepsilon^M_i\}$, but one can define another basis by putting

$$e^M_i = \mathbf{D}_i ,$$

i.e.

$$e^M_i = S^{\,j}_i(x)\partial_j : \ e^M_i \hat{F} = \mathbf{D}_i F(x) . \tag{4.17a}$$

$\{e^M_1 \ldots e^M_N\}$ is another local frame of $T_{\mathcal{M}}(M)$ since $S(x)$ is regular. One then has[*]

$$v_M \hat{F} = V^i(x) e^M_i (\hat{F}(M)) = V^i(x) \mathbf{D}_i F(x) . \tag{4.17b}$$

---

[*]To formulae (4.17), we should add another relation (9.4) $v_M\ (\hat{F}(M)) = \underline{d}\hat{F}(M)\ v_M = v^i\ (x)\ \partial_i\ F(x) = V^i\ (x)\ \mathbf{D}_i\ F(x) .$

There are two sets of components of v_M at our disposal, the $v^i(x)$ components of (4.9) and the present multiplet $\{V^i(x)\}$.

This is a point which should not be overlooked: in Sec. 1 we considered a point transformation (using the language of Level 1) corresponding to a change of chart (change of the homeomorphism φ); in its turn such a change induces a change of frame in $T_{\mathcal{M}}(M)$ and finally a change of the components of v_M. The interpretation of (4.17) is different: there, we keep the same chart (same φ) but we parametrize $T_{\mathcal{M}}(M)$ by a new frame: $\{\varepsilon_i^M\} \to \{e_i^M\}$. Although both concepts can be linked together, they clearly lead to distinct procedures. Returning to formula (4.14) which can be written as

$$[e_i^M, e_j^M] = c_{ij}^p e_p \ , \tag{4.18}$$

a relation which shows that the vectors $\{e_k^M\}$ of the basis build up a Lie algebra. It is also simple to verify that the set $\{v_M\}$ may be endowed with a Lie algebra structure. For a proof we calculate the commutator of two vectors v_M and w_M. Let $\hat{F}$ be a C^∞ function. We may introduce a new C^∞ function $F^{(w)}$ by

$$\hat{F}^{(w)}(M) = w_M \hat{F} \ . \tag{4.19}$$

We then calculate

$$\begin{aligned}
v_M w_M \hat{F} = v_M \hat{F}^{(w)} &= v^j(x)\partial_j\{w^i(x)\partial_i F(x)\} \\
&= C_j^{(w)} v^j(x) + v^j(x)w^i(x)\frac{\partial^2 F(x)}{\partial x^i \partial x^j} \ ,
\end{aligned} \tag{4.20a}$$

where

$$C_j^{(w)} = \frac{\partial w^i(x)}{\partial x^j}\frac{\partial F(x)}{\partial x^i} \ . \tag{4.20b}$$

Clearly the product $v_M w_M$ is not an internal operation of the algebra of the tangent vectors because of the second term of the right-hand side of (4.20a). But consider now the commutator

$$\begin{aligned}
[v_M, w_M]\hat{F} &= v_M w_M \hat{F} - w_M v_M \hat{F} \\
&= C_j^{(w)} v^j(x) + v^j(x)w^i(x)\frac{\partial^2 F(x)}{\partial x^i \partial x^j} \\
&\quad - \left(C_j^{(v)}(x)w^j(x) + w^j(x)v^i(x)\frac{\partial^2 F(x)}{\partial x^i \partial x^j} \right) \\
&= C_j^{(w)}(x)v^j(x) - C_j^{(v)}(x)w^j(x) \ .
\end{aligned}$$

The last equality can be brought into the form

$$\frac{\partial w^i(x)}{\partial x^j}\frac{\partial F(x)}{\partial x^i}v^j(x) - \frac{\partial v^i(x)}{\partial x^j}\frac{\partial F(x)}{\partial x^i}w^j(x)$$
$$= (v^j(x)\partial_j w^i(x) - w^j(x)\partial_j v^i(x))\varepsilon_i^M \hat{F} \ ,$$

and we may finally write

$$[v_M, w_M] = (v^j(x)\partial_j w^i(x) = w^j(x)\partial_j v^i(x))\varepsilon_i^M \ . \tag{4.21}$$

This formula clearly shows that the set $\{v_M\}$ endowed with the internal operation [,] builds up a Lie algebra.

One then verifies

$$[\alpha v, w] = \alpha[v, w] \ , \quad [v, w + u] = [v, w] + [v, u] \ ,$$
$$[v, w] = -[w, v] \tag{4.22}$$

and the **Jacobi identity**

$$[v_1, [v_2, v_3]] + [v_2, [v_3, v_1]] + [v_3, [v_1, v_2]] = 0 \ . \tag{4.23}$$

We also notice, by the way, that for any C^∞ function $\hat{F}(M)$, the expression $\hat{F}(M)u_M$ is a vector and so is $v_M \hat{F}(M)u_M$. It will be necessary to consider later on (Sec. 18, Level 2) this type of vectors. The expression of their components in local coordinates is a very simple affair:

$$\hat{F}(M)u_M \hat{H} = F(x)u^i(x)\partial_i H(x) \ , \tag{4.24}$$

$$\{v_M \hat{F} u_M\}\hat{H} = v^i(x)\partial_i\{F(x)u^j(x)\partial_j H(x)\} \ . \tag{4.25}$$

We finally define the components of a p-contravariant tensor: we need only to consider the vector space product of p-vector spaces,

$$T^p_{\mathcal{M}}(M) = T_{\mathcal{M}}(M) \otimes T_{\mathcal{M}}(M) \otimes \ldots \ .$$

One of its frames is given by the set of N^p tensor products:

$$\{e^M_{i_1} \otimes e^M_{i_2} \otimes \ldots e^M_{i_p} : i_k = 1 \ldots N\} \ , \quad M \in \mathcal{M}_N \ .$$

Then any element of $T^p_{\mathcal{M}}(M)$ vector space will be of the form

$$A_M = A^{i_1 \ldots i_p}(x)e^M_{i_1} \otimes \ldots e^M_{ip} \ . \tag{4.26}$$

It will be seen in the next paragraph (see (5.7)) that these components $A^{i_1 \cdots i_p}$ of A_M follow the transformation law given by (4.14), Level 1.

5. *Change of Chart*

A change of chart, i.e. a change of the homeomorphism, induces for the different mathematical objects we have been studying certain laws of transformation which define their respective variances.

We begin by considering any C^∞ function $\hat{F}$, its transformation law under a change of chart $\varphi \to \varphi'$ having already been given:

$$\hat{F}(M) = F(x) = F'(x') , \tag{5.1}$$

where $x = \varphi(M), x' = \varphi'(M)$ and $F(x)$ and $F'(x)$ are the respective images of $\hat{F}(M)$ by φ and φ'.

Taking into account the relation

$$x' = \varphi'(M) = \varphi' \circ \varphi^{-1}(x) = x'(x) ,$$

the derivatives chain rule allows to write

$$\frac{\partial F(x)}{\partial x^i} = \frac{\partial x'^k}{\partial x^i} \frac{\partial F'(x')}{\partial x'^k} . \tag{5.2}$$

This is a formula we shall use in establishing the transformation law for the components of a vector v_M. Let us consider the two natural bases $\{\varepsilon_i^M = \partial_i\}$ and $\{\varepsilon'^M_i = \partial'_i\}$ corresponding to the two charts $x = \varphi(M), x' = \varphi'(M)$:

$$v_M \hat{F} = v^i(x)\varepsilon_i^M \hat{F} = v'^i(x')\varepsilon'^M_i \hat{F}' , \tag{5.3}$$

or

$$v^i(x)\partial_i F(x) = v'^k(x')\partial'_k F'(x') . \tag{5.4}$$

Its left-hand side can be brought into the form

$$v^i(x)\partial_i F(x) = v^i(x)\frac{\partial x'^k}{\partial x^i}\partial'_k F'(x') , \tag{5.5}$$

or

$$v^i(x)\frac{\partial x'^k}{\partial x^i}\varepsilon'^M_k \hat{F}' = v'^k(x')\varepsilon'^M_k \hat{F}' , \tag{5.6}$$

and finally

$$v'^k(x') = \frac{\partial x'^k}{\partial x^i} v^i(x) \, , \tag{5.7}$$

a formula identical to formula (2.11) of Level 1, which was the starting point of the taxonomic definition of a contravariant vector field.

Jointly with formula (5.7), one may also write the transformation law of ε_i^M:

$$\begin{aligned} \varepsilon'^M_i \hat{F} &= \partial'_i F'(x') = \frac{\partial x^k}{\partial x'^i} \partial_k F'(x'(x)) \\ &= \frac{\partial x^k}{\partial x'^i} \varepsilon_k^M \hat{F}' \, , \end{aligned} \tag{5.8}$$

i.e.,

$$\varepsilon'^M_i = \frac{\partial x^k}{\partial x'^i} \varepsilon_k^M \, . \tag{5.9}$$

One can use the term **contragredient transformations** to characterize the (5.7) and (5.9) laws which deal with natural basis. Similar formulae can be given also for arbitrary basis, following (4.17) we consider the two bases $\{e_i^M\}$ and $\{e'^M_i\}$

$$e_i^M = S_i^j(x)\varepsilon_j^M \, , \quad e'^M_i = S'^j_i(x)\varepsilon'^M_j \, , \tag{5.10}$$

where S and S' are regular $N \times N$ arbitrary matrices. We multiply (5.9) by S'^j_i to obtain

$$e'^M_i = S'^j_i \varepsilon'^M_j = S'^j_i \frac{\partial x^k}{\partial x'^j} (S^{-1})^l_k e_l^M \, , \tag{5.11a}$$

or

$$e'^M_i = \left(S' \frac{\mathbf{D}(x)}{\mathbf{D}(x')} S^{-1} \right)^l_i e_l^M \, . \tag{5.11b}$$

6. *Map of a Manifold $\mathcal{M}_N$ into a Manifold $\mathcal{M}'_{N'}$*

In Sec. 11 of Level 1, we did consider the problem of the deformation of a manifold. Our comments were intuitive and also lacking in precision; we want now to cast the same question in the framework of the intrinsic formulation. But before any development, a word of caution about notations seems "à propos": up to now in our present approach, we have dealt with a point M of a single definite manifold $\mathcal{M}_N$ and several homeomorphisms between $\mathcal{M}_N$ and $\mathbb{R}^N$ defining several systems of coordinates (charts) denoted by $x, x', \ldots$. In the present paragraph and in the following one, we shall consider two manifolds $\mathcal{M}_N$ and $\mathcal{M}'_{N'}$: their respective points will be denoted by $M, P, \ldots$ and $M', P', \ldots$, and their coordinates will be given by the N-uplets $x, y, \ldots$ and $x', y' \ldots$. If we are going over to a new chart, the respective coordinates of the same points will be respectively $\xi, \eta, \ldots$ and $\xi', \eta', \ldots$. Let ϕ be a map $\mathcal{M}_N \to \mathcal{M}'_{N'}$ such that

$$M' = \phi(M) , \qquad (6.1)$$

with $M \in \mathcal{M}_N$ and $M' \in \mathcal{M}'_{N'}$, both manifolds being supposed to be differentiable. Locally, if one considers two charts respectively of $\mathcal{M}_N$ and $\mathcal{M}'_{N'}$ defined by the respective homeomorphisms φ and ψ such that

$$x = \varphi(M) , \quad x' = \psi(M') ,$$

one has

$$x' = \psi(M') = \psi \circ \phi(M) = \psi \circ \phi \circ \varphi^{-1}(x) = \overline{\phi}(x) , \qquad (6.2a)$$

or in matrix form

$$x'^\lambda = \overline{\phi}^\lambda(x^1 \ldots x^N) = x'^\lambda(x^1 \ldots x^N) . \qquad (6.2b)$$

The existence of $\partial x^\lambda / \partial x^k, \lambda = 1 \ldots N', k = 1 \ldots N$, implies the existence of the N' by N Jacobian matrix $\mathbf{D}(x')/\mathbf{D}(x)$ and ϕ will be called **a differentiable map**. It will also be shown later on (see (6.8)) that such a concept is intrinsic, even if it appears here as a local one.

The map ϕ induces a map $D\phi_M^{M'}$ between the two respective planes to $\mathcal{M}_N$ and $\mathcal{M}'_{N'}$ at the points $M \in \mathcal{M}_N$ and $M' \in \mathcal{M}'_{N'}$. This map is linear as it will be seen in formula (6.4).

Then any vector $v_M \in T_{\mathcal{M}}(M)$ is transformed into a vector $v'_{M'} \in T_{\mathcal{M}}(M')$:

$$v'_M = D\phi_M^{M'} v_M , \qquad (6.3)$$

$D\phi_M^{M'}$ is called the **differential of ϕ** at $M \in \mathcal{M}_N$. We also notice that the notation $D\phi_M^{M'}$ is explicit, though cumbersome, and in all cases where no ambiguity is to be feared we shall omit the indices.

We define $D\phi_M^{M'}$ by asking whether, if Γ is a curve of $\mathcal{M}_N$ and $\Gamma' \subset \mathcal{M}'_{N'}$ is its transform by ϕ, the transform of the tangent $v_{M'}$ to Γ at M is the tangent $v'_{M'}$ to Γ' at M' or in other words whether for any function $\hat{G}'$ defined on $\mathcal{M}'_{N'}$, one has

$$v'_{M'}\hat{G}'(M') = v_M\{\hat{G}' \circ \phi\}(M) , \qquad (6.4)$$

where all the arguments have been carefully displayed. To have a local expression of (6.4), we introduce the frames $\{\varepsilon^M_j, j = 1 \ldots N\}$ for $\mathcal{M}_N$ and $\{\varepsilon'^{M'}_\nu : \nu = 1 \ldots N'\}$ for $\mathcal{M}'_{N'}$. The left-hand side of (6.4) can be locally written as

$$v'_{M'}\hat{G}'(M') = v'^\lambda(x')\varepsilon'^{M'}_\lambda\hat{G}'(M') = v'^\lambda(x')\frac{\partial G'(x')}{\partial x'^\lambda} \qquad (6.5)$$

and its right-hand side as

$$v_M\{\hat{G}' \circ \phi\}(M) = v^k(x)\partial_k G'(x') = v^k(x)\frac{\partial x'^\nu}{\partial x^k}\partial'_\nu G'(x') . \qquad (6.6)$$

Taking into account that $\hat{G}'$ is arbitrary and C^∞, we identify (6.5) and (6.6) to obtain

$$v'^\nu(x') = \frac{\partial x'^\nu}{\partial x^k}v^k(x) , \qquad (6.7a)$$

a formula which reads in matrix form:

$$v'(x') = \frac{\mathbf{D}(x')}{\mathbf{D}(x)}v(x) , \qquad (6.7b)$$

where v' and v should be considered as column matrices with N' and N elements.

It may also be noticed that the former considerations can be intuitively summarized in a few words: between the infinitesimal dx and dx' elements of the tangent spaces to $\mathcal{M}_N$ and $\mathcal{M}'_{N'}$, there exists the relation

$$dx' = \frac{\mathbf{D}(x')}{\mathbf{D}(x)}dx .$$

In connection with that last formula, let us add a few words on the intrinsic character of the concept of differentiability which has been presented locally. We consider two diffeomorphisms φ and φ' for $\mathcal{M}_N$ and two other ψ and ψ' for $\mathcal{M}'_{N'}$ such that for $M \in \mathcal{M}_N$ and $M' \in \mathcal{M}'_{N'}$, one has

$$x = \varphi(M) , \quad \xi = \varphi'(M) ; \quad x' = \psi(M') , \quad \xi' = \psi'(M') .$$

The chain rule for Jacobian matrices reads

$$\frac{\mathbf{D}(\xi')}{\mathbf{D}(x')} = \frac{\mathbf{D}(\xi)}{\mathbf{D}(x)}\frac{\mathbf{D}(\xi')}{\mathbf{D}(\xi)}\frac{\mathbf{D}(x)}{\mathbf{D}(x')} \ . \tag{6.8}$$

But from their very definition, the matrices of the right-hand side do exist, hence the Jacobian matrix of the left-hand side should also exist whatever the four diffeomorphisms $\varphi, \varphi'; \psi, \psi'$. Then the definition of $D\phi_M^{M'}$ given precisely by $\mathbf{D}(\xi')/\mathbf{D}(x')$ is independent of any choice of chart, i.e. is intrinsic. We remark also that the rank of a diffeomorphism defined by the **rank** of the corresponding Jacobian matrix is also an intrinsic notion.

As a special case, let us consider two manifolds $\mathcal{M}_N$ and $\mathcal{M}'_{N'}$ with the same dimensions $N = N'$: any differentiable map between neighbourhoods of $\mathcal{M}_N$ is called a **local transformation** of $\mathcal{M}_N$. If the local transformation is an isometry (Riemannian manifold), one speaks of **motion** on $\mathcal{M}_N$.

It is also worthwhile to answer the following question: up to now, we have considered a differentiable map $D\phi : T_\mathcal{M} \to T_{\mathcal{M}'}$; we want also to consider differentiable maps $T_\mathcal{M}^{\otimes p} \to T_\mathcal{M}^{\otimes p}$. Their definition depends essentially on the techniques of the tensor products: the corresponding map $D\phi^{\otimes p}$ is locally given by

$$(\text{mat } D\phi^{\otimes p})_{i_1 \dots i_p}^{\lambda_1 \dots \lambda_p} = \frac{\partial x'^{\lambda_1}}{\partial x^{i_1}} \cdots \frac{\partial x'^{\lambda_p}}{\partial x^{i_p}} \ . \tag{6.9}$$

As a conclusion to the present paragraph, let us introduce two notions which are familiar to many physicists: the first is the concept of a **passive transformation**. A point $M \in \mathcal{M}_N$ can be located in two charts $(U_\mathcal{M}(M), \varphi)$ and $(U_\mathcal{M}(M), \varphi')$ by its respective coordinates $x = \varphi(M), x' = \varphi'(M)$: one goes from x to x' by a passive transformation which also induces in the corresponding tangent vector space a relation between the coordinates $v^i(x)$ and $v'^i(x')$ of any given vector $v_{M'}$ as seen in Sec. 5.

An **active transformation** represents a map $\mathcal{M}_N \to \mathcal{M}'_{N'}$ or a map $\mathcal{M}_N \to \mathcal{M}_N$. If such a map is furthermore differentiable, one can define $D\phi$, a linear map between the corresponding tangent spaces, as in (6.3).

7. *Active Transformation and Lie Derivative*

We shall investigate now active transformations and their local properties (see Sec. 11, Level 1), such that they map a neighborhood of a given point of $\mathcal{M}_N$ into a neighborhood of another point of $\mathcal{M}_N$. We will suppose that they are one-to-one correspondences (bijective maps) and that, together with their inverses, they are differentiable: they will be called (again) **diffeomorphisms**. It is needless to add that such concepts are essential in the theory of continuous media, fluid dynamics,... . We now introduce a family of diffeomorphisms $\{\phi_\tau, \tau \in \mathcal{I} \subset \mathbb{R}\}$, parametrized by the real number τ, and such that any point $M \in \mathcal{M}_N$ is mapped into the point $M' \in \mathcal{M}_N$ and so are their corresponding neighbourhoods. We will denote M' by M_τ or $M(\tau)$:

$$M' = M(\tau) = M_\tau = \phi_\tau(M) \tag{7.1a}$$

and require that

$$M(0) = M_0 = \phi_0(M) = M . \tag{7.1b}$$

In other words, ϕ_τ becomes the identity for $\tau = 0$. It may happen that a single chart covers M, M_τ and their neighbourhoods, but, in general, the coordinates of M and M_τ are given by two homeomorphisms φ and ψ:

$$x = \varphi(M) , \quad \xi_\tau = \psi(M_\tau) ,$$

then

$$\xi_\tau = \psi(M_\tau) = \psi \circ \phi_\tau \circ \varphi^{-1}(x) = \Psi_\tau(x) ,$$
$$\Psi_\tau = \psi \circ \phi_\tau \circ \varphi^{-1} . \tag{7.2}$$

In other words, the multiplet ξ is a C^∞ function of x. From now on, we shall denote the coordinates of M_τ by $\xi(x, \tau)$, instead of $\xi(\tau)$, then

$$\xi^k(x, \tau) = \xi^k(x^1 \ldots x^N, \tau) = \Psi_\tau(x^1 \ldots x^N) . \tag{7.3}$$

These formulae show that if one keeps the point M fixed and let τ vary, one gets a curve Γ and $\mathcal{M}_N$ going through M; its tangent vector $Y_{M(\tau)}$ at the point M_τ has

$$Y^k(x, \tau) = \frac{\partial \xi^k(x, \tau)}{\partial \tau} \tag{7.4a}$$

as components (see (4.10)). In particular, at $M = M(0)$ the tangent to Γ at the point M,

$$Y^k(x) = \left. \frac{\partial \xi^k(x, \tau)}{\partial \tau} \right|_{\tau=0} , \tag{7.4b}$$

and to any local transformation defined by one of the diffeomorphisms of the family $\{\phi_\tau, \tau \in I \subset \mathbb{R}\}$, we may associate a vector field $Y_{M(\tau)}$ which, as we will see in Secs. 25, 26, generates a one-parameter pseudo group.

We now define the Lie derivative L_Y of a vector v_M along a vector Y_M given by (7.4b):

$$L_{Y_M} v_M = \lim_{\tau=0} \cdot [D\phi_\tau^{-1} v_{M_\tau} - v_M] \tag{7.5}$$

and we will use the equivalent notation $(\mathcal{L}_Y v)_M$. The term $D\{\phi_\tau^{-1}\}$ is the differential of ϕ_τ^{-1}, i.e. the $\xi \to x$ transformation. Our next task is to show that the definition (7.5) is identical to the definition (11.7) of Level 1. We remark that the definition (7.4b) of Y_M^k leads to the infinitesimal transformation $x \to \xi(x, \tau)$:

$$\xi^k(x, \tau) = x^k + Y^k(x)d\tau = x^k + \delta x^k(x, \tau) . \tag{7.6}$$

Let us recall the following formulae (see (3.14) and (3.15), Level 1)

$$\frac{\mathbf{D}(\xi)}{\mathbf{D}(x)} = I + \frac{\mathbf{D}(\delta x)}{\mathbf{D}(x)} , \tag{7.7}$$

$$\frac{\mathbf{D}(x)}{\mathbf{D}(\xi)} = \left(\frac{\mathbf{D}(\xi)}{\mathbf{D}(x)}\right)^{-1} , \tag{7.8}$$

$$\frac{\mathbf{D}(x)}{\mathbf{D}(\xi)} = I - \frac{\mathbf{D}(\delta x)}{\mathbf{D}(x)} , \tag{7.9}$$

where the symbols x, ξ, v', v denote the following column matrices

$$x = \begin{pmatrix} x^1 \\ \vdots \\ x^N \end{pmatrix} , \quad \xi = \begin{pmatrix} \xi^1(x, \tau) \\ \vdots \\ \xi^N(x, \tau) \end{pmatrix} ; \quad v(x) = \begin{pmatrix} v^1(x) \\ \vdots \\ v^N(x) \end{pmatrix} ,$$

$$v'(x) = \begin{pmatrix} v'^1(x) \\ \vdots \\ v'^N(x) \end{pmatrix} . \tag{7.10}$$

We may now use these formulae to bring (7.6) into the form

$$(L_Y v)(x) = \lim_{\tau=0} \cdot \frac{1}{\tau} \left[\left(I - \frac{\mathbf{D}(\delta x)}{\mathbf{D}(x)}\right) v(\xi) - v(x) \right] . \tag{7.11}$$

We also have

$$v'(\xi) = \frac{\mathbf{D}(\xi)}{\mathbf{D}(x)} v(x) = \left(I + \frac{\mathbf{D}(\delta(x))}{\mathbf{D}(x)}\right) v(x) , \qquad (7.12)$$

which can be written as

$$v'(\xi) - v(x) = \frac{\mathbf{D}(\delta x)}{\mathbf{D}(x)} v(x) . \qquad (7.13)$$

We bring all these results into (7.10) and get

$$\begin{aligned}
(L_Y(v))(x) &= \lim_{\tau=0} \cdot \frac{1}{\tau} \left[v(\xi) - v(x) - \frac{\mathbf{D}(\delta x)}{\mathbf{D}(x)}\right] v(x) \\
&= \lim_{\tau=0} \cdot \frac{1}{\tau} [v(\xi) - v'(\xi)] = -\lim_{\tau=0} \cdot \frac{1}{\tau} \delta v(x) ,
\end{aligned} \qquad (7.14)$$

the local expression $(L_Y v)_x$ of $(L_Y v)_M$, which is minus the quotient of the form variation $\delta v(x)$ of $v(x)$ by τ for $\tau = 0$. The identity of the intrinsic definition of the Lie derivative with the one we gave in (11.7), Level 1 is now clearly established.

We consider again formula (11.7) of Level 1 which can be written using the present notations:

$$(L_Y v)^i = Y^k \delta_k v^i - v^k \partial_k Y^i , \qquad (7.15)$$

where the right-hand side was called the Poisson bracket of Y and v. We now want to show it is also the commutator $[Y_M, v_M]$ of the vectors Y_M and v_M. In the natural frame $\{\varepsilon_k^M = \partial_k, k = 1 \ldots N\}$ (see (4.9)) the components of v_M and Y_M are

$$v_M = v^i(x)\varepsilon_i^M , \quad Y_M = Y^k(x)\varepsilon_k^M , \qquad (7.16)$$

and, for any arbitrary function $\hat{F}$, one may write

$$\begin{aligned}
[Y_M, v_M]\hat{F} &= [Y^k \partial_k, v^i \partial_i] F(x) \\
&= Y^k \partial_k \{v^i \partial_i F\} - v^i \partial_i \{Y_k \partial_k F\} .
\end{aligned}$$

Then by calculations similar to the ones used in formulae (4.19) up to (4.21), one obtains

$$[Y_M, v_M]\hat{F} = (Y^k \partial_k v_i - v^k \partial_k Y^i)\partial_i F , \qquad (7.17)$$

and, since $\hat{F}$ is any arbitrary C^∞ function,

$$[Y_M, v_M] = (Y^k \partial_k v^i - v^k \partial_k Y^i)\varepsilon_i^M , \qquad (7.18)$$

and also

$$L_Y v_M = [Y_M, v_M] . \qquad (7.19a)$$

From now on, we shall omit the index M to Y_M and write simply Y. A special case is the following one:

$$L_Y \varepsilon_k^M = (\partial_k Y^i(x))\varepsilon_i^M . \qquad (7.19b)$$

We further remark that (7.19a) leads to

$$L_Y v_M = -L_v Y_M , \qquad (7.20a)$$

and that from Jacobi's identity, one obtains

$$[L_X, L_Y]v_M = L_X L_Y v_M - L_Y L_X v_M$$
$$= L_{[X,Y]} v_M . \qquad (7.20b)$$

Let us quote some useful formulae: it is clear that L_Y is linear:

$$L_Y\{u_M + v_M\} = L_Y u_M + L_Y v_M \qquad (7.21)$$

and that the formula (11.6) of Level 1 can be written as

$$L_Y \hat{F}(M) = Y^k \partial_k F(x) . \qquad (7.22)$$

For a C^∞ function $\hat{F}$, let us prove that

$$L_Y\{\hat{F}(M)v_M\} = \hat{F}(M)L_Y v_M + Y_M(\hat{F})v_M . \qquad (7.23)$$

Indeed, from the definition (7.5) of a Lie derivation, one has

$$L_Y\{\hat{F}(M)v_M\} = \lim_{\tau=0} \cdot \frac{1}{\tau}[D\phi_{\tau M}^{-1 M_\tau}\{\hat{F}(\phi_\tau(M)\}v_{\phi_\tau(M)} - \hat{F}(M)v_M] .$$

But since $D\phi_\tau^{-1}$ is a linear map,

$$L_Y\{\hat{F}(M)v_M\} = \lim_{\tau=0} \cdot \frac{1}{\tau}[(\hat{F}(\phi_\tau(M))D\phi_{\tau M}^{-1 M_\tau} v_{\phi_\tau}(M) - \hat{F}(\phi_\tau(M)v_M)$$
$$+ \hat{F}(\phi_\tau(M))v_M - \hat{F}(M)v_M] .$$

The first bracket of the right-hand side is $\hat{F}(M)L_Y v_M$ (since $\phi_0(M) = M$)); the remaining terms give rise to the following limit:

$$\lim_{\tau=0} \cdot \frac{1}{\tau}[\hat{F}(\phi_\tau(M)) - \hat{F}(M)]v_M = \lim_{\tau=0} \frac{1}{\tau}[F(x + \tau Y) - F(x)]v_M$$
$$= (Y^k(x)\partial_k F(x))v_M = (Y_M(\hat{F}))v_M$$

and, as a conclusion, (7.23) is proved. Notice that, using the concept of exterior derivatives (see Sec. 9, formula (5)), the last term $Y_M(\hat{F})$ of (7.23) can be replaced by $\underline{d}F(Y_M)$.

We still have to give the definition of the Lie derivative of any contravariant tensor T_M. In analogy with (7.5), we will adopt the definition

$$L_Y T_M = \lim_{\tau=0} \cdot \frac{1}{\tau}[D\phi_{\tau M}^{-1M_\tau} T_{\phi_\tau}(M) - T_M] . \tag{7.24}$$

We may also give two other formulae similar to (7.21), (7.22): for any two tensors T_M and S_M of the same type one has

$$L_Y\{T_M + S_M\} = L_Y T_M + L_Y S_M , \tag{7.25a}$$

$$L_Y\{T_M \otimes S_M\} = (L_Y T_M) \otimes S_M + T_M \otimes L_Y S_M . \tag{7.25b}$$

The proof of the first formula is obvious; that of the second formula requires some heavy, but simple, calculations. We notice first that for a tensor product $T_M \otimes S_M$ one has the following formula:

$$D\phi_{\tau M}^{M'}\{T_M \otimes S_M\} = D\phi_{\tau M}^{M'}T_M \otimes D\phi_{\tau M}^{M'}S_M .$$

In the calculations that follow, we will simply write $D\phi_\tau$ instead of the more explicit notation above and we shall calculate $L_Y\{T_M \otimes S_M\}$ following (7.24):

$$L_Y\{T_M \otimes S_M\} = \lim_{\tau=0} \cdot \frac{1}{\tau}[D\phi_\tau^{-1}\{T_{\phi_\tau(M)} \otimes S_{\phi_\tau(M)} - T_M \otimes S_M\}]$$
$$= \lim_{\tau=0} \cdot \frac{1}{\tau}[D\phi_\tau^{-1}\{T_{\phi_\tau(M)} \otimes S_{\phi_\tau(M)}\} - T_M \otimes D\phi_\tau^{-1}S_{\phi_\tau(M)}$$
$$+ T_M \otimes D\phi_\tau^{-1}S_{\phi_\tau(M)} - T_M \otimes S_M]$$
$$= \lim_{\tau=0} \cdot \frac{1}{\tau}[(D\phi_\tau^{-1}T_{\phi_\tau(M)} - T_M) \otimes D\phi_\tau^{-1}S_{\phi_\tau(M)}$$
$$+ T_M \otimes (D\phi_\tau^{-1}S_{\phi_\tau(M)} - S_M)] . \tag{7.26}$$

We now use (7.25) and perform some simple transformations to get

$$L_Y\{T_M \otimes S_M\} = \lim_{\tau=0} \cdot\frac{1}{\tau}(D\phi_\tau^{-1}T_{\phi_\tau(M)} - T_M) \otimes \lim_{\tau=0} \cdot(D\phi_\tau^{-1}S_{\phi_\tau(M)})$$
$$+ T_M \otimes \lim_{\tau=0} \cdot\frac{1}{\tau}(D\phi_\tau^{-1}S_{\phi_\tau(M)} - S_M) \ ,$$

which readily proves (7.25b).

As an application of (7.25), we may calculate the Lie derivative of the second rank contravariant tensor

$$A_M = A^{ij}(x)\varepsilon_i^M \otimes \varepsilon_j^M \ .$$

One has

$$L_Y A_M = (L_Y A^{ij}(x))\varepsilon_i^M \otimes \varepsilon_j^M + A^{ij}(x)(L_y\varepsilon_i^M) \otimes \varepsilon_j^M$$
$$+ A^{ij}(x)\varepsilon_i^M L_\xi\varepsilon_j^M \ .$$

We now take into account formulae (7.23) and (7.19b) and after some simple transformations have

$$L_Y\{A^{ij}(x)\varepsilon_i^M \otimes \varepsilon_j^M\} = [Y^m\partial_m A^{ij} - (\partial_k A^i)A^{kj}$$
$$- (\partial_k Y^j)A^{ik}]\varepsilon_i^M \otimes \varepsilon_j^M \ . \qquad (7.27)$$

We may notice, by the way, that as in formula (11.8) of Level 1 one can replace everywhere in the right-hand side ordinary partial derivatives with covariant ones.

C. Cotangent Vector Space
8. *Some Reminders*

E_N being a vector space, let us consider the linear map $\Omega : E_N \to \mathbb{R}$ such that for any $x \in E_N$,

$$\Omega(x) \in \mathbb{R} \ . \qquad (8.1)$$

Ω is then called a **linear form** and we may endow the set of linear forms with a vector space structure to constitute the **dual space** E_N^* of E_N.

Let $\{e_i, i = 1, \dots N\}$ be a basis of E_N and for a vector $x \in E_N$

$$x = x^k e_k : \ x^k \in \mathbb{R} \qquad (8.2a)$$

consider N maps $\{\omega^1 \ldots \omega^N\}$ such that

$$x^k = \omega^k(x) . \qquad (8.2b)$$

The ω^k are linear forms, since for $\lambda \in \mathbb{R}$ and x and y elements of E_N, one has

$$\omega^k(\lambda x) = \lambda x^k = \lambda \omega^k(x) ,$$
$$\omega^k(x+y) = x^k + y^k = \omega^k(x) + \omega^k(y) , \qquad (8.3)$$

and, choosing $x = e_j$,

$$\omega^k(e_j) = \delta_j^k . \qquad (8.4)$$

They are linearly independent and build up a basis for E_N^*. Furthermore, for any $f^* \in E_N^*$ and $x \in E_N$,

$$f^*(x) = f^*(x^i e_i) = x^i f^*(e_i) = x^i f_i^* : f_i^* \in \mathbb{R} . \qquad (8.5)$$

The real numbers f_i^* are the components of f^*; indeed, the last equality of (8.5) may be written as

$$f^*(x) = f_i^* \omega^i(x) , \qquad (8.6a)$$

or, in other words,

$$f^* = \omega^i f_i^* . \qquad (8.6b)$$

The physicist may have recognized by now that a linear form is a "**bra**" $\langle|$ of Dirac and a vector a "**ket**" $|\rangle$.

Consider now a change of basis $\{e_1 \ldots e_N\} \rightarrow \{e'_1 \ldots e'_N\}$ through the regular matrix A:

$$e'_k = A_k^j e_j : A_k^j \in \mathbb{R} . \qquad (8.7)$$

Let a be the corresponding matrix for the change $\{\omega^1 \ldots \omega^N\} \rightarrow \{\omega'^1 \ldots \omega'^N\}$:

$$\omega'^i = a^i{}_j \omega^j . \qquad (8.8)$$

Writing

$$\delta_k^i = \omega'^i(e'_k) = \omega'^i(A^j{}_k e_j) = a^i_j \omega^j(A^l_k e_l) = a^i{}_j A^j{}_k , \qquad (8.9a)$$

one gets

$$a = A^{-1} . \qquad (8.9b)$$

The transformation law for any form f^* can be obtained from

$$f^* = \omega^i f_i^* = \omega'^i f_i'^*$$

and one gets, using (8.7),

$$f_i'^* = (A^{-1})_i^j f_j^* \ . \tag{8.10}$$

Let us add to the former elementary considerations some reminders on tensor products; the **tensor product** $x \otimes y$ is defined by

$$x \otimes y = x^i e_i \otimes y^j e_j = x^i y^j e_i \otimes e_j \ , \tag{8.11}$$

where $\{e_j \otimes e_j : i, j = 1 \ldots N\}$ is a basis of $E_N \otimes E_N$. One may introduce **symmetrized** and **antisymmetrized products** of two vectors x and y, namely,

$$(x \otimes y)_s = \frac{1}{2}(x^i y^j + x^j y^i)e_i \otimes e_j \ ,$$

$$(x \otimes y)_{as} = \frac{1}{2}(x^i y^j - x^j y^i)e_i \otimes e_j = x^i y^j \frac{1}{2}(e_i \otimes e_j - e_j \otimes e_i) \tag{8.12}$$

$$= x^i y^j e_i \wedge e_j = x \wedge y = -y \wedge x \ .$$

All preceding considerations apply as well to E_N^*: the tensor product $f_1^* \otimes f_2^*$ may be defined as follows:

$$f^* \otimes f^{*\prime}(x \otimes y) = f^*(x)f^{*\prime}(y) \ . \tag{8.13a}$$

We may also introduce a basis $\{\omega^k\}$ of $E^*{}_N$ and write

$$f^* \otimes f^{*\prime} = f_i^* f^{*\prime}{}_j \omega^i \otimes \omega^j \ . \tag{8.13b}$$

These results can be readily generalized to a tensor product of n-forms. For the definition of the antisymmetric tensor product of n-forms, we write

$$f_1^* \wedge f_2^* \wedge \ldots f_n^* = \frac{1}{n!} \sum_{i_k}^{P} \eta_{i_1 \ldots i_n} f_{i_1}^* \otimes f_{i_2} \otimes \ldots f_{i_n}^* \ ,$$

$$i_k = 1 \ldots n \ , \tag{8.14}$$

where Σ^P represents the sum of all possible permutations of the indices $i_1 \ldots i_N$, and $\eta_{i_1 \ldots i_n}$ the sign of each permutation ($\pm$ if the number of transpositions is even or odd) allowing to go over from $i_1 \ldots i_n$ to $1 \ldots n$.

Finally, a **p-form** α is an element of $E_N^* \otimes E_N^* \otimes \ldots (p$ factors) such that

$$\alpha = \alpha_{i_1 \ldots i_p} \omega^{i_1} \wedge \omega^{i_2} \wedge \ldots \omega^{i_p} \tag{8.15}$$

and the coefficients $\alpha_{i_1 \ldots i_p}$ are real numbers (see also Sec. 9). One may also point out that for two forms of degrees p and q one has

$$\alpha \wedge \beta = (-1)^{pq} \beta \wedge \alpha , \tag{8.16a}$$

and, if $\alpha = \beta$,

$$\alpha \wedge \alpha = (-1)^{p^2} \alpha \wedge \alpha , \tag{8.16b}$$

where if $p = 2k + 1$ (odd), p^2 is also odd and $\alpha \wedge \alpha = 0$.

As a conclusion, let us make a last remark: consider the 2-form $\omega^{i_1} \wedge \omega^{i_2}$ in E_2: it can take only 3 values $0, \omega^1 \wedge \omega^2, -\omega^1 \wedge \omega^2$, and so can the 2-form $\varepsilon^{i_1 i_2} \omega^1 \wedge \omega^2 (\varepsilon^{i_1 i_2}$ is the numerical Levi-Civita tensor in E_2^*, i.e. the completely antisymmetric tensor of second order in E_2^*), then

$$\omega^{i_1} \wedge \omega^{i_2} = \varepsilon^{i_1 i_2} \omega^1 \wedge \omega^2 \tag{8.17a}$$

and, by a simple recursive method, one can show that in E_N^*,

$$\omega^{i_1} \wedge \omega^{i_2} \wedge \ldots \omega^{i_N} = \varepsilon^{i_1 i_2 \ldots i_N} \omega^1 \wedge \omega^2 \wedge \ldots \omega^N , \tag{8.17b}$$

$\varepsilon^{i_1 \ldots i_N}$ being again the Levi-Civita tensor (see Problem 1.7).

Remarks: We summarize the foundations of the tensor product theory. Consider two vector spaces E_N and $E_{N'}$ with respective bases $\{e_1 \ldots e_N\}$ and $\{e_1' \ldots e_{N'}'\}$, and let us define the space $E_N \times E_{N'}$ as the set of couples $\{x, y\}$ with $x \in E_N, y \in E_{N'}$ and also the map $E_N \times E_{N'} \to E_N \otimes E_{N'}$, a vector space of NN' dimensions, such that

$$x \otimes y \in E_N \otimes E_{N'} ,$$

and let us require for the preceding map the following properties:

(a) x, x' and y, y' being respectively elements of E_N and $E_{N'}$, one has

$$x \otimes (y + y') = x \otimes y + x \otimes y' ,$$
$$(x + x') \otimes y = x \otimes y + x' \otimes y .$$

(b) For any $\alpha \in \mathbb{R}$,

$$(\alpha x) \otimes y = x \otimes (\alpha y) = \alpha(x \otimes y) \ .$$

(c) One of the possible bases for $E_N \otimes E_{N'}$ is $\{e_i \otimes e'_j : i = 1 \ldots N, j = 1 \ldots N'\}$.

9. *Cotangent Vector Space • Covectors and Covariant Tensors • Exterior Derivation*

Consider the dual $T^*_{\mathcal{M}}(M)$ of the tangent vector space $T_{\mathcal{M}}(M) : T^*_{\mathcal{M}}(M)$ is the **cotangent vector space** at $M \in \mathcal{M}_N$; its elements, denoted by α_M, are called **linear 1-forms** (or simply 1-forms), **covariant vectors** or **covectors**. Let $\{\omega^1_M \ldots \omega^N_M\}$ be a moving frame (basis at M) of $T^*_{\mathcal{M}}(M)$ such that

$$\omega^i_M(e^M_k) = \delta^i_k \ .$$

The set of the ω's is called the **dual basis** to $\{e^M_1 \ldots e^M_n\}$ of $T_{\mathcal{M}}(M)$. Any element α_M of $T^*_{\mathcal{M}}$ is then a linear combination

$$\alpha_M = \alpha_i(x)\omega^i_M \ . \tag{9.1a}$$

The $\{x^i\}$ are the coordinates of M in a certain chart characterized by the homeomorphism $\varphi : x = \varphi(M)$. Then for any $v_M \in T_{\mathcal{M}}(M)$,

$$\alpha_M(v_M) = \alpha_i(x)\omega^i_M(v^j(x)e^M_j) = \alpha_i(x)v^i(x) \ . \tag{9.1b}$$

By the method of the space products presented in the preceding paragraph, we may define a vector space

$$T^*_{\otimes p} = T^*_{\mathcal{M}}(M) \otimes T_{\mathcal{M}}(M) \otimes \ldots \ ,$$

the factor being p in number and the dimension of the space being pN. One of its possible bases will be given by the N^P vectors

$$\{\omega^{i_1}_M \otimes \omega^{i_2}_M \otimes \ldots \omega^{i_p}_M : \ i_k = 1 \ldots N, k = 1 \ldots p\}$$

and a p-covariant tensor A_M will be written as

$$A_M = A_{i_1 \ldots i_p}(x)\omega^{i_1}_M \otimes \ldots \omega^{i_p}_M \ . \tag{9.2}$$

One may easily obtain the transformation law of $a_{i_1\ldots i_p}(x)$ (which in the vocabulary used at Level 1 was called a p-covariant tensor field) in perfect agreement with the law (4.3) of Level 1.

Among all p-covariant tensors, we shall distinguish the antisymmetric ones:

$$\alpha_M = \alpha_{i_1\ldots i_p}(x)\omega_M^{i_1} \wedge \ldots \omega_M^{i_p} \ . \tag{9.3}$$

Such tensors will be called **differential p-forms**. Let us make three comments about this last definition: we notice first that no symmetric part of $\alpha_{i_1\ldots i_p}(x)$ can contribute to (9.3a) because of the complete antisymmetry of $\omega^{i_1} \wedge \ldots \omega^{i_p}$. Next, remark that starting from a set of components $\alpha_{i_1\ldots i_p}$ with no preferred symmetry, it is straightforward to build up a set of antisymmetric components. Indeed, we may change at will the names of the dummy indices $i_1 \ldots i_p$ and obtain $p!$ other expressions of α_M:

$$\alpha_M = \alpha_{i_1 i_2 \ldots i_p}\omega^{i_1} \wedge \omega^{i_2} \wedge \ldots \omega^{i_p} = \alpha_{i_2 i_1 i_3 \ldots i_p}\omega^{i_2} \wedge \omega^{i_1} \wedge \ldots \omega^{i_p}$$
$$= \ldots \tag{9.3a}$$

and we may write

$$\alpha_M = \frac{1}{p!}(\alpha_{i_1 i_2 \ldots i_p} - \alpha_{i_2 i_1 \ldots i_p} + \ldots)\omega^{i_1} \wedge \omega^{i_2} \wedge \ldots \omega^{i_p} \ . \tag{9.3b}$$

One may use several notations to denote the bracket; we shall use either $\alpha_{(i_1\ldots i_p)}$ or $p!\alpha_{[i_1\ldots i_p]}$. The third remark concerns the set $\{\omega^{i_1} \wedge \ldots \omega^{i_p}\}$: it is not a basis for the vector space of differential p-forms, which will be denoted by $\wedge^p T^*_{\mathcal{M}}(M)$, since $\omega^i \wedge \omega^j = -\omega^j \wedge \omega^i$. But we may extract a basis by considering the subset

$$\{\omega^{i_1} \wedge \omega^{i_2} \wedge \ldots \omega^{i_p} : i_1 < i_2 < i_3 < \ldots i_p\} \ .$$

Then, the factor $1/p!$ does not appear anymore in α_M and one has

$$\alpha_M = \sum_{i_1 < i_2 < \ldots i_p} \alpha_{i_1} \ldots i_p \omega^{i_1} \wedge \ldots \omega^{i_p} = \sum_{i_1 < i_2 < \ldots i_p} \alpha_{(i_1\ldots i_p)}\omega^{i_1} \wedge \ldots \omega^{i_p} \ . \tag{9.3c}$$

When such a restriction is imposed in the summation, the components of α_M are called **strict components** or **independent components** (by physicists!).

We want now to introduce an operation which will play a central part in our forthcoming considerations: such an operation is the **exterior derivation** denoted by $\underline{d}$, in contradistinction to the usual symbol d of ordinary differentiation.

The map $\underline{d}$ is defined as follows:

$$\{\text{space of the differential } p\text{-forms}\}$$
$$\rightarrow \{\text{space of the differential } (p+1)\text{-forms}\}$$

with the following properties.

(1) A C^∞ function is a 0-form (also a 0-tensor).

(2) If α, β, γ are three p-forms such that $\alpha = \beta + \gamma$, then

$$\alpha = \beta + \gamma: \quad \underline{d}\alpha = \underline{d}\beta + \underline{d}\gamma \ .$$

(3) $$\underline{d}^2 = 0 \ .$$

(4) $$\underline{d}\{\alpha \wedge \beta\} = (\underline{d}\alpha) \wedge \beta + (-1)^p \alpha \wedge \underline{d}\beta$$

if α is a p-form.

(5) Let $\hat{F}$ be a C^∞ function (0-form), then $\underline{d}\hat{F}_M$ is a 1-form (following (1)) and for any vector $v_M \in T_{\mathcal{M}}(M)$, one has

$$\underline{d}\hat{F}v_M = v_M \hat{F} \ .$$

Postulate (5) may be analyzed using (4.8) and (4.17b); indeed,

$$\underline{d}\hat{F}v_M = v^i(x)\partial_i F(x) = V^i(x)\mathbf{D}_i F(x) \ . \tag{9.4}$$

We may choose for $F(x)$ the coordinate x^i of the point M in the natural basis $\{\varepsilon_j^M\}$; then

$$\underline{d}x^i \varepsilon_j^M = \varepsilon_j^M(x^i) = \delta_j^i \ . \tag{9.5a}$$

As a consequence the set $\{\underline{d}x^1 \ldots \underline{d}x^N\}$ is the dual basis of the natural basis $\{\varepsilon_1^M \ldots \varepsilon_N^M\}$. The 1-form α_M given by (9.1) can also be written as

$$\alpha_M = \alpha_i^{(n)}(x)\underline{d}x^i \ , \tag{9.5b}$$

where $\alpha_i^{(n)}(x)$ are the natural components of α_M (the index (n) will be omitted below).

The denomination of the differential form is justified, since for any C^∞ function $\hat{F}(M)$ one can write

$$\underline{d}\hat{F}_M = \partial_i F(x)\underline{d}x^i \; , \tag{9.6}$$

since

$$\underline{d}\hat{F}_M(v_M) = \partial_i F(x)\underline{d}x^i(v^j \varepsilon_j^M) = v^i(x)\partial_i F(x) \; . \tag{9.7}$$

The physicist has to distinguish the vector dx, element of E_N with components dx^j in the natural basis, and the differential 1-forms $\underline{d}x^j$. Furthermore,

$$\underline{d}x^j(dx) = \underline{d}x^j(dx^i \varepsilon_i^M) = dx^j \; . \tag{9.8}$$

To the preceding rules, we add another rule:

(6) For a C^∞ function $\hat{F}$ and a differential form α_M, we shall define $F(M) \wedge \alpha_M$ as

$$\hat{F}_M \wedge \alpha_M = \hat{F}(M)\alpha_M \; ,$$

and following rule (4) we also have

$$\underline{d}\{\hat{F}_M \wedge \alpha_M\} = \underline{d}\hat{F}_M \wedge \alpha_M + \hat{F}_M \wedge \underline{d}\alpha_M \; . \tag{9.9}$$

See also (9.24).

In Sec. 4, we did introduce the two frames $\{\varepsilon_j^M\}$ and $\{e_j^M\}$ of $T_{\mathcal{M}}(M)$ with the relation given by (4.17):

$$e_i^M = S_i^j(x)\varepsilon_j^M \; . \tag{9.10}$$

We may also consider the respective dual bases $\{\underline{d}x^j\}$ and $\{\omega_M^j\}$ such that, following (8.10),

$$\omega_M^i = (S^{-1})_j^i \underline{d}x^j \; . \tag{9.11}$$

In the natural and local frames of $T_{\mathcal{M}}(M)$ for a function $\hat{F}$ of (9.5), one gets

$$\underline{d}\hat{F}_M = \partial_i F(x)\underline{d}x^i = \partial i F(x)S_j^i \omega_M^j = (\mathbf{D}_j F(x))\omega_M^j \; , \tag{9.12}$$

where we took into account the definition of the pfaffian derivative (4.11),

$$\mathbf{D}_i = S_i^j(x)\partial_j \; . \tag{9.13}$$

One goes over from (9.5) to (9.12) by replacing ∂_j by $\mathbf{D}_j$; the structure coefficients $c^m{}_{ij}$ were introduced by (4.14):

$$[\mathbf{D}_i \mathbf{D}_j] = c^m{}_{ij}(x)\mathbf{D}_m \; . \tag{9.14}$$

We shall find again these coefficients in the **Cartan-Maurer formula**:

$$\underline{d}\omega_M^k = -\frac{1}{2}c^k{}_{ij}\omega_M^i \wedge \omega_M^j \ , \tag{9.15}$$

which we shall now prove. We thus observe, first of all, that since ω_M^k is a 1-form, $\underline{d}\omega_M^k$ is a 2-form and may be expressed by a linear combination of $\omega_M^i \wedge \omega_M^j$. We still have to calculate explicitly the coefficients c_{ij}^k of (9.15) in order to show that they are indeed the structure coefficients which appear in (9.14). In all our calculations, the matrix S^{-1} will occur very often, and we shall denote it by s:

$$Ss = sS = 1 \ . \tag{9.16}$$

We then have

$$\begin{aligned}
\underline{d}\omega_M^k(e_i^M \wedge e_j^M) &= \underline{d}\{s_n^k(x)\underline{d}x^n\}(e_i^M \wedge e_j^M) \\
&= \partial_m s_n^k(x)\underline{d}x^m \wedge \underline{d}x^n(e_i^M \wedge e_j^M) \ .
\end{aligned}$$

If we express e_i^M, e_j^M as functions of the ε_i^M, we may also write

$$\underline{d}\omega_M^k\{e_i^M \wedge e_j^M\} = (\partial_m s_n^k(x))S_i^l(x)S_j^p(x)\underline{d}x^m \wedge \underline{d}x^n(\varepsilon_l^M \wedge \varepsilon_p^M) \ ,$$

or

$$\begin{aligned}
\underline{d}x^m \wedge \underline{d}x^n(\varepsilon_l^M \wedge \varepsilon_p^M) &= \frac{1}{2}(\underline{d}x^m \wedge \underline{d}x^n - \underline{d}x^n \wedge \underline{d}x^m)(\varepsilon_l^M \wedge \varepsilon_p^M) \\
&= \frac{1}{2}(\delta_l^m\delta_p^n - \delta_p^m\delta_l^n) \ ,
\end{aligned}$$

and finally

$$\underline{d}\omega_M^k(e_i^M e_j^M) = \frac{1}{2}S_i^l(x)S_j^p(x)(\partial_l s_p^k(x) - \partial_p s_l^k(x)) \ . \tag{9.17}$$

Let us differentiate (9.16) with respect to x^p:

$$\partial_p\{Ss\} = (\partial_p S)s + S\partial_p s = 0 \ ,$$

hence

$$S(x)\partial_p s(x) = -(\partial_p S(x))s(x) \ ,$$

or

$$S_i^l(x)\partial_p s_l^k(x) = -(\partial_p S_l^i(x))s_l^k(x) ,$$
$$S_j^p(x)\partial_l s_p^k(x) = -(\partial_l S_j^p(x))s_p^k(x) .$$

Bringing into (9.17) these last results, one obtains

$$\underline{d}\omega_M^k(e_i^M \wedge e_j^M) = -\frac{1}{2}c_{ij}^k = -\frac{1}{2}c_{lm}^k\omega_M^l \wedge \omega_M^m(e_i^M \wedge e_j^M) ,$$

hence

$$\underline{d}\omega_M^k = -\frac{1}{2}c_{lm}^k(x)\omega_M^l \wedge \omega_M^m , \qquad (9.18)$$

precisely the Cartain-Maurer formula.

We will now treat another problem: how does the volume element of a multiple integral behave under a change of variables? We will have, using exterior differentiations, a more systematic approach than the one used in classical analysis. Let us first consider the volume element $dx^1 dx^2$ of a double integral and the change of variables:

$$x^1 = f(u,v) , \quad x^2 = g(u,v) . \qquad (9.19)$$

It is well-known that one cannot write $dx^1 dx^2 = df dg$; let us show that one can replace that notation by the following:

$$\underline{d}x^1 \wedge \underline{d}x^2 = \underline{d}f \wedge \underline{d}g . \qquad (9.20)$$

One has indeed

$$\begin{aligned}
\underline{d}x^1 \wedge \underline{d}x^2 &= \left(\frac{\partial f}{\partial u}\underline{d}u + \frac{\partial f}{\partial v}\underline{d}v\right) \wedge \left(\frac{\partial g}{\partial u}\underline{d}u + \frac{\partial g}{\partial v}\underline{d}v\right) \\
&= \frac{\partial f}{\partial u}\frac{\partial g}{\partial v}\underline{d}u \wedge \underline{d}v + \frac{\partial f}{\partial v}\frac{\partial g}{\partial u}\underline{d}v \wedge \underline{d}u \\
&= \frac{\partial(f,g)}{\partial(u,v)}\underline{d}u \wedge \underline{d}v .
\end{aligned} \qquad (9.21)$$

The Jacobian $\partial(f,g)/\partial(u,v)$ automatically appears if one uses the notations of the exterior algebra. The generalization to N variables can be readily made: let us consider a change of coordinates $x \to x'$ such that

$$x^k = f^k(x'^1 \ldots x'^N) : \ k = 1 \ldots N , \qquad (9.22)$$

then

$$\underline{dx}^1 \wedge \ldots \underline{dx}^N = \frac{\partial f^1}{\partial x'^{i_1}} dx'^{i_1} \wedge \frac{\partial f^2}{\partial x'^{i_2}} \underline{dx}^{i_2} \wedge \ldots \frac{\partial f^N}{\partial x'^N} \underline{dx}'^N$$

$$= \frac{\partial f^1}{\partial x'^{i_1}} \ldots \frac{\partial f^N}{\partial x'^{i_N}} \underline{dx}'^{i_1} \wedge \ldots \underline{dx}'^{i_N} \tag{9.23}$$

$$= \frac{\partial(f^1 \ldots f^N)}{\partial(x'^1 \ldots x'^N)} \underline{dx}'^1 \wedge \ldots \underline{dx}'^N \ ,$$

as one verifies after some transformations.

We conclude the present paragraph by proving some formulae of constant use in the calculations of exterior derivatives. We go back to formula (9.4) and notice that this formula may take the form

$$\underline{d}\hat{F}_M v_M = (\mathbf{D}_i F(x)) V^i(x) = \mathbf{D}_i F(x) \omega_M^i(v_M) \ ,$$

hence

$$\underline{d}\hat{F}_M = (\mathbf{D}_i F(x)) \omega_M^i \ , \tag{9.24}$$

being the generalization of (9.5).

Another useful formula deals with the exterior differential of any differential form. Let us begin with a 1-form α_M,

$$\alpha_M = \alpha_i(x) \omega_M^i \ .$$

The definition of $\underline{d}$ and formulae (9.9), (9.15) and (9.24) lead to

$$\underline{d}\alpha_M = \underline{d}\{\alpha_i(x) \omega_M^i\} = (\underline{d}\alpha_i(x)) \wedge \omega_M^i + \alpha_i(x) \underline{d}\omega_M^i$$

$$= \mathbf{D}_j \alpha_i(x) \omega_M^j \wedge \omega_M^i - \frac{1}{2} c_{kl}^i(x) \alpha_i(x) \omega_M^k \wedge \omega_M^l \tag{9.25}$$

$$= \left(\mathbf{D}_k \alpha_l(x) - \frac{1}{2} \alpha_i(x) c_{kl}^i(x) \right) \omega_M^k \wedge \omega_M^l \ .$$

If, instead of ω^i, one uses the natural basis $\underline{dx}^i$, the former formula is reduced to

$$\underline{d}\alpha_M = \partial_i \alpha_j(x) \underline{dx}^i \wedge \underline{dx}^j \tag{9.26}$$

and, for a differential p-form,

$$\alpha_M = \alpha_{j_1 \ldots j_p}(x) \underline{dx}^{j_1} \wedge \ldots \underline{dx}^{j_p} \ , \tag{9.27}$$

one has

$$\underline{d}\alpha_M = \partial_i \alpha_{j_1 \ldots j_p}(x) \underline{dx}^i \wedge \underline{dx}^{j_1} \wedge \ldots \underline{dx}^{j_p} \ . \tag{9.28}$$

One last linguistic distinction: any differential p-form such as (9.27) should be called a differential p-form in its natural basis: this is a cumbersome denomination and we shorten it by saying simply p-form when there is no misunderstanding to be feared.

10. *Change of Chart Map of a Manifold $\mathcal{M}_N$ into a Manifold $\mathcal{M}'_{N'}$*
Lie Derivative

The corresponding problems for $T_{\mathcal{M}}(M)$ were studied in paragraphs 5 and 6; we now want to solve them for $T^*_{\mathcal{M}}(M)$.

The solution of the problem of the **change of chart** is straightforward: let α_M be a differential 1-form and v_M any vector of $\mathcal{M}_N$ at M. Considering two charts of the same neighborhood of $\mathcal{M}_N$ corresponding to the homeomorphisms $\varphi' \neq \varphi$, one has

$$\alpha_{\varphi(M)}(v_{\varphi(M)}) = \alpha_{\varphi'(M)}(v_{\varphi'(M)}) \ . \tag{10.1a}$$

Locally, this formula with $x = \varphi(M), x' = \varphi'(M)$ means that

$$\alpha_j(x)\underline{d}x^j(v_M) = \alpha'_i(x')\underline{d}x'^i(v_M) \ . \tag{10.1b}$$

But to go from x to x', we need N local transformations

$$x'^i = f^i(x^1 \ldots x^N) = x'^i(x^1 \ldots x^N) \ , \tag{10.2}$$

so that

$$\underline{d}x'^i = \partial_j f(x)\underline{d}x^j \ . \tag{10.3}$$

We now bring that result into (10.1) to obtain

$$\alpha_j(x)\underline{d}x^j(v_M) = \alpha'_i(x')(\partial_j f^i(x))\underline{d}x^j(v_M) \tag{10.4}$$

and, by identification of both sides,

$$\alpha_j(x) = \frac{\partial x'^i}{\partial x^j}\alpha'_i(x') \ , \tag{10.5}$$

which agrees with formula (4.2) of Level 1:

$$\alpha'_i(x') = \frac{\partial x^j}{\partial x'^i}\alpha_j(x) \ . \tag{10.6}$$

The second problem concerns the **map of a manifold $\mathcal{M}_N$ into another $\mathcal{M}'_{N'}$**: it requires some attention. Let $P \in \mathcal{M}_N, P' \in \mathcal{M}'_{N'}$ and consider their respective neighborhoods $U_{\mathcal{M}}(P)$ and $U_{\mathcal{M}'}(P')$ and let ϕ be a diffeomorphism between these two neighborhoods. Then to any point $M \in U_{\mathcal{M}}(P)$, there corresponds $M' \in U_{\mathcal{M}'}(P')$ such that

$$M' = \phi(M) \ .$$

Let us consider the tangent and cotangent vector spaces at P and P' and the following map induced by ϕ

$$T_M(P) \overset{D\phi_M^{M'}}{\to} T_{M'}(P') : \quad v'_{M'} = D\phi_M^{M'} v_M \ ,$$
$$T_M^*(P) \overset{D\phi_{M'}^{*M}}{\leftarrow} T_{M'}^*(P') : \quad \alpha_M = D\phi_{M'}^{*M} \alpha'_{M'} \ , \tag{10.7}$$

where $D\phi$ has been defined in (6.4) and $D\phi^*$ (denoted very often also by ϕ^*) will be defined below. The reader should carefully note that the directions of the arrows in (10.7) are different for the two maps.

We shall suppose that the transformation law for a differential 1-form α_M obeys the following rule:

$$\alpha'_{M'}(D\phi_M^{M'} v_M) = (D\phi_{M'}^{*M} \alpha'_{M'}) v_M \ , \tag{10.8}$$

which can be simply written as

$$\alpha'_{M'}(v'_{M'}) = \alpha_M(v_M) \ , \tag{10.9}$$

where

$$v'_{M'} = D\phi_M^{M'} v_M \ , \quad \alpha_M = D\phi_{M'}^{*M} \alpha'_{M'} \ . \tag{10.10}$$

We now use local coordinates to transform (10.9) into

$$\alpha'_{M'}(v'^{\lambda}(x')\varepsilon'^P_{\lambda}) = \alpha_M(v^k(x)\varepsilon^M_k) : \quad \begin{matrix} k = 1 \ldots N \\ \lambda = 1 \ldots N' \end{matrix} \tag{10.11}$$

and introduce into (10.11) the components $\alpha'_{\lambda}(x')$ and $\alpha_k(x)$ of $\alpha'_{M'}$ and α_M respectively to obtain

$$\alpha'_{\lambda}(x')v'^{\lambda}(x') = \alpha_k(x)v^k(x) \ . \tag{10.12}$$

But $v'^{\lambda}(x')$ can be expressed by means of (6.19) as a function of $v^k(x)$, then

$$\alpha'_{\lambda}(x')\frac{\partial x'^{\lambda}}{\partial x^k}v^k(x) = \alpha_k(x)v^k(x) \ , \tag{10.13}$$

i.e.

$$\alpha_k(x) = \frac{\partial x'^{\lambda}}{\partial x^k}\alpha'_{\lambda}(x') \ . \tag{10.14}$$

We summarize now the local expressions of some of the formulae given above. We notice first that the relation $M' = \phi(M)$ has the local expression

$$x'^\lambda = \phi^\lambda(x^1 \ldots x^N) = x'^\lambda(x^1 \ldots x^N) \ , \qquad (10.15)$$

then the first of the formula (10.7) takes the form

$$D\phi_m^{M'} v_M = v'_{M'} = v'^\lambda(x')\varepsilon'^P_\lambda = \frac{\partial x'^\lambda}{\partial x^k} v^k(x)\varepsilon'^P_\lambda \ , \qquad (10.16)$$

where $\lambda = 1 \ldots N'$ and $k = 1 \ldots N$. The second formula (10.7) can be written as

$$D\phi_{M'}^{*M} \alpha'_{M'} = \alpha_M = \alpha_k(x)\underline{d}x^k = \frac{\partial x'^\lambda}{\partial x^k} \alpha'_\lambda(x)\underline{d}x^k \qquad (10.17a)$$

and we may use a matrix notation

$$D\phi_{M'}^{*M} \alpha'_{M'} = \alpha'(x')\frac{\mathbf{D}(x')}{\mathbf{D}(x)}\underline{d}x \ , \qquad (10.17b)$$

provided that $\alpha'(x')$ is considered as a row matrix while $\underline{d}x$ is a column matrix with $\underline{d}x^k$ as elements.

We now come to the definition of the Lie derivative of a covariant tensor. As above $D\phi^*$ will be considered as a linear diffeomorphism between $T^*_{\mathcal{M}'}(P')$ and $T^*_{\mathcal{M}}(P)$ and we shall suppose that $\mathcal{M}_N = \mathcal{M}'_{N'}$. Then for any covariant tensor $\underline{T}_{M'}$, we shall define

$$L_Y \underline{T}_M = \lim_{\tau=0} \cdot \frac{1}{\tau}[D\phi_{\tau M'}^{*M} \underline{T}_{M'} - \underline{T}_M] \ , \qquad (10.18)$$

where $M' = \phi_\tau(M)$ as in (7.1).

We remark that L_Y is linear, obeys the Leibniz rule (as in (7.21) and (7.22)) and is a derivative on the algebra of the differentiable covariant tensors. As an example, we may consider the case $\underline{T}_M = \alpha_M$, a differential 1-form: we recall formulae (7.13), (7.14). The coordinates $\xi^k(x, \tau)$ are

$$\xi_k(x, \tau) = x^k + Y^k(x)d\tau$$

and

$$\xi(x, 0) = x \ , \quad Y^k(x) = \frac{\partial \xi^k(x, \tau)}{\partial \tau}\bigg|_{\tau=0} \ .$$

Using (10.17), one obtains

$$D\phi^{*M}_{\tau M'}\alpha_{M'} = (\partial_k \xi^j(x + Y(x)d\tau))\alpha_j(x + Y(x)d\tau)\underline{d}x^k ,$$

the arguments of ξ_j and α^j being $x + Y(x)d\tau$. We then expand them up to first order in $d\tau$ and obtain

$$\begin{aligned}
&(\partial_k\{\xi^j(x,0) + Y^i(x)d\tau\partial_i\xi^j(x,\tau)|_{\tau=0})(\alpha_j(x) + Y^l(x)d\tau\partial_l\alpha_j(x))\underline{d}x^k \\
&= (\delta^j_k + \partial_k Y^i(x)d\tau\delta^j_i)(\alpha_j(x) + Y^l(x)d\tau\partial_l\alpha_j(x))\underline{d}x^k \\
&= (\alpha_k(x) + Y^l(x)\partial_l\alpha_k(x)d\tau + (\partial_k Y^j(x))\alpha_j(x))\underline{d}x^k .
\end{aligned}$$

We now use (10.18) to get

$$L_Y\alpha_M = [Y^i(x)\partial_i\alpha_k(x) + (\partial_k Y^i(x))\alpha_i(x)]\underline{d}x^k , \qquad (10.19)$$

a formula which agrees completely with (11.9) of Level 1. In the special case

$$\alpha_M = \underline{d}x^j = \delta^j_l\underline{d}x^l ,$$

one gets

$$L_Y\underline{d}x^j = \partial_i Y^j(x)\underline{d}x^i = \underline{d}y^j(x) . \qquad (10.20)$$

The calculation of the Lie derivative of any covariant tensor can now be undertaken along the lines which led to formula (7.26): the result completely agrees with (11.10) of Level 1. We leave these calculations to the reader.

Let us present two useful formulae. The first one,

$$L_Y\{\alpha_M \wedge \beta_M\} = (L\alpha_M) \wedge \beta_M + \alpha_M \wedge L_Y\beta_M , \qquad (10.21)$$

can be proved by antisymmetrization (see (7.21)). The second concerns the Lie derivative of any differential p-form:

$$\begin{aligned}
\mathcal{L}_Y\alpha_M = (Y^j(x)\partial_j\alpha_{i_1\ldots i_p}(x) + \alpha_{ji_2\ldots i_p}\partial_{i_1}Y^j(x) + \ldots \\
+ \alpha_{i_1\ldots i_{p-1}j}(x)\partial_{i_p}Y^j(x))\underline{d}x^{i_1} \wedge \ldots \underline{d}x^{i_p} . \qquad (10.22)
\end{aligned}$$

In the present paragraph we considered the matrix elements of $D\phi^*$, map of $T^*_{\mathcal{M}}(P)$ into $T^*_{\mathcal{M}'}(P')$, and wish to find the corresponding representation of the map $(D\phi^*)^{\otimes p}$ of $(T^*_{\mathcal{M}})^{\otimes p}$ into $(T^*_{\mathcal{M}})^{\otimes p}$. We may use considerations very similar to the ones used in Sec. 6. One finds

$$(\text{mat}(D\phi^*)^{\otimes p})^{\lambda_1\ldots\lambda_p}_{i_1\ldots i_p} = \frac{\partial x'^{\lambda_1}}{\partial x^{i_1}} \cdots \frac{\partial x'^{\lambda_p}}{\partial x^{i_p}} , \qquad (10.23)$$

with $\lambda_r = 1 \ldots N'$ and $j_k = 1 \ldots N$. This is a remark which will be useful in Sec. 14 on the integration of differential forms.

11. *Total Differential, Closed and Exact Differential Forms*

Physicists make frequent uses of the concept of total differential $df(x)$ of a function $f(x^1 \ldots x^N)$; they define a total differential as the infinitesimal increment of the function f when each of its arguments is submitted to an infinitesimal increment $dx^k, k = 1 \ldots N$.

But physics needs "linear forms". For example, in thermodynamics the elementary quantity of heat δQ exchanged with the external medium by a certain mass of divariant fluid of volume v and pressure p when these variables become $v + dv$ and $p + dp$ is given by

$$\delta Q = h(p, v)dp + l(p, v)dv . \tag{11.1}$$

The elementary heat is denoted by δQ and not dQ just to remind the reader that, although an infinitesimal, δQ is not a total differential. If $T(t)$ is the absolute temperature of the fluid, $\delta Q/T = dS$ is the total differential of the entropy and T is the **integrating factor** of the linear form δQ. See also Problem 2.6.

1-Linear forms are useful tools in physics:

$$\delta \alpha = \alpha_i(x^1 \ldots x^N)dx^i , \tag{11.2}$$

and it is well-known in classical analysis that if the **integrability conditions** hold

$$\partial_j \alpha_i(x) = \partial_i \alpha_j(x) : \; i \neq j , \tag{11.3}$$

then $\delta \alpha$ is a total differential $d\alpha$.

The mathematician is more careful with the corresponding concept of differential 1-forms: the concept of infinitesimal quantity is no more admitted in mathematics and the increments dx^j become differential 1-forms $\underline{d}x^j$ such that if

$$dx = dx^j \varepsilon_j^M , \tag{11.4a}$$

then

$$\underline{d}x^k(dx) = dx^k . \tag{11.4b}$$

Let us now consider a 1-form

$$\alpha_M = \alpha_i(x^1 \ldots x^N)\underline{d}x^i , \tag{11.5}$$

which will be a **closed 1-form** if

$$\underline{d}\alpha_M = 0 \ . \tag{11.6}$$

An example of closed forms is the 1-form $\underline{d}\hat{F}_M$, where $\hat{F}$ is a C^∞ function.

Take now a closed 1-form and ask the question: is it possible to find a 0-form $\hat{F}$ such that

$$\alpha_M = \underline{d}\hat{F}_M \ . \tag{11.7}$$

The answer to that question is generally "no", except if $\mathcal{M}_N = \mathbb{R}^N$, then (11.6) has (11.7) as a consequence, the converse being obvious. This is the **Poincaré lemma**: for its proof the reader may consult [Bibl. 9], p. 215. When (11.7) is satisfied, one says that α_M is an **exact 1-form**. In such a case, formula (11.7) leads to

$$\alpha_M = \alpha_i(x)\underline{d}x^i = d\hat{F}_M$$

and, as a consequence,

$$\alpha_i(x) = \partial_i F(x) \ , \tag{11.8}$$

a formula from which one can derive again the integrability conditions (11.3). These conclusions can be generalized very simply for any p-form and the Poincaré lemma remains also valid. We shall find later on (Sec. 17) some application to Betti numbers.

Because of their physical applications, we may dwell upon these last points: a vector field of forces is said to be **conservative** iff there exists a function $V(x)$, the **potential**, such that

$$X_k(x) = -\partial_k V(x) \ , \tag{11.9}$$

where x are space coordinates (or generalized coordinates). Note the time t is not an argument of V.

Equations (11.3) and (11.8) are the translation of that remark and its consequences in the language of differential geometry: consider indeed the differential 1-form

$$\alpha_M = X_k(x)\underline{d}x^k \tag{11.10}$$

and suppose α_M to be closed, then

$$\begin{aligned}
0 = \underline{d}\alpha_M &= \partial_j X_k(x)\underline{d}x^j \wedge dx^k \\
&= \sum_{j<k}(\partial_j X_k(x) - \partial_k X_j(x))\underline{d}x^j \wedge \underline{d}x^k \ ,
\end{aligned}$$

implying that

$$\partial_j X_k(x) - \partial_k X_j(x) = 0 \ , \tag{11.11}$$

which are precisely the integrability conditions (11.3).

If $\mathcal{M}_N$ is the vector space $\mathbb{R}^N$, then from the Poincaré lemma α_M is also exact:

$$\alpha_M = -d\hat{V}(M) = -\partial_k V(x)\underline{dx}^k \ , \tag{11.12}$$

which implies (11.9). In $\mathbb{R}^3$ both conditions (11.9) and (11.11) read

$$\mathbf{X}(\mathbf{x}) = -\mathbf{\nabla}V(\mathbf{x}) \ ,$$
$$\mathbf{\nabla} \wedge \mathbf{X}(\mathbf{x}) = \mathrm{curl}\mathbf{X}(\mathbf{x}) = 0 \ .$$

Another important question concerns the conditions under which the flux of a vector field is conservative. Such a concept plays a central part in electrostatics and in the theory of Newtonian potential. The reader will find in Sec. 16, Part (B) some comments on that problem.

Remark: Notice that the Poincaré lemma is no more valid if the form α_M is singular at certain points of $\mathbb{R}^N$: although α_M becomes differentiable on $\mathbb{R}^N$ without the set of such points, such a manifold is no more homeomorphic to $\mathbb{R}^N$.

A classical example is provided by the form

$$\alpha_M = -\frac{x^2}{(x^1)^2 + (x^2)^2}\underline{dx}^1 + \frac{x^1}{(x^1)^2 + (x^2)^2}\underline{dx}^2 = \underline{d}\left\{\mathrm{arc\ tg}\,\frac{x^2}{x^1}\right\} \ ,$$

differentiable in $\mathbb{R}^2 - \{0\}$, but not exact.

D. Integration of Differential Forms

The development of important chapters of physics require the use of the transformation formulae of multiple integrals; such formulae are called **Green's** and **Stokes'** formulae. Potential theory, electrostatics, deformable media theories,... would still be in their infancy without the use of such formulae. But it should be remarked that a rigorous presentation of these formulae is outside the limits of classical analysis; it needs the intensive use of the machinery of modern differential geometry. In return, such an elaborate presentation brings to light fundamental analogies between these formulae which are particular cases of a single formula – **Stokes' formula**.

12. *Some Classical Reminders*

We consider the affine space $\mathcal{E}_3$ with an orthonormal frame $\{0, \mathbf{e}_1, \mathbf{e}_2, \mathbf{e}_3\}$ and a vector field

$$\mathbf{X}(\mathbf{x}) = \begin{pmatrix} X^1(\mathbf{x}) \\ X^2(\mathbf{x}) \\ X^3(\mathbf{x}) \end{pmatrix} \in E_3 \ ,$$

components with continuous derivatives within a compact $D \subset E_3$.

(a) *Stokes' formula:* Let S be a non-closed surface defined in D and having the curve $\partial S = C$ as boundary. Examples of such surfaces are given in all elementary textbooks. The surface $S \subset D$ is supposed to have a single unitary normal $\mathbf{N}$ at each of its points and the orientation of the normal defines the orientation along the contour C following Maxwell's rule.

Then **Stokes' formula** can be written as

$$\int_S \operatorname{curl}\mathbf{X}(\mathbf{x}) \cdot \mathbf{N}(\mathbf{x}) dS(\mathbf{x}) = \int_S \boldsymbol{\nabla} \wedge \mathbf{X} \cdot \mathbf{N} dS = \oint_C \mathbf{X}(\mathbf{x}) \cdot d\mathbf{x} \ , \qquad (12.1)$$

i.e., the flux of the curl of $\mathbf{X}$ through S is equal to the line integral of $\mathbf{X}$ along C. There are several ways for the evaluation of the left-hand side of (12.1). One could use the expression of dS in parametric coordinates as given at Level 0 and triangulize S into elementary parallelograms (see any textbook in classical analysis). Let us only remind the reader that using the frame defined at the beginning of the present paragraph, the surface integral of the left-hand side of (12.1) can be expressed as a sum of three double integrals:

$$\int_{S_2} (\partial_2 X_3 - \partial_3 X_2) dx^2 dx^3 + \int_{S_1} (\partial_3 X_1 - \partial_1 X_3) dx^3 dx^1$$
$$+ \int_{S_3} (\partial_1 X_2 - \partial_2 X_1) dx^2 dx^3 \ , \qquad (12.2)$$

where S_1, S_2, S_3 are the orthogonal projections of S in the corresponding coordinate planes. The right-hand side of (12.1) is a simple line integral which requires some comments (see also (15.11)).

As a consequence, let us consider two surfaces S and S' with a single normal $\mathbf{N}$ at each of their points and having C as a common boundary; the flux of curl $\mathbf{X}$ is the same for these two surfaces and is equal to

$$\oint_C \mathbf{X} \cdot d\mathbf{x} \ . \qquad (12.3)$$

That real number represents, following the considerations of Sec. 2, Level 0, a topological invariant of S (which remains constant when S is continuously deformed).

(b) *Green's formula:* Let V be a volume included in D with a closed boundary $\partial V = S$. We suppose that S divides D in two domains: one interior and the other one exterior to S, we also suppose that there is a single normal at each point of S and that this normal is oriented toward the exterior of S. Then *Green's formula* reads

$$\int_V \boldsymbol{\nabla} \cdot \mathbf{X}(\mathbf{x}) d_3 x = \int_V \operatorname{div} \mathbf{X} d_3 x = \oint_S \mathbf{X}(\mathbf{x}) \cdot \mathbf{N}(\mathbf{x}) dS(x) , \qquad (12.4)$$

where $d_3 x = dx^1 dx^2 dx^3$ is the classical volume element in orthogonal coordinates. One also has the alternative form

$$\int_V (\partial_1 X^1 + \partial_2 X^2 + \partial_3 X^3) dx^1 dx^2 dx^3$$
$$= \int_{S_1} X^1 dx^2 dx^3 + \int_{S_2} X^2 dx^3 dx^1 + \int_{S_3} X^3 dx^1 dx^2 , \qquad (12.5)$$

where S_1, S_2, S_3 have already been defined in Part (A). Formula (12.5) states that the volume integral of div $\mathbf{X}$ is equal to the flux of $\mathbf{X}$ through the boundary S of V. We notice that (12.4) implies that the flux of $\mathbf{X}$ is conservative if $\nabla \cdot \mathbf{X} = 0$. As a reminder, let us quote the following relations

$$\operatorname{div} \operatorname{curl} \mathbf{X} = 0 , \quad \operatorname{div} \mathbf{X} = 0 \Leftrightarrow \mathbf{X} = \operatorname{curl} \mathbf{A} ,$$
$$\operatorname{curl} \operatorname{grad} F = 0 . \qquad (12.6)$$

13. *Integration of a Differential Form*

We want to define the integral of a differential p-form on a manifold $\mathcal{M}_p$ of the same dimension. We also notice that one integrates only covariant tensors. Let us first **define the orientation of a frame**: suppose x and x' to be the coordinates of $M \in \mathcal{M}_p$ in two frames Σ and Σ'; the two frames will have the same orientation if $\partial(x')/\partial(x) > 0$, otherwise their orientations are opposed. Then, let α_M be a p-form defined on $\mathcal{M}_p$ and have the compact K as support. In order to define

$$\int_{\mathcal{M}_p} \alpha_M ,$$

we begin by considering the special p-form

$$\omega_M = F(x^1 \ldots x^p)\underline{d}x^1 \wedge \underline{d}x^2 \wedge \ldots \underline{d}x^p \; , \tag{13.1}$$

where $F(x)$ is the function $\hat{F}(M)$ with K as support and we define

$$\int_{\mathcal{M}_p} \omega_M = \int_{-\infty}^{+\infty} \ldots \int_{-\infty}^{+\infty} F(x^1 \ldots x^p)dx^1 \ldots dx^p \; , \tag{13.2}$$

where the right-hand side is a p-uplet integral of classical analysis. We may then come back to α_M by remarking that in local coordinates α_M is a sum of terms identical to (13.2) (a more careful interpretation uses a partition of unity, see [Bibl. 9], p. 203). Furthermore, that definition does not depend on the chosen chart, provided the change of variables in $\mathbb{R}^p$ does not change the chosen orientation, since we saw indeed in (9.23) that the Jacobian is automatically introduced for any change of variables.

14. *p-Rectangles, Chains and Boundaries*

The generalization of Green's and Stokes' formulae require the introduction of the concept of a boundary which belongs to topological algebra. Such a concept was intuitively introduced in Sec. 12; we want to extend it to an $\mathcal{M}_p$ manifold.

In order to perform such a generalization, we will consider a map of $\mathcal{M}_p$ into a **p-rectangle** defined in $\mathbb{R}^p$, i.e., a set of points with $\{x^1 \ldots x^p\}$ as coordinates obeying the system of inequalities

$$a^i \le x^i \le b^i \; : \; i = 1 \ldots p \; . \tag{14.1}$$

We notice straightaway that following (13.2) any differential p-form can be integrated on a p-rectangle.

We now introduce the concept of elementary p-chain. It is defined by the following pair: an oriented p-rectangle $\mathcal{P}$ via the numbering of its coordinates and a differentiable map $\psi : \mathcal{P} \to \mathcal{M}_p$, the corresponding map $D\psi^*$ being

$$T_{\mathcal{P}}{}^* \overset{D\psi^*}{\leftarrow} T_{\mathcal{M}}^* \; . \tag{14.2}$$

If $\alpha_M \in T^*\mathcal{M}$ is a 1-form, then $D^*\psi\alpha_M \in \mathcal{P}$ and, following (13.2),

$$\int_{\psi(\mathcal{P})} \alpha_M = \int_c \alpha_M = \int_{\mathcal{P}} D\psi^*_{\alpha_M} \; , \tag{14.3a}$$

a formula which means that the evaluation of the right-hand side integral amounts to the calculation of a p-multiple integral.

If α_M is a p-form, a straightforward generalization leads to

$$\int_{\mathcal{M}_p} \alpha_M = \int_c \alpha_M = \int_{\mathcal{P}} D\psi^{*\otimes p}\alpha_M , \qquad (14.3b)$$

where we have taken formula (10.23) into account. Using a simpler language and supposing the preceding integrands to be defined in the natural dual basis $\{\underline{dx}^j\}$, we look for a transformation such that the domain of integration $\mathcal{M}_p$ is transformed in the p-rectangle $\mathcal{P}$ and we applied (13.2).

Before coming to Stokes' theorem, let us complete our comments on the chains: two elementary p-chains c and c' are equal or opposed if, for any $\alpha_M \in T_{\mathcal{M}}^*$,

$$\int_c \alpha_M = \int_{c'} \alpha_M \quad \text{or} \quad \int_c \alpha_M = -\int_{c'} \alpha_M . \qquad (14.4)$$

Let us consider a finite set of elementary p-chains $\{c_i\}$, the **p-chain C** will be denoted by

$$C = \sum k_i c_i \qquad (14.5)$$

with k_i real numbers, then we define

$$\int_C \alpha_M = \sum k_i \int_{c_i} \alpha_M . \qquad (14.6)$$

Finally, if ψ is a diffeomorphism, the (elementary) p-chain is called an (elementary) **p-domain of integration**.

We now define the concept of boundary: any **face** of the p-rectangle $\mathcal{P}$ will be defined by the corresponding $(p-1)$-rectangle obtained by choosing one of the coordinates – say x^j – equal to a^j or b^j. There are $2p$ faces to a p-rectangle, their set constitutes the **boundary** $\partial\mathcal{P}$ of the p-rectangle.

How does one define the orientation of a face? Let us define the one corresponding to the choice of the coordinate x^j.

If j is even: if $x^j = a^i$ the orientation of the face is the same as the orientation of the frame at that point, if $x^j = b^j$ the orientation is opposite.

If j is odd: the choice of the orientation is the opposite of the one for j even. For an example, see next paragraph. The boundary of a p-rectangle is the set of the $(p-1)$-rectangles (number $2p$), and for a chain

$$\partial C = \sum k_i \partial c_i . \qquad (14.7)$$

15. *Stokes' Theorem*

As we will see below, Stokes' theorem constitutes a generalization of Green's and Stokes' formulae (see Sec. 12). Let us first of all, state Stokes' theorem: C is a chain in $\mathcal{M}_N$ and α_M a differential form (class C^1); one can write

$$\int_C \underline{d}\alpha_M = \int_{\partial C} \alpha_M \; ; \tag{15.1}$$

one also uses the following form

$$\langle C, \underline{d}\alpha_M \rangle = \langle \partial c, \alpha_M \rangle \; , \tag{15.2}$$

defined as a transcription of (15.1).

We shall prove the theorem for a 2-rectangle, its extension to a p-rectangle, although quite long, does not necessitate the introduction of new concepts. We consider then a 2-rectangle

$$a^1 \leq x^1 \leq b^1 \; , \quad a^2 \leq x^2 \leq b^2 \; , \tag{15.3}$$

in the orthonormal frame $\{0, \mathbf{e}_1, \mathbf{e}_2\}$:

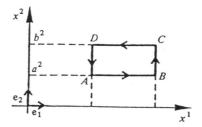

Let us define its orientation: we notice first that its boundary $\partial\mathcal{P}$ is, following (14.7), the rectangle $ABCD$:

$$
\begin{aligned}
AB &: x^2 = a^2, a^1 \leq x^1 \leq b^1 \\
BC &: x^1 = b^1, a^2 \leq x^2 \leq b^2 \\
CD &: x^2 = b^2, a^1 \leq x^1 \leq b^1 \\
DA &: x^1 = a^1, a^2 \leq x^2 \leq b^2 \; .
\end{aligned}
\tag{15.4}
$$

Their orientation is given by the rule (14.7):

$$
\begin{aligned}
AB &: x^2 = a^2, j \text{ even: its orientation is the one of } \{0, \mathbf{e}_1, \mathbf{e}_2\} \\
BC &: x^1 = a^1, j \text{ odd: its orientation is the one of } \{0, \mathbf{e}_1, \mathbf{e}_2\} \\
CD &: x^2 = b^2, j \text{ even: its orientation is opposed to } \{0, \mathbf{e}_1, \mathbf{e}_2\} \\
DA &: x^1 = a^1, j \text{ even: its orientation is opposed to } \{0, \mathbf{e}_1, \mathbf{e}_2\}
\end{aligned}
\tag{15.5}
$$

and the orientations given in the tableau correspond to the ones of the figure. Another way to obtain the orientation is to suppose that the normal N to the plane (e_1, e_2) is toward the observer and use Maxwell's rule.

We now come to the explicit expression of the integral of (15.1): we shall follow the rules given at the end of the preceding paragraph. Consider the 1-form

$$\alpha_M = \alpha_1(x^1, x^2)\underline{d}x^1 + \alpha_2(x^1, x^2)\underline{d}x^2 \tag{15.6}$$

and let us calculate

$$\underline{d}\alpha_M = (\partial_1\alpha_2 - \partial_2\alpha_1)\underline{d}x^1 \wedge \underline{d}x^2 . \tag{15.7}$$

Following formula (13.2), one has

$$\int_C \alpha_M = \iint_{\mathcal{P}} \partial_1\alpha_2 dx^1 dx^2 - \int_{\mathcal{P}} \partial_2\alpha_1 dx^1 dx^2 , \tag{15.8}$$

and, performing the integration on one of the variables, one gets

$$\int_C \underline{d}\alpha_M = \int_{a^2}^{b^2} (\alpha_2(b^1, x) - \alpha_2(a^1, x))dx - \int_{a_1}^{b_1} (\alpha_1(x, b^2) - \alpha_1(x, a^2))dx . \tag{15.9}$$

Finally, the integral of α_M along the boundary ∂C amounts to the calculation of the integral along $ABCD$,

$$\int_{\partial C} \alpha_m = \int_{ABCD} \alpha_1(x^1, x^2)dx^1 + \int_{ABCD} \alpha_2(x^1, x^2)dx^2 , \tag{15.10}$$

when the first integral of the right-hand side gives two integrals which do not necessarily vanish (AB and CD) and the second integral of the same side two other integrals (BC and DA). The result so obtained corresponds exactly to the expression (15.9). Stokes' theorem is then proved for a 2-rectangle provided one applies rule (15.7) (see also Sec. 16). In classical analysis, (15.9) and (15.10) build up the well-known **Riemann formula**:

$$\int_{\partial C} P(x, y)dx + Q(x, y)dy = \iint_C \left(\frac{\partial Q}{\partial x} - \frac{\partial P}{\partial y}\right) dx dy , \tag{15.11}$$

where ∂C is the boundary of the planar surface enclosed by the curve C.

We may also remark that the map ψ defined above is reduced to an identity for a p-rectangle; thus Stokes' theorem, as formulated above, is proved. Its extension to $\mathcal{M}_p$ is now clear since one only needs to find a map $\psi : \mathcal{M}_p \to \mathcal{P}$.

Several special techniques are elaborated by physicists, mainly in $\mathcal{E}_3$. If there are singularities of the field, one defines surface curl or divergence. One may notice a last consequence of Stokes' formula. One has

$$\partial^2 C = 0 \tag{15.12}$$

for a chain C. Indeed, for any form α_M let us integrate over $\partial^2 C = \partial\partial C$, then following (15.2)

$$\langle \partial\partial C, \alpha_M \rangle = \langle \partial C, \underline{d}\alpha_M \rangle = \langle C, \underline{d}^2\alpha_M \rangle = 0 , \tag{15.13}$$

and (15.12) is proved.

16. *Stokes' and Green's Formulae Revisited*

We now go back to Sec. 12 and show that Stokes' and Green's formulae are special cases of the general Stokes' formula given in the previous paragraph.

(a) *Stokes' formula:* We start from the differential 1-form

$$\alpha_M = \alpha_i(x)\underline{d}x^i \tag{16.1}$$

and using (9.9) we calculate

$$\begin{aligned}
\underline{d}\alpha_M &= \partial_j \alpha_i \underline{d}x^j \wedge \underline{d}x^i = \frac{1}{2}(\partial_j \alpha_i - \partial_i \alpha_j)\underline{d}x^j \wedge \underline{d}x^i \\
&= \sum_{j<i} \int_S (\partial_j \alpha_i - \partial_i \alpha_j)\underline{d}x^j \wedge \underline{d}x^i ;
\end{aligned} \tag{16.2}$$

then following (15.1) one gets the classical Stokes' formula.

$$\int_S \sum_{j<i}(\partial_j \alpha_i - \partial_j \alpha_i)\underline{d}x^j \wedge \underline{d}x^i = \int_{\partial S} \alpha_i \underline{d}x^i ,$$

$$i, j = 1, 2, 3 . \tag{16.3}$$

Same remarks as in Sec. 14 apply.

(b) *Green's formula:* The calculations are a little more complicated than in (a); we consider the 2-form

$$\alpha_M = \frac{1}{2}\varepsilon^i{}_{jk}\alpha_i(x)\underline{d}x^j \wedge \underline{d}x^k , \tag{16.4}$$

where the indices i, j, k take the values 1,2,3 and ε^i_{jk} is the Ricci tensor of the third order. Suppose that the summation on the index i is performed, then

$$
\begin{aligned}
\alpha_M &= \frac{1}{2}(\varepsilon^1{}_{jk}\alpha_1(\mathbf{x}) + \varepsilon^2{}_{jk}\alpha_2(\mathbf{x}) + \varepsilon^3{}_{jk}\alpha_3(\mathbf{x}))\underline{d}x^j \wedge \underline{d}x^k \\
&= \frac{1}{2}(\varepsilon^1{}_{23}\alpha_1\underline{d}x^2 \wedge \underline{d}x^3 + \varepsilon^1{}_{23}\alpha_1\underline{d}x^3 \wedge \underline{d}x^2 \\
&\quad + \varepsilon^2{}_{13}\alpha_2\underline{d}x^1 \wedge \underline{d}x^3 + \varepsilon^2{}_{31}\alpha_2\underline{d}x^3 \wedge \underline{d}x^1 \\
&\quad + \varepsilon^3{}_{12}\alpha_3\underline{d}x^1 \wedge \underline{d}x^2 + \varepsilon^3{}_{21}\alpha_3\underline{d}x^2 \wedge \underline{d}x^1) \\
&= \alpha_1(\mathbf{x})\underline{d}x^1 \wedge dx^3 + \alpha_2(\mathbf{x})\underline{d}x^3 \wedge \underline{d}x^1 + \alpha_3(\mathbf{x})\underline{d}x^1 \wedge \underline{d}x^2 ,
\end{aligned}
\tag{16.5}
$$

and, in the vocabulary of Euclidean vector analysis, this is the flux of the vector $\boldsymbol{\alpha}(\mathbf{x})$, i.e. $\boldsymbol{\alpha}(\mathbf{x}) \cdot \mathbf{N}(\mathbf{x})dS$.

We also may calculate

$$
\underline{d}\alpha_M = \frac{1}{2}\varepsilon^i{}_{jk}\partial_l\alpha_i(\mathbf{x})\underline{d}x^l \wedge \underline{d}x^j \wedge \underline{d}x^k ,
\tag{16.6}
$$

but, following (8.16),

$$
\underline{d}x^l \wedge \underline{d}x^j \wedge \underline{d}x^k = \varepsilon^{ljk}\underline{d}x^1 \wedge \underline{d}x^2 \wedge \underline{d}x^3 ,
$$

and finally

$$
\begin{aligned}
\underline{d}\alpha_M &= \frac{1}{2}\varepsilon^i{}_{jk}\varepsilon^{ljk}\partial_l\alpha_i(\mathbf{x})\underline{d}x^1 \wedge \underline{d}x^2 \wedge \underline{d}x^3 \\
&= \delta^{il}\partial_l\alpha_i(\mathbf{x})\underline{d}x^1 \wedge \underline{d}x^2 \wedge \underline{d}x^3 = \sum_i \partial_i\alpha_i(\mathbf{x})\underline{d}x^1 \wedge \underline{d}x^2 \wedge \underline{d}x^3 .
\end{aligned}
\tag{16.7}
$$

The last equality is the divergence of the vector $\boldsymbol{\alpha}(\mathbf{x})$ (see Problem 1.7). We find again Green's formula. Formula (16.7) throws new light on the general concept of flux: let us indeed consider the special case $\mathcal{M}_N = \mathbb{R}^N$; if we require that α_M is a closed 2-form, i.e. $\underline{d}\alpha_M = 0$, then, from Poincaré lemma, α_M is also exact

$$
\alpha_M = \underline{d}\hat{\phi}(M)
$$

and the function $\hat{\phi}$ is the flux. Its conservation can be expressed through the use of Stokes' theorem.

17. *Cohomology. Betti's Numbers*

We are now going to study some topological consequences of the theory of differential forms.

Let $\mathcal{M}_N$ be a differential manifold and consider the following two sets:

$Z^p(\mathcal{M}_N)$ is the set of all closed p-forms ($\underline{d}\alpha_M = 0$),

$B^p(\mathcal{M}_N)$ is the set of all exact p-forms ($\alpha_M = \underline{d}\hat{F}(M)$).

We may give to each of these sets a vector space structure since the sum of two closed (exact) forms is again a closed (exact) form and if α_M is closed (exact), so is the form $\lambda\alpha_M, \lambda \subset \mathbb{R}$.

Remarking that $B^p(\mathcal{M}_N) \subset Z^p(\mathcal{M}_N)$, one may define the quotient vector space

$$H^p(\mathcal{M}_N) = Z^p/B^p .$$

The elements of $H^p(\mathcal{M}_N)$ are then equivalence classes defined as follows: if α_M and β_M are two closed p-forms (elements of $Z^p(\mathcal{M}_N)$) such that

$$\alpha_M - \beta_M \in B^p , \tag{17.1}$$

then the class of all β_M obeying (17.1) is an equivalence class with α_M as representative. We may also notice that (7.1) can also be written as

$$\alpha_M - \beta_M = \underline{d}\gamma_M , \tag{17.2}$$

where γ_M is a certain $(p-1)$-form. The α_M and β_M forms are said to be **homologous** and the vector space H^p represents the (de Rham) **cohomology**. One shows furthermore that if $\mathcal{M}_N$ is compact, $H^p(\mathcal{M}_N)$ has a finite dimension denoted by b_p, **Betti's number** of $\mathcal{M}_N$ of order p.

The $Z^0(\mathcal{M}_N)$ calls for some comments: following its definition, it is the set of all 0-form $\hat{F}(M) = F(x)$ such that $\underline{d}\hat{F}_M = 0$, locally that statement is equivalent to the vanishing of all partial derivatives $\partial_i F(x)$. Then $H^p(\mathcal{M}_N)$ has for elements all the functions F which are constants on all connex parts of $\mathcal{M}_N$. One also may remark that $B^0(\mathcal{M}_N)$ is empty, i.e. $b^0 = 0$.

One may also consider the special case $\mathcal{M}_N = \mathbb{R}^N$: following Poincaré's lemma, any closed differential form is an exact one, then $Z^p(\mathbb{R}^N) = B^p(\mathbb{R}^N)$. The dimension of $H^0(\mathbb{R}^N)$ is thus 0 (see (17.2)) and all the corresponding Betti's numbers vanish.

One calls **coboundary** an exact differential form and **cocyle** a closed one. These additions to the vocabulary are not gratuitous, one may indeed look for similarities between the relations $\underline{d}^2 = 0$ and $\partial^2 = 0$.

Betti's number lead to the definition of the **Euler-Poincaré characteristic** introduced at Level 0, Sec. 2. One considers the alternate sum

$$\chi(\mathcal{M}_N) = \sum_{1}^{N} (-1)^q B_q \ , \tag{17.3}$$

and $\chi(\mathcal{M}_N)$ can be shown to be identical to the number $\chi(S)$ introduced at Level 0.

As a conclusion of the preceding remarks, we may state that the enumeration of certain classes of differential forms on $\mathcal{M}_N$ can lead to topological properties of that manifold and this is somewhat an unexpected result!

We may also add that the theories of differential geometry on cohomology and homology have registered remarkable progress in recent times: they constitute a really important chapter in that domain (generalization of several of the concepts developed in Sec. 2, Level 0). Nevertheless, such developments are clearly outside the program of the present textbook; the interested reader may consult [Bibl. 12], Vol. 3 and also monographs quoted in the bibliography of Chap. 4.

Up to now, none of the theorems we have been studying can be used for the characterization of a curved surface: no criterion has been stated which will allow us to distinguish between a curved and a planar surface. As far as physics is concerned, such a distinction is of overwhelming importance. This is therefore one of the points which we will consider in the next paragraph.

Although rigorous and intrinsic, our comments will lack intuitive content. Such an approach was undertaken in paragraphs 3 of Level 0 and 5,6,7 of Level 1.

E. Theory of Linear Connections
18. *Linear Connections, Covariant (or Absolute) Derivative*

On a manifold $\mathcal{M}_N$, we consider a point $M \in \mathcal{M}_N$, the tangent vector $T_{\mathcal{M}}(M)$ and the cotangent vector spaces $T_{\mathcal{M}}^*(M)$.

A linear connection is a map

$$T_{\mathcal{M}}(M) \xrightarrow{\nabla^M} T_{\mathcal{M}}(M) \otimes T_{\mathcal{M}}^*(M) \ , \tag{18.1}$$

by which any vector $v_M \in T_{\mathcal{M}}$ is transformed into $\nabla^M v_M$, a tensor once contravariant and once covariant. To complete the definition of ∇^M, we shall state six postulates: we will study below the first three (formulae

(18.2), (18.3), (18.5)), and later on the last three (formulae (18.29), (18.36), (18.37)). We state:

(1) For any two vectors v_M and w_M, one has

$$\nabla^M \{v_M + w_M\} = \nabla^M v_M + \nabla^M w_M \ . \tag{18.2}$$

(2) Let $\hat{F}$ be a C^∞ function on $\mathcal{M}_N$ (or equivalently a 0-form, a 0-covariant, or a 0-contravariant tensor field), then $\hat{F}(M)v_M$ is clearly a vector (of $T_\mathcal{M}(M)$) and

$$\nabla^M \{\hat{F}(M)v_M\} = v_M \otimes \underline{d}\hat{F}(M) + \hat{F}(M)\nabla^M v_M \ , \tag{18.3}$$

a relation similar to the Leibniz rule.

(3) Let $\{e_i{}^M\}$ be any frame at $M \in \mathcal{M}_N$; then for any vector v_M, one has

$$v_M = v^i(x)e_i^M \ : \ x = \varphi(M) \tag{18.4}$$

and, taking (4.17) into account, we assume that

$$(\nabla^M \hat{F})(v_M) = v_M(\hat{F}) = \underline{d}\hat{F}(M) = v^i(x)\partial_i F(x) = V^i(x)\mathbf{D}_i F(x) \ . \tag{18.5}$$

For the time being, using postulates (1) and (2), we may bring ∇^M into an interesting form

$$\nabla^M v_M = \nabla^M \{v^i(x)e_i^M\} = e_i^M \otimes \underline{d}v^i(x) + v^i(x)\nabla^M e_i^M \ , \tag{18.6}$$

a formula which shows that $\nabla^M v_M$ will be defined once the $\nabla^M e_i{}^M$ are known. But these last objects are elements of $T_\mathcal{M}(M) \otimes T^*{}_\mathcal{M}(M)$; then if the set $\{\omega^i_M\}$ is the dual frame of $\{e_i^M\}$, the set $\{e_k^M \otimes \omega^j_M\}$ is a frame of that same space. We define the **affine connection coefficient** $\gamma^k{}_{ij}(x)$ by

$$\nabla^M e_i^M = \gamma^k{}_{ji}(x)e_k^M \otimes \omega^j_M \ . \tag{18.7}$$

In particular, if the frame $\{e_k^M\}$ is the natural one $\{\varepsilon_k^M\}$ as defined in (4.8) then the $\gamma^k{}_{ij}(x)$ will be denoted by $\Gamma^k{}_{ij}(x)$ already introduced at (5.5), Level 1. Finally $\nabla^M v_M$ may be written as

$$\nabla^M v_M = e_i^M \otimes \underline{d}v^i(x) + e_k^M \otimes \Omega^k{}_j(M)v^j(x) \ . \tag{18.8}$$

We can also introduce the **connection matrix** whose elements $\Omega^k{}_j(M)$:

$$\Omega^k{}_i(M) = \gamma^k_{ji}(x)\omega^j_M \ , \tag{18.9}$$

are 1-forms and, instead of (18.6), we may write $\nabla^M v_M$ using these new notations as

$$\nabla^M v_M = e_i^M \otimes \underline{d}v^i(x) + e_k^M \otimes \Omega_j^k(M)v^j(x) \ . \qquad (18.10)$$

Another notation, already introduced in Level 1, is also very much in use; one introduces the functions $\nabla_j v^i(x)$ of x depending on the two indices i, j such that

$$\nabla^M v_M = \nabla_j v^i(x)e_i^M \otimes \omega_M^j \qquad (18.11)$$

and, since in (9.32) we saw that

$$\underline{d}v^i(x) = \mathbf{D}_j v^i(x)\omega_M^j \ , \qquad (18.12)$$

the expression (18.8) can be written as

$$\nabla^M v_M = (\mathbf{D}_j v^i(x) + \gamma^i{}_{kj}(x)v^k(x))e_i^M \otimes \omega_M^j \ . \qquad (18.13)$$

Comparing (18.13) and (18.11), one obtains the expression of the set $\{\nabla_j v^i(x)\}$ as components of $\nabla^M v_M$ in the basis $\{e_i^M \otimes \omega_M^j\}$:

$$\nabla_j v^i(x) = \mathbf{D}_j v^i(x) + \gamma^i{}_{kj}(x)v^k(x) \ . \qquad (18.14)$$

In the natural basis $\{\varepsilon_i^M \otimes \underline{d}x^j\}$, we have

$$(\nabla_j v_{(n)}^i(x))_n = \partial_j v_{(n)}^i(x) + \Gamma^i_{jk}(x)v_{(n)}^k(x) \ , \qquad (18.15)$$

where the $v_{(n)}^i(x)$ are the natural components of v_M. A formula which is to be compared with formula (5.16), Level 1 (with a slight change of notations, since one sums over the second index k, and not over the first index as in Level 1). The reader will notice a different interpretation of ∇_i from the one given at Level 1: there ∇_j was a differential operator acting on the vector $A^i(x)$ and transforming it into a tensor field once contravariant and once covariant, while in the present case $\nabla_j v^i$ is presented as a whole, a function of x, component of $\nabla^M v_M$.

The map ∇^M is also called a **covariant derivation**; its present intrinsic definition generalizes the one given at Level 1 which was valid only for the basis $\varepsilon_i^M \otimes \underline{d}x^j$. This is also an important point for the physicist. We may now define the **covariant derivative along a vector** u_M: since for any vector $v_M, \nabla^M v_M$ is an element of the space $T_{\mathcal{M}}(M) \otimes T^*_{\mathcal{M}}(M)$, we may calculate its value for any vector u_M by the formula

$$(\nabla^M v_M)(u_M) = \nabla_j v^i(x) e_i^M \otimes \omega_M^j(u_M) \ . \tag{18.16}$$

But as $\{\omega_M^j\}$ is the dual basis of $\{e^M{}_j\}$ we can write

$$(\nabla^M v_M)(u_M) = u^j(x) \nabla_j v^i(x) e_i^M \ , \tag{18.17}$$

and this is the definition of

$$\nabla_u^M v_M = (\nabla^M v_M)(u_M) = u^j(x) \nabla_j v^i(x) e_i^M \ , \tag{18.18}$$

which is another vector, element $T_{\mathcal{M}}(M)$. For $u_M = e_k^M$,

$$\nabla_{e_k}^M v_M = \delta_k^j \nabla_j v^i(x) e_i^M = \nabla_k v^i(x) e_i^M \ , \tag{18.19}$$

a formula which shows that $\nabla_k v^i(x)$ is the i-th component of the vector $\nabla_{e_k}^M$. If we now take

$$u_M = e_k^M \ , \quad v_M = e_i^M \ ,$$

then

$$\nabla_{e_k}^M e_i^M = \gamma^j{}_{ki}(x) e_j^M \tag{18.20}$$

and $\gamma^j_{ki}(x)$ is the j-th component of $\nabla_{e_k}^M e^M{}_i$.

We may also quote two other important properties of $\nabla_u^M v_M$:

(a) Let u_M, v_M, w_M be vectors and λ, μ two real numbers, then

$$\begin{aligned}\nabla_{\lambda u + \mu v}^M w_M &= \lambda \nabla_u^M w_M + \mu \nabla_v^M w_M \\ &= (\nabla^M w_M)(\lambda u_M + \mu v_M) = \nabla_{\lambda u}^M w_M + \nabla_{\mu v}^M w_M \ , \end{aligned} \tag{18.21}$$

an interesting relation.

(b) Let $\hat{F}$ be a C^∞ function, then

$$\nabla_u^M \{\hat{F}(M) v_M\} = v_M \underline{d}\hat{F}(M) u_M + \hat{F}(M)(\nabla^M v_M)(u_M) \ , \tag{18.22}$$

where $\underline{d}\hat{F}(M)(u_M)$ is a number (see (18.24)). But according to formula (18.3), one has

$$\nabla_u^M \{\hat{F}(M) v_M\} = (u_M \hat{F}(M)) v_M + \hat{F}(M) \nabla_u^M v_M \ , \tag{18.23}$$

and, taking (9.4) into account, we arrive at the formula

$$\nabla_u^M \{\hat{F}(M)v_M\} = (u^i(x)\partial_i F(x))v_M + \hat{F}(M)\nabla_u^M v_M \ . \tag{18.24}$$

It is also worthwhile to notice that (18.22) is a special case of the relation

$$(\nabla_{\hat{F}_M u}^M)(\hat{G}(M)v_M) = \hat{F}(M)u_M \hat{G}(M) + \hat{F}(M)\hat{G}(M)\nabla_u^M v_M \ . \tag{18.25}$$

Indeed, a straightforward application of (18.25) leads to

$$\nabla_{\hat{F}_M u}^M v_M = \hat{F}(M)\nabla_u^M v_M \ . \tag{18.26}$$

The two maps ∇^M and ∇_u^M act on elements of $T_{\mathcal{M}}(M)$; we want to extend that action to elements of $T_{\mathcal{M}}^*(M)$, i.e. to 1-form α_M. We therefore define $\nabla_v^M \alpha_M$ as the map

$$T_{\mathcal{M}}^*(M) \xrightarrow{\nabla_v^M \alpha_M} T_{\mathcal{M}}^*(M) \ , \tag{18.27}$$

and consequently $\nabla_v^M \alpha_M(u_M)$ is a real function of $M \in \mathcal{M}_N$.

We then state rule (4).

(4) We suppose that $\nabla_v^M \alpha_M(u_M)$ follows a Liebniz-type law, i.e.

$$\nabla_v^M \{\alpha_M(u_M)\} = (\nabla_v^M \alpha_M)u_M + \alpha_M(\nabla_v^M u_M) \ , \tag{18.28}$$

a relation which leads to the following definition of the 1-form $\nabla_v^M \alpha_M$:

$$(\nabla_v^M \alpha_M)u_M = \nabla_v^M \{\alpha_M(u_M)\} - \alpha_m(\nabla_v^M u_M) \ . \tag{18.29}$$

We then proceed as in (18.11) and define the components $\nabla_j \alpha_i(x)$ (a set of function of x depending on the indices j, i) of $\nabla_v^M \alpha_M$:

$$\nabla_u^M \alpha_M = v^j(x)\nabla_j \alpha_i(x)\omega_M^i \ . \tag{18.30}$$

We now put $u_M = e_k^M$ in (18.29), k being chosen once for all,

$$\begin{aligned} (\nabla_v^M \alpha_M)e_k^M &= v^j(x)\nabla_j \alpha_i(x)\omega_M^i(e_k^M) \\ &= v^j(x)\nabla_j \alpha_k(x) \ . \end{aligned} \tag{18.31}$$

Going back to (18.29) one has, following (18.20),

$$\begin{aligned} (\nabla_v^M \alpha_M)e_k^M &= \nabla_v^M \{\alpha_M(e_k^M)\} - \alpha_M(v^j(x)\gamma_{jk}^p(x)e_p^M) \\ &= \nabla_v^M \alpha_k(x) - v^j(x)\gamma_{jk}^p(x)\alpha_p(x) \ . \end{aligned} \tag{18.32}$$

The first term of the last line can be read as

$$\nabla_v^M \alpha_k(x) = (\nabla^M \alpha_k(x))v_M = v^j(x)\mathbf{D}_j\alpha_k(x) \ ,$$

with the help of (18.5). We notice finally that the field $\{v^j(x)\}$ can be arbitrarily chosen; comparing then (18.31) to (18.32), one gets

$$\nabla_j\alpha_k(x) = \mathbf{D}_j\alpha_k(x) - \gamma_{jk}^p(x)\alpha_p(x) \tag{18.33}$$

and instead of (18.30) one may also write

$$\nabla_v^M \alpha_M = v^j(x)(\mathbf{D}_j\alpha_k(x) - \gamma^i{}_{kj}(x)\alpha_i(x))\omega_M^k \ . \tag{18.34}$$

We may also choose $\alpha_M = \omega_M^a$, a being chosen once for all; then $\alpha_i(x) = \delta_i^a$ and

$$\nabla_v^M \omega^a = -v^j(x)\gamma^i{}_{jk}(x)\delta_i^a\omega_M^k = -v^j(x)\gamma^a{}_{jk}(x)\omega_M^k \ . \tag{18.35}$$

We want to extend all the previous results to **any kind** of tensor at M; we shall need two further rules (5) and (6).

(5) For two tensors T_M and S_M,

$$\nabla_v^M \{T_M + S_M\} = \nabla_v^M T_M + \nabla_v^M S_M \ . \tag{18.36}$$

(6) One also admits a Leibniz-type rule

$$\nabla_v^M \{T_M \otimes S_M\} = (\nabla_v^M T_M) \otimes S_M + T_M \otimes \nabla_v^M S_M \ . \tag{18.37}$$

As an example, we may apply these rules to a doubly covariant tensor at M:

$$\begin{aligned}
\nabla_v^M \{A_{ij}(x)\omega_M^i \otimes \omega_M^j\} = v^l(x)(\mathbf{D}_l A_{ij}(x) &- A_{kj}\gamma_{li}^k(x) \\
&- A_{ik}(x)\gamma^k{}_{lj}(x))\omega_M^i \otimes \omega_M^j \ ,
\end{aligned} \tag{18.38}$$

where we have used rule (3) (18.5) and rule (6) (18.37). It should be noted that the bracket gives the expression of $\nabla_l A_{ij}(x)$ in the terminology of Level 1 in an arbitrary basis, instead of the natural one.

19. *Curvature Tensor*

Let us first of all state two relations which will be of importance in our considerations on the curvature tensor: we consider three vectors u_M, v_M, w_M being elements of $T_{\mathcal{M}}(M)$, and three C^∞ functions $\hat{F}, \hat{G}, \hat{H}$. Since following (4.24) and (4.25), such expressions as $\hat{F}(M)u_M\hat{G}(M)v_M$ are well-defined, we may calculate the commutator

$$
\begin{aligned}
[\hat{F}(M)u_M, \hat{G}(M)v_M]\hat{H}(M) &= \hat{F}(M)u_M\{\hat{G}(M)v_M\hat{H}(M)\} \\
&\quad - \hat{G}(M)v_M\{\hat{F}(M)u_M\hat{H}(M)\} \\
&= \hat{F}(M)\hat{G}(M)[u_M, v_M]\hat{H}(M) + \hat{F}(M)u_M\hat{G}(M)v_M\hat{H}(M) \\
&\quad - \hat{G}(M)v_M\hat{F}(M)u_M\hat{H}(M)
\end{aligned}
\tag{19.1}
$$

and, since $\hat{H}$ is any C^∞ function

$$
\begin{aligned}
&[\hat{F}(M)u_M, \hat{G}(M)v_M] \\
&= \hat{F}(M)\hat{G}(M)[u_M, v_M] + \hat{F}(M)(u_M\hat{G}(M))v_M - \hat{G}(M)(v_M\hat{F}(M))u_M \ .
\end{aligned}
\tag{19.2}
$$

Such a formula is very clumsy indeed and so would be all the following formulae. We shall simplify their appearance by adopting some conventions: "we will everywhere omit the indication of $M \in \mathcal{M}_N$, we shall then write e_j, ω^j instead of e_j^M, ω_M^j; instead of $\hat{F}(M), \hat{G}(M)$ we shall simply write $F, G, \ldots$, using $F(x), G(x)$ to denote the values of these functions at $x = \varphi(M)$, where φ is the homeomorphism defining a chart".

In spite of some slight possible ambiguity in the meaning of the formulae, they appear notably simplified and more compact and readable; for instance, (19.2) will be written as follows:

$$
[Fu, Gv]FG[u, v] + F(uG)v - G(vF)u \ .
\tag{19.3}
$$

For the definition of the **curvature tensor** we shall proceed as in Sec. 7, Level 1, avoiding if possible, any local reasoning. The geometric considerations which appear in that paragraph 7 are still valid and important since they allow to go beyond the axiomatic and formal framework of the considerations below. Let us define $R(u, v)$:

$$
R(u, v) = [\nabla_u, \nabla_v] - \nabla_{[u,v]} \ ,
\tag{19.4a}
$$

which is antisymmetric:

$$
R(u, v) = -R(v, u) \ .
\tag{19.4b}
$$

Clearly from its very definition it is a map

$$T_{\mathcal{M}}(M) \longrightarrow T_{\mathcal{M}}(M) \ ,$$

i.e. any vector $w \in T_{\mathcal{M}}(M)$ is transformed into another vector $R(u,v)w$. Let F, G, H be three C^{∞} functions and w some other vector; we want to show that

$$R(Fu, Gv)Hw = (FGH)R(u,v)w \ . \tag{19.5}$$

Let us begin by proving that

$$R(u,v)Hw = (H)R(u,v)w \ , \tag{19.6}$$

or, indeed,

$$\begin{aligned} R(u,v)Hw &= [\nabla_u, \nabla_v]Hw - \nabla_{[u,v]}Hw \\ &= \nabla_u\{\nabla_v Hw\} - \nabla_v\{\nabla_u Hw\} - \nabla_{[u,v]}Hw \ . \end{aligned} \tag{19.7}$$

We now transform the two brackets of the preceding formula using (18.23); one gets

$$\begin{aligned} R(u,v)Hw &= \nabla_u\{(v(H)w + H\nabla_v w)\} - \nabla_v\{(u(H)w) \\ & H\nabla_u w\} - ([u,v](H))w - H\nabla_{[u,v]}w \ , \end{aligned} \tag{19.8}$$

and, using once more the same rule, one obtains

$$\begin{aligned} R(u,v)Hw &= u(v(H))w + v(H)\nabla_u w + u(H)\nabla_v w \\ & H\nabla_u\{\nabla_v u\} - v(uH)w - u(H)\nabla_v w \\ & - v(H)\nabla_u w - H\nabla_v\{\nabla_u w\} - [u,v]H - H\nabla_{[u,v]}w \ . \end{aligned} \tag{19.9}$$

Cancelling certain terms and collecting together some others we finally get (19.6). One may also show that

$$R(Fu,v)w = FR(u,v)w \ , \tag{19.10}$$

or, indeed,

$$R(Fu,v)w = [\nabla_{Fu}, \nabla_v]w - \nabla_{[Fu,v]}w \ . \tag{19.11}$$

As an exercise, the reader may derive, from the last relation, formula (19.10), using (18.26) and (18.21). The antisymmetry of $R(u,v)$ with respect to u and v leads to another formula similar to (19.10) for $R(u, Gv)$

and, after some simple considerations, to the general formula (19.5). This is a formula which will lead us to the determination of the components of $R(u, v)$. Let t be some vector at $M : t \in T_{\mathcal{M}}(M)$ and $\alpha \in T^*_{\mathcal{M}}(M)$ a 1-form; then from its very definition $\alpha(R(u, v)t)$ is a real number. Let $\{e_k\}$ and $\{\omega^k\}$ be, respectively, the frames of $T_{\mathcal{M}}(M)$ and $T^*_{\mathcal{M}}(M)$, then

$$\alpha(R(u, v)t) = \alpha_i(x)\omega^i\{R(u^j(x)e_j, v^k(x)e_k)t^l(x)e_l\} , \qquad (19.12)$$

where $\alpha_i(x), u^j(x), v^j(x)$ are real functions of $x = \varphi(M), \varphi$ being the homeomorphism associated with the chart we are considering. Following (9.5), one may write

$$\alpha(R(u, v)t) = \omega^i[R(e_j, e_k)e_l]\alpha_i(x)u^j(x)v^k(x)t^l(x) \qquad (19.13)$$

and define the tableau of functions of x:

$$R^i{}_{jkl}(x) = \omega^i\{R(e_j, e_k)e_l\} . \qquad (19.14)$$

We now show that $R^i{}_{jkl}(x)$ are the components of the curvature tensor already given at Level 1. Let us replace $R(u, v)$ in (19.14) by its definition (19.4):

$$\begin{aligned}R^i{}_{jkl}(x) &= \omega^i\{([\nabla_{e_j}, \nabla_{e_k}] - \nabla_{[e_j, e_k]})e_l\} \\ &= \omega^i\{([\nabla_{e_j}, \nabla_{e_k}] - c^m_{jk}\nabla_{e_m})e_l\} ,\end{aligned} \qquad (19.15)$$

where we have introduced the structure coefficients (4.18) and used (18.26) to get (19.15). The use of formulae (18.19) up to (18.27) and the introduction of Pfaff derivative (see (4.17)) lead to

$$\begin{aligned}R^i{}_{jkl}(x) =&\omega^i\{(e_j\{\gamma^p{}_{kl}(x)\}e_p + \gamma^p{}_{kl}(x)\nabla_{e_j}e_p \\ &- e_k\{\gamma^p{}_{kl}(x)\}e_p - \gamma^p{}_{jl}(x)\nabla_k e_p - c^m{}_{jk}(x)\gamma^p{}_{ml}(x)e_p\} \\ =&\omega^i\{(\mathbf{D}_j\gamma^p{}_{kl}(x))e_p + \gamma^p{}_{kl}(x)\gamma^q{}_{jp}(x)e_q \\ &- (\mathbf{D}_k\gamma^p{}_{jl}(x)e_p - \gamma^p{}_{jl}(x)\gamma^q{}_{kp}(x)e_q - c^m{}_{jk}(x)\gamma^p{}_{ml}(x)e_q\}\end{aligned}$$
$$(19.16)$$

and, since $\omega^m(e_n) = \delta^m_n$, one also has

$$\begin{aligned}R^i{}_{jkl}(x) =& \mathbf{D}_j\gamma^i{}_{kl}(x) - \mathbf{D}_k\gamma^i{}_{jl}(x) + \gamma^p{}_{kl}(x)\gamma^i{}_{jp}(x) \\ &- \gamma^p{}_{jl}(x)\gamma^i{}_{kp}(x) - c^m{}_{jk}(x)\gamma^i{}_{ml}(x) .\end{aligned} \qquad (19.17)$$

In the natural bases $\{\varepsilon_k^M\}$ and $\{\underline{d}x^k\}$, one obtains a formula already given at Level 1:

$$R^i{}_{jkl}(x) = \underline{d}x^i\{[\nabla_{\epsilon_j}, \nabla_{\epsilon_k}]\varepsilon_l\}$$
$$= \partial_j \Gamma^i_{kl}(x) - \partial_k \Gamma^i_{jl}(x) + \Gamma^p{}_{kl}(x)\Gamma^i{}_{jp}(x) - \Gamma^p{}_{jl}(x)\Gamma^i{}_{kp}(x) .$$
$$(19.18)$$

Remark: Given the importance of formulae (19.5), (19.12), (19.13) and (19.14), we want to be more explicit and introduce all the missing arguments.

Formula (19.5) should be read as follows:

$$R(\hat{F}(M)u_M, \hat{G}(M)v_M)\hat{H}(M)w_M = \hat{F}(M)\hat{G}(M)\hat{H}(M)R(u_M, v_M)w_M .$$
$$(19.5')$$

Let us then consider (19.12),

$$\alpha_M\{R(u_M, v_M)t_M\} = \alpha_i(x)\omega_M^i\{R(u^j(x)e_j^M, v^k(x)e_k^M)\}t^l(x)e_l^M . \quad (19.12')$$

The use of (19.5') leads to

$$\alpha_m\{R(u_M, v_M)t_M\} = \omega_M^i\{R(e_j^M, e_k^M)\}\alpha_i(x)u^j(x)v^k(x)t^l(x) \quad (19.13')$$

and the right-hand side of (19.13) reads

$$R^i{}_{jkl}(x)\alpha_i(x)u^j(x)v^k(x)t^l(x) . \quad (19.14')$$

Every argument is now explicitly displayed; the expressions are quite cumbersome.

20. *Torsion Tensor*

One also considers the **torsion tensor** defined as follows:

$$\mathcal{T}(u, v) = \nabla_u v - \nabla_v u - [u, v] , \quad (20.1a)$$

$$\mathcal{T}(u, v) = -\mathcal{T}(v, u) . \quad (20.1b)$$

One can prove a formula similar to (19.5):

$$\mathcal{T}(Fu, Gv) = (FG)\mathcal{T}(u, v) . \quad (20.2)$$

Its proof is easy. From its very definition the torsion is a vector; let $u \in T_{\mathcal{M}}(M), v \in T_{\mathcal{M}}(M), \alpha \in T^*_{\mathcal{M}}(M)$ and consider the real number

$$\alpha\{T(u,v)\} .$$

Let $\{e_k\}$ and $\{\omega^k\}$ be the respective frames of $T_{\mathcal{M}}(M)$ and $T^*_{\mathcal{M}}(M)$; one has

$$\alpha\{T(u,v)\} = \alpha_i(x)\omega^i\{T(u^j(x)e_j, v^k(x)e_k)\} , \qquad (20.3)$$

where α_i, u^i, v^i are functions of x, components of α, u, v. As a consequence of (20.2),

$$\alpha\{T(u,v)\} = \omega^i\{T(e_j, e_k)\}\alpha_i(x)u^j(x)v^k(x) . \qquad (20.4)$$

We introduce the components of the torsion tensor

$$T^i{}_{jk}(x) = \omega^i\{T(e_j, e_k)\} , \qquad (20.5)$$

components which clearly can be defined (see (20.1)) by the relation

$$T^i{}_{jk}(x) = \omega^i\{\nabla_{e_j} e_k - \nabla_{e_k} e_j - [e_j, e_k]\} . \qquad (20.6)$$

Taking (18.20) into account, one also has

$$T^i_{jk}(x) = \gamma^i_{jk}(x) - \gamma^i_{kj}(x) - c^i_{jk}(x) . \qquad (20.7)$$

Suppose that instead of $\{e_k\}, \{\omega^k\}$ we consider the natural frames $\{\varepsilon_k\}$, $\{\underline{dx}^k\}$; then

$$T^i_{jk}(x) = \underline{dx}^i\{\nabla_{\varepsilon_j}\varepsilon_k - \nabla_{\varepsilon_k}\varepsilon_j\} = \Gamma^i_{kj}(x) - \Gamma^i_{jk}(x) , \qquad (20.8)$$

since all the c^i_{jk} vanish and the components of the torsion tensor are twice the antisymmetric part of the affine connection coefficients. Formula (20.8) can also be found in Sec. 7, Level 1.

21. *Curvature and Torsion 2-Forms*

To the curvature and torsion tensors there correspond two 2-forms, respectively, the **curvature 2-form** and the **torsion 2-form** which will be defined for arbitrary frames $\{e_k\}$ and $\{\omega^k\}$ of $T_{\mathcal{M}}(M)$ and $T^*_{\mathcal{M}}(M)$.

Let us begin by considering the curvature 2-form

$$\mathcal{R}^i_j(M) = \frac{1}{2}R^i{}_{jkl}w^k \wedge w^l \qquad (21.1)$$

and show that $\mathcal{R}^i_j$ obeys the relation

$$\mathcal{R}^i_j(M) = \underline{d}\Omega^i_j(M) + \Omega^i_p(M) \wedge \Omega^p_j(M) , \tag{21.2}$$

where Ω^i_j are the elements of the connection matrix (18.9):

$$\Omega^i_j(M) = \gamma^i{}_{lj}(x)\omega^l . \tag{21.3}$$

In order to prove (21.2), let us first calculate, following (9.9) and (9.15),

$$\underline{d}\Omega^i_j = \frac{1}{2}(\mathbf{D}_k\gamma^i_{jl}(x) - \mathbf{D}_l\gamma^i{}_{kj}(x) - \gamma^i_{jp}(x)c^p{}_{kl}(x))\omega^k \wedge \omega^l , \tag{21.4}$$

and then

$$\Omega^i_p \wedge \Omega^p_j = \frac{1}{2}(\gamma^i{}_{kp}\gamma^p{}_{lj} - \gamma^i{}_{lp}\gamma^p{}_{kj})\omega^k \wedge \omega^l . \tag{21.5}$$

Let us add these two expressions in order to get the right-hand side of (21.1); taking into account the components $R^i{}_{jkl}$ as given by (19.17), we obtain (21.2).

We may also define the torsion 2-form

$$T^i(M) = \frac{1}{2}T^i{}_{jk}(x)\omega^j \wedge \omega^k \tag{21.6}$$

and, taking the expression of T^i_{jk} given by (20.7), we can show that

$$T^i(M) = \underline{d}\omega^i + \Omega^i_j \wedge \omega^j . \tag{21.7}$$

One may use (21.2) and (21.7) to prove the Bianchi identities, important in general relativity; they were proved in Problem 1.11, see also [Bibl. 8], p. 240.

F. Riemannian Geometry

The considerations on Riemannian geometry as they were given in Sec. 8, Level 1 were purely local, i.e., relative to the choice of a given basis: we want now to present intrinsically the same matter. As a consequence the previous results of Level 1 and the ones of Sec. 3, Level 0 will appear in a new light.

22. *The ds^2 of a Differentiable Manifold $\mathcal{M}_N$*

A **metric** on $\mathcal{M}_N$ will be introduced by a continuous twice covariant tensor field such that

(a) $\underline{G}_M$ is symmetric,

(b) for any $M \in \mathcal{M}_N$, the real number $\underline{G}_M(u_M \otimes v_M)$ will vanish if and only if for all vectors $u_M \in T_{\mathcal{M}}(M)$ one has $v_M = 0$ (and vice-versa); $\underline{G}_M$ is then called a **nondegenerate tensor field**. Any manifold $\mathcal{M}_N$ endowed with a metric obeying the two conditions (a) and (b) is called a **Riemannian manifold**.

We consider now a frame $\{e_k^M\}$ and its dual basis $\{\omega_M^k\}$ in $T_{\mathcal{M}}(M)$ and $T^*_{\mathcal{M}}(M)$ respectively, then

$$\underline{G}_M = g_{ij}(x)\omega_M^i \otimes \omega_M^j \in T^*_{\mathcal{M}}(M) \otimes T^*_{\mathcal{M}}(M) \,, \tag{22.1}$$
$$g_{ij}(x) = g_{ji}(x) \,,$$

and, for any pair of vectors $\{u_M, v_M\}$,

$$\underline{G}_M(u_M \otimes v_M) = g_{ij}(x)\omega^i(u_M) \otimes \omega^j(v_M) = g_{ij}(x)u^i(x)v^i(x) \,. \tag{22.2}$$

This is a bilinear symmetry form which will be denoted $\langle u, v \rangle$ and the vector space $T_{\mathcal{M}}(M)$ will be endowed with the scalar product $\langle u, v \rangle$.

The expression of $\underline{G}_M$ in the natural basis $\{\underline{d}x^k\}$ of $T^*_{\mathcal{M}}(M)$ is denoted $\underline{G}_M^{(n)}$ and reads

$$\underline{G}_M^{(n)} = g_{ij}(x)\underline{d}x^i \otimes \underline{d}x^j \,. \tag{22.3}$$

Let dx be a vector with components (natural) dx^k,

$$dx = dx^k \varepsilon_k^M \,,$$

and consider

$$G_M^{(n)}(dx \otimes dx) = g_{ij}(x)\underline{d}x^i \otimes \underline{d}x^j (dx \otimes dx) = g_{ij}(x)\underline{d}x^i(dx) \tag{22.4a}$$
$$\otimes \underline{d}x^j(dx) = g_{ij}(x)dx^i dx^j \,.$$

This is a function of x which was previously called ds^2:

$$ds^2 = \underline{G}_M^{(n)}(dx \otimes dx) \,. \tag{22.4b}$$

Just to avoid a superabundance of notations, we shall use $\underline{ds}^2$ to denote $\underline{G}_M$: in that new notation, $\underline{ds}^2$ is now twice covariant tensor field and (22.4a) will be written

$$\underline{ds}^2(dx \otimes dx) = g_{ij}(x)dx^i dx^j \,. \tag{22.4c}$$

The scalar product $\langle , \rangle$, introduced above, associates to any vector u the number

$$\|u_M\|_g^2 = g_{ij}(x)u^i(x)u^j(x) \ .$$

$\|u_M\|_g$ is not a norm and it may be complex. If for any $u_M \neq 0$,

$$\|u_M\|_g^2 > 0 \ ,$$

then the corresponding manifold is **proper Riemannian**. If not, it is **pseudo Riemannian** and the metric is said to be **indefinite**.

For a pseudo Riemmannian manifold (as the one of special relativity), if for $u_M \neq 0$, one has

$$\|u_m\|_g = 0 \ ,$$

then the vector is said to be **null (isotropic or light-like)**. The set of all null vectors builds up the **isotropy cone** at $M \in \mathcal{M}_N$ in $T_{\mathcal{M}}(M)$.

The rules which were given at Level 0 (Sec. 3) and Level 1 (Sec. 8) for the raising and lowering of the indices are clearly valid and now justified.

If in the expression of ds^2 it happens that the $g_{ij}(x)$ are constants and more particularly if

$$g_{ij}(x) = \pm\delta_{ij} \ ,$$

the manifold is said to be **flat**. If

$$g_{ij}(x) = \delta_{ij} \ ,$$

the manifold is said to be **Euclidean**.

23. *Connection on a Riemannian Manifold*

From now on, we will use the notation $\underline{ds}^2$ (instead of $\underline{G}_M$) : $\underline{ds}^2$ is thus a twice covariant tensor field. Given a Riemannian manifold we want to show that the following two conditions imply a unique, well-defined connection on the manifold:

(1) the torsion tensor in the manifold vanishes,

(2) the covariant derivative of the tensor $\underline{ds}^2$ vanishes.

Let us begin with the condition (2): for any vector v_M,

$$\nabla_v^M \underline{ds}^2 = \nabla_v^M \{g_{ij}(x)\omega_M^i \otimes \omega_M^j\} = 0 \ , \qquad (23.1)$$

taking into account (18.38) and the symmetry of the $g_{ij}(x)$, one has locally

$$\mathbf{D}_l g_{ij}(x) = g_{ik}(x)\gamma_{lj}^k(x) + g_{jk}\gamma^k{}_{li}(x) \ . \qquad (23.2)$$

Let us now consider condition (2); using (21.7) we obtain a second relation between $\underline{d}\omega^i$ and the matrix elements Ω^i_j (1-forms) of the connection matrix

$$\mathcal{T}^i = 0 = \underline{d}\omega^i + \Omega^i_j \wedge \omega^j \ , \tag{23.3}$$

where the argument M has been omitted. In local coordinates, Eq. (20.7) for the tensor $c^i_{jk}(x)$ can be used to obtain

$$\mathcal{T}^i_{jk}(x) = \gamma^i_{jk}(x) - \gamma^i{}_{kj}(x) - c^i_{jk}(x) = 0 \ . \tag{23.4a}$$

The introduction of the generalized Christoffel symbol

$$[ij,l] = \frac{1}{2}(\mathbf{D}_i g_{jl}(x) + \mathbf{D}_j g_{li}(x) - \mathbf{D}_l g_{ij}(x)) \tag{23.4b}$$

will allow us to show that the connection on a Riemannian manifold is uniquely determined by the γ-symbols following (18.14) and (18.33).

Indeed, let us go back to (23.2): a circular permutation of the indices l, i, j leads to two new equations. Let us add the two first equations and substract the last one from their sum. By calculations similar to the ones developed in (9.6), (9.7), (9.8) of Level 1 and the use of (23.4a), one obtains after certain simplifications

$$2[ij,l] = g_{ik}(x)c^k{}_{lj}(x) - g_{kj}(2\gamma^k{}_{il}(x) - c^k_{li}(x)) + g_{il}(x)c^k{}_{ij}(x) \ , \tag{23.5}$$

which can be written as

$$[ij,l] = \frac{1}{2}(g_{ik}(x)c^k_{il}(x) + g_{kj}(x)c^k{}_{li}(x) + g_{ki}(x)c^k_{ij}(x)) + g_{kj}(x)\gamma^k{}_{il}(x) \ . \tag{23.6}$$

The term $g_{kj}\gamma^k{}_{il}$ is the last term in (23.6): by multiplication with g^{kp}, we obtain the γ-symbols:

$$\gamma^p{}_{il} = g^{pj}[ij,l] - \frac{1}{2}(g^{pj}g_{ik}c^k{}_{lj} + g^{pj}g_{kl}c^k{}_{ij} + c^p_{li}) \tag{23.7}$$

and hence the connection on the Riemannian manifold. In the natural frame $\{\varepsilon^M_k\}$, the c^i_{kj} vanish and (23.7) reduces to

$$\Gamma^p{}_{ij} = g^{pj}[ij,l] \ . \tag{23.8}$$

In that special case, the $\mathbf{D}_i$ are simply ∂_i and $[ij,l]_n$ is given by either (3.22), Level 0 or (9.8), Level 1. We want to conclude the present paragraph with some considerations which will be important later on. Let us consider formula (23.2) once more. We recall two previous formulae: in (9.24) we said that for any C^∞ function

$$\underline{d}\hat{F}(M) = \mathbf{D}_i F(x)\omega_M^i \ ,$$

and in (18.9) we defined the connection matrix

$$\Omega_i^k(M) = \gamma^k{}_{ji}(x)\omega_M^j \ .$$

Using these two formulae, we multiply both sides of (23.2) by ω_M^l, and get

$$\underline{d}g_{ij} = \Omega_{ij} + \Omega_{ji} \ , \qquad\qquad (23.9)$$

where

$$\Omega_{ij} = g_{ik}\Omega_j^k = g_{ik}\gamma^k{}_{lj}\omega^l \ .$$

Together with (23.9), we recall also the Cartan-Maurer formula (9.25)

$$\underline{d}\omega_M^k = -\frac{1}{2}c^k{}_{ij}\omega_M^i \wedge \omega^j{}_M \ , \qquad\qquad (23.10)$$

where the structure coefficients are defined by (9.14) (see also (4.14) and (4.18)). In the special case where $g_{ij}(x)$ are constants, $[ij, l]$ as defined by (23.4) vanishes and (23.7) is reduced to

$$\gamma_{il}^p = \frac{1}{2}[g^{pj}c_{ijl} + g^{pj}c_{lji} - c_{li}^k] \ , \qquad\qquad (23.11a)$$

where

$$c_{ijl} = g_{ik}c^k{}_{jl} \ .$$

Still simpler is the case of a flat manifold, with $\varepsilon_p = \pm 1$; its metric tensor is

$$g_{pj} = \varepsilon_p \delta_p^j \ , \qquad g^{pj} = \varepsilon_p \delta_p^j$$

and

$$\Omega_{pk} = g_{pk}\gamma^p{}_{jl}\omega_M^l = \frac{\varepsilon_p}{2}(c_{kpl} + c_{lpk} - c_{pkl})\omega_M^l \ . \qquad\qquad (23.11b)$$

All these formulae will be used in Problem 2.7 for the calculation of the Γ-symbols and the Riemann tensor.

As a last remark, we want to investigate the relation between exterior and covariant derivations in a Riemannian space. We first consider a function $\hat{F}$; one has then clearly

$$\underline{d}\hat{F}(M) = \partial_k F(x)\underline{d}x^k = \nabla_k F(x)\underline{d}x^k . \tag{23.12}$$

Consider now a 1-form α_M, one has

$$\underline{d}\alpha M = \partial_j \alpha_k(x)\underline{d}x^j \wedge \underline{d}x^k . \tag{23.13a}$$

But following (5.5) of Level 1, one may also write

$$\nabla^M \alpha_M = (\partial_j \alpha_k(x) - \Gamma^s{}_{kj}\alpha_s)\underline{d}x^j \wedge \underline{d}x^k , \tag{23.13b}$$

where the last term in the bracket vanishes, i.e.,

$$\Gamma^s{}_{kj}\alpha_s \underline{d}x^j \wedge \underline{d}x^k = 0 ,$$

since Γ_{kj} is symmetric in k, j and $\underline{d}x^k \wedge \underline{d}x^j$ antisymmetric. Comparing (23.13a) and (23.13b), one gets

$$\underline{d}\alpha_M = \nabla_j \alpha_k(x)\underline{d}x^j \wedge \underline{d}x^k . \tag{23.14}$$

Note that formula (23.12) can be recast into an intrinsic form. For any vector u_M, one has

$$\underline{d}\hat{F}_M(u_M) = \nabla_k F(x)\underline{d}x^k(u^i e_i^M) = u^k \nabla_k F(x) = \nabla_u^M \hat{F} . \tag{23.15}$$

It has been stressed, more than once, that differential geometry is not necessarily based on a metric: metric concepts may be introduced after a topological study of differentiable manifolds. In the next paragraph, the generalization of geodesics into autoparallel curves will be a perfect illustration of that point.

24. *Geodesics and Autoparallel Curves*

Let a curve Γ be given on a manifold $\mathcal{M}_N$; locally, in a given chart, its image is the curve C,

$$C : x = x(\lambda) , \tag{24.1}$$

where λ belongs to the real interval $\mathcal{I}$ defining the curve C. Let $\{e_k^M\}$ be a basis of $T_\mathcal{M}(M)$; the vector

$$t_M = \frac{dx^k}{ds}e_k^M = \dot{x}^k e_k^M \tag{24.2}$$

is tangent to Γ at M. On the other hand, the $\underline{ds}^2$ of $\mathcal{M}_N$ is given by (22.2)

$$\underline{ds}^2(t_m \otimes t_M) = g_{ij}\dot{x}^i\dot{x}^j \ . \tag{24.3}$$

We may then consider two important cases.

(A) $\mathcal{M}_N$ *is a proper Riemannian manifold*: then $ds^2 > 0$ and, if λ_I and λ_II are two values of λ in (24.1) such that $\lambda_\mathrm{I} \in \mathcal{I}, \lambda_\mathrm{II} \in \mathcal{I}$, then the length of the corresponding arc of C is

$$s(\lambda_\mathrm{I}, \lambda_\mathrm{II}) = \int_{\lambda_\mathrm{I}}^{\lambda_\mathrm{II}} \sqrt{g_{ij}(x(\lambda))\dot{x}^i\dot{x}^j}\, d\lambda \tag{24.4}$$

and the square root is well-defined. As in Level 0, Sec. 4 the geodesics will be defined as the extremals of $s(\lambda_\mathrm{I}, \lambda_\mathrm{II})$. They obey the Euler-Lagrange system of differential equations (see (13.2), Level 1):

$$\ddot{x}^k + \Gamma^k{}_{ij}(x)\dot{x}^i\dot{x}^j + \dot{x}^k A(x(\lambda)) = 0 \tag{24.5}$$

with

$$A(x(\lambda)) = \ddot{s}(\lambda)/\dot{s}(\lambda) \ ,$$

where (see (14.2) Level 1)

$$s(\lambda) = \int_{\lambda_\mathrm{I}}^{\lambda} \sqrt{g_{ij}(x(\lambda))\dot{x}^i\dot{x}^j}\, d\lambda \ , \tag{24.6}$$

and the Γ-coefficients are given by (22.8).

(B) $\mathcal{M}_N$ *is pseudo-Riemannian*: then the square root appearing in (24.5) has no meaning. Geodesics will be defined as the extremals of

$$J(\lambda_\mathrm{I}, \lambda_\mathrm{II}) = \int_{\lambda_\mathrm{I}}^{\lambda_\mathrm{II}} g_{ij}(x(\lambda))\dot{x}^i\dot{x}^j\, d\lambda \tag{24.7a}$$

and a simple calculation shows that the Euler-Lagrange equations read

$$\ddot{x}^k + \Gamma^k{}_{ij}(x)\dot{x}^i\dot{x}^j = 0 \ . \tag{24.7b}$$

Such a system of differential equations represent the autoparallel curves which were studied in Sec. 6, Level 1 and one can see that autoparallel curves are geodesics (and vice versa) if the parameter λ is chosen to be s, since the factor A in (24.5) vanishes.

All the results of Sec. 3, Level 1 clearly remain valid.

G. Vector Fields and Lie Groups

25. *Vector Fields and One-Parameter Pseudo Groups*

The present paragraph will be divided into two parts:

Part (A): In a given affine space $\mathcal{E}_N$, we shall study the trajectories of a vector field $X(x) : x \in \mathcal{E}_N, X(x) \in \mathcal{E}_N$.

Part (B): The same problem will be intrinsically considered. We will be led to the study of one-parameter groups and pseudo groups. We begin with:

Part (A): When a physicist wants to study a vector field, an electric, a magnetic field or a field of forces, he begins his study by looking at the **trajectories of the field**. They are called **lines of forces** where forces are involved, **lines of current** when one considers the flow of the velocities of a given fluid or an electrical current in a conducting medium. By definition, they represent curves whose tangents in each of their points $x = \{x^1 \ldots x^N\}$ are along the direction $X(x)$. If $x = x(t), (t \in \mathcal{I} \subset \mathbb{R})$, is the equation of one of these curves, then

$$\frac{dx(t)}{dt} = \boldsymbol{X}(x(t)) , \qquad (25.1a)$$

or

$$\frac{dx^1(t)}{dt} = X^1(x(t)) \ldots , \qquad \frac{dx^N(t)}{dt} = X^N(x(t)) , \qquad (25.1b)$$

a system of $N - 1$ equations which also reads

$$\frac{dx^1}{X^1(x)} = \cdots \frac{dx^N}{X^N(x)} . \qquad (25.2)$$

Systems (25.1) and (25.2) are also said to describe a **flow of trajectories**. Under Lipschitz-type conditions for instance, we consider a domain $\Delta \subset \mathcal{E}_N$ where all $X^k(x) \neq 0$; then if ξ is any point of Δ, one can assert that one and only one trajectory goes through ξ:

$$x_\xi = x_\xi(t) : \ x_\xi(0) = \xi .$$

This is a **Cauchy problem** or an **initial value problem**. Such a problem may be formulated differently: one considers solutions of (25.2) depending on $N - 1$ arbitrary parameters $A_1 \ldots A_{N-1}$ such that

$$x^k = x^k(t; A_1 \ldots A_N) \ . \tag{25.3}$$

These $N - 1$ functions of t describe a certain curve in $\mathcal{E}_N$ and one can fix the constants A_k such that for $t = 0, x = \xi$. But such a curve can also be considered as the intersection of $N - 1$ hypersurfaces

$$F_k(x^1 \ldots x^N) = B_k \ ,$$

where the B_k are constants. Indeed, for such an hypersurface, one has

$$0 = \frac{dF_k(x(t))}{dt} = \frac{\partial F_k}{\partial x^j} \frac{dx^j}{dt} = X^j(t) \frac{\partial F_k}{\partial x^j} \ , \tag{25.4}$$

taking (25.1) into account. One is then able to associate with (25.1) a partial derivative equation

$$X^j(x) \partial_j F(x^1 \ldots x^N) = 0 \ . \tag{25.5}$$

$N - 1$ linearly independent solutions of (25.5) in the domain Δ, called **first integrals** of (25.1), define the trajectory under study. One shows in classical analysis that (25.1) and (25.5) are two differential systems which are equivalent. Physicists often use another formalism to describe solutions of (25.2): the **evolution operator** formalism. Let us describe the trajectory obeying the initial condition $x_\xi(0) = \xi$ as follows:

$$\xi \rightarrow x_\xi(t) : \ x_\xi(t) = \mathcal{U}_X(t)\xi \ , \quad \mathcal{U}_X(0) = I \ , \tag{25.6}$$

or, in other words, to any point $\xi \in \Delta$, there corresponds by means of the map $\mathcal{U}_X(t)$ a trajectory $x = x_\xi(t)$ of the field $X(x)$. Take then $x_\xi(t + s)$, where t and s are real numbers; this is a solution (one also says **integral** of (25.1)) and, using the definition (25.6), one has

$$x_\xi(t + s) = \mathcal{U}_X(t + s)\xi : \ x_\xi(t + s)|_{t+s=0} = \xi \ . \tag{25.7}$$

On the other hand, the equation

$$x_\xi(t + s)|_{t=0} = \mathcal{U}_X(t + s)\xi|_{t=0} = \mathcal{U}_X(s)\xi \ , \tag{25.8}$$

leads to

$$x_\xi(t+s) = \mathcal{U}_X(t)\mathcal{U}_X(s)\xi \ . \tag{25.9a}$$

Comparing (25.7) and (25.9), it finally turns out that

$$\mathcal{U}_X(t+s) = \mathcal{U}_X(t) \circ \mathcal{U}_X(s) \ . \tag{25.9b}$$

This is a group multiplication law and the $\{\mathcal{U}_X(t)\}$, when t describes $\mathcal{I}$, are said to form a **pseudo-group** (see (25.14)). The former considerations suggest another notation for the evolution operator; we may introduce, indeed, a particular notation, the **exponential map** by the definition (see for further developments formulae (26.38), (26.41)):

$$\mathcal{U}_X(t) = \exp tX \ , \tag{25.10}$$

since one has

$$\exp(t+s)X = \exp tX \circ \exp sX \ .$$

We may conclude those preliminary considerations with some comments on the exponential of a matrix which is another exponential map but distinct to the one we just presented. Let Ω be an N by N real and regular matrix with elements Ω_j^i, supposed to be independent of the parameter t. Several physical problems lead to the system of differential equations

$$\frac{dx(t)}{dt} = \Omega x(t) \ , \tag{25.11a}$$

where $x(t)$ is a vector of components $\{x^1(t)\ldots x^N(t)\}$. Such an equation is equivalent to the differential system

$$\frac{dx^i(t)}{dt} = \Omega_j^i x^j(t) \ . \tag{25.11b}$$

The solution of (25.11) is really simple; we first calculate the set of derivatives $\{d^k(x)/dt^k\}$ at $t = 0$,

$$\frac{d^k x^i(t)}{dt^k}\bigg|_{t=0} = \Omega_{l_1}^i \Omega_{l_2}^{l_1} \ldots \Omega_j^{l_k} x^j(t)|_{t=0} = (\Omega^k)_j^i \xi^j \ ,$$

and then build up the Taylor series at $t = 0$:

$$x^i(t) = \sum \frac{t^n}{n!} \frac{d^n x^i(t)}{dt^n}\bigg|_{t=0} = \sum \frac{t^n}{n!}(\Omega^n)_j^i \xi^j$$
$$= (e^t\Omega)_j^i \xi^j = U_{\Omega j}{}^i \xi^j$$
$$\Rightarrow x(t) = e^{t\Omega\xi} = U_\Omega \xi \ .$$

The map U_Ω is an example of the evolution operator and it also obeys the equation

$$\frac{dU_\Omega}{dt} = \Omega U_\Omega(t) \ . \tag{25.12a}$$

Equation (25.11) means that to each point $\xi \in \mathcal{E}_N$ there goes an associated trajectory $x = x(t)$. The set of the tangents $\{X(t)\}$ at this trajectory builds up a field (of the electromagnetic or mechanical type, for instance). Such a field, which is also called a **flow**, is defined by

$$X(t) = \frac{dx^{(}t)}{dt} = \Omega e^{t\Omega\xi} = \Omega x(t) \ . \tag{25.12b}$$

The set of matrices $\{U_\Omega\}$ builds up a group, which, as we will see in the next paragraph, is a one-parameter pseudo Lie group of parameter t.

There is an obvious generalization of (25.11): if we suppose that Ω depends also on the parameter t, then the solution of

$$\frac{dx(t)}{dt} = \Omega(t)x(t) \tag{25.13a}$$

cannot be represented by an exponential as above, although one can still define an evolution operator $U_\Omega(t, t_0)$ such that

$$\frac{dU_\Omega(t, t_0)}{dt} = \Omega(t)U_\Omega(t, t_0) : \ U_\Omega(t_0, t_0) = 1 \ . \tag{25.13b}$$

Such a set of matrices builds up a pseudo group with the multiplication law

$$U_\Omega(t, t')U_\Omega(t', t_0) = U_\Omega(t, t_0) \ . \tag{25.14}$$

Part (B): We intend to recast the results of Part (A) in an intrinsic form. But before any development, let us state two slight changes of notation which will shorten our formulae: instead of (2.6), Level 2,

$$\hat{\mathcal{F}}(\lambda) = M(\lambda) \ ,$$

describing the curve $\Gamma \subset \mathcal{M}_N$, we shall use

$$\hat{F}_t = M_t \ , \quad t \in \mathcal{I} \subset \mathbb{R} \ ,$$

where $\hat{F}_t$ is a map $\mathbb{R} \to \mathcal{M}_N$. The image $C \subset \mathbb{R}^N$ of Γ in some chart defined by a given diffeomorphism φ is given by

$$x = x(t) = \varphi(\hat{\mathcal{F}}_t) = \varphi \circ \hat{\mathcal{F}}(t) = \mathcal{F}(t) \ , \tag{25.15a}$$

where $\mathcal{F}(t)$ symbolizes the NC^∞ functions $\mathcal{F}^k(t)$.

The tangent to $\Gamma \subset \mathcal{M}_N$ at the point M_t is given by (4.10):

$$X_{M_t} = \frac{dx^i(t)}{dt}\varepsilon_i^{M_t} , \qquad (25.15b)$$

where the notation X_{M_t} replaces the previous notation $\tau_M(\lambda)$.

Since X_{M_t} is a vector, one may write

$$X_{M_t} = X^k(x(t))\varepsilon_k^{M_t} \qquad (25.15c)$$

in the natural basis $\{\varepsilon_k^{M_t}\}$, after the comparison of the two equations (25.15). One can also choose to represent (25.15b) by $d\hat{F}_t/dt$, then the previous expressions can be replaced by a single relation

$$\frac{d\hat{\mathcal{F}}_t}{dt} = X_{M_t} \qquad (25.15d)$$

and Eq. (25.1) becomes its local form.

The evolution operator can still be written as

$$M_t = \hat{\mathcal{F}}_t = (\exp t X_M) M_0 , \qquad (25.16)$$

where $M_0 \in \mathcal{M}_N$. This is an **exponential map** for the vector field X_{M_t}: it defines the **flow** of the vector X_M. The set of exponential maps is again a pseudo group (local)*. One last remark: the operator i_χ of Problem 2.3 may be used to bring the corresponding equation associated with (25.16) into an intrinsic form, since that equation can be written as

$$i_\chi \underline{d}\hat{F} = 0 \qquad (25.17)$$

and $\hat{F}(M)$ is a first integral of (25.2).

26. *Lie Groups and Lie Groups of Transformations*

In this field also, physicists and mathematicians adopt fundamentally different strategies. Starting from the knowledge he acquired while working with translations, rotations and Lorentz transformations, the physicist favours a method which leads directly to the crux of his problem. The

---

*For its local intepretation, see formulae (26.38), (26.41).

mathematician is more ambitious: he analyzes more precisely and deeply the general problem under study and asks questions at different levels and gets, sometimes, involved, but always rigorous answers. Group theory and Lie group theory have become by now an invaluable research tool in atomic physics and in theory of elementary particles: they rightly deserve some attention.

(A) *The physicist's point of view** is essentially local. We may consider an affine space $\mathcal{E}_N$ and its point $M \in \mathcal{E}_{N'}$, the associated vector space $\mathbb{R}^N$, whose elements are N-uplets $x = \{x^1 \ldots x^N\}$ and a homeomorphism φ:

$$\mathcal{E}_N \xrightarrow{\varphi} \mathbb{R}^N \; ,$$
$$M \in \mathcal{E}_N : \; x = \varphi(M) \; , \tag{26.1}$$

a second vector space $\mathbb{R}^r$ of elements $a = \{a^1 \ldots a^r\}$, and finally a set of differentiable maps $F_{(a)} : \mathcal{E}_N \to \mathcal{E}_N$, each element of the set being indexed by r-uplets:

$$a = \{a^1 \ldots a^r\} \; .$$

We state that the set $\{\hat{F}_{(a)}\}$ builds up a group with the laws:

I $\qquad\qquad\qquad\qquad \hat{F}_{(0)} = I \; ,$

II $\qquad\qquad\qquad\qquad \forall \hat{F}_{(a)} : \; \exists \hat{F}_{(a)}^{-1} \; . \tag{26.2}$

To these rules, we have to add the multiplication law of the group

III $\qquad\qquad\qquad\qquad \hat{F}_{(a')} \circ \hat{F}_{(a)} = \hat{F}_{(a'')} \; ,$

where a, a', a'' are three r-uplets obeying the relation

III' $\qquad\qquad a'' = \Theta(a, a') : \; a''^\lambda = \Theta^\lambda(a^1 \ldots a^r, a'' \ldots a'^r) \; ,$

where Θ^i is a well-defined C^∞ function of a, a', characteristics of the group. At this point, we shall make some useful comments: from the very definition of $\hat{F}_{(a)}$, to any point $M \in \mathcal{E}_N$ we associate another point $M_{(a)}$ such that

$$M_{(a)} = \hat{F}_{(a)}(M) : \; M_{(a^1 \ldots a^r)} = \hat{F}_{(a' \ldots a^r)}(M) \; . \tag{26.3}$$

Then φ being the homeomorphism characteristic of the chart under consideration, one has locally

$$\varphi(M_{(a)}) = \varphi \circ \hat{F}_{(a)}(M) = \varphi \circ \hat{F}_{(a)} \circ \varphi^{-1}(x) \; , \tag{26.4a}$$

*Among many published textbooks, let us quote two: H. Bacry, *Lectures on Group Theory and Particle Theory*, Gordon and Breach Publishers and also a classic in this domain: M. Hammermesh, *Group Theory and Its Application to Physical Problems*, Pergamon Press.

or, in other terms,

$$x_{(a)} = F_{(a)}(x) : \quad x^j_{(a^1 \dots a_r)} = F^j_{(a^1 \dots a^r)}(x^1 \dots x^N)$$
$$j = 1 \dots N \ . \tag{26.4b}$$

Since the map $\hat{F}_{(a)}$ is a differentiable one, $F_{(a)}(x)$ is also differentiable with respect to the N variable x^k, we shall suppose that it is also differentiable with respect to the r parameter a^λ: under such conditions, physicists call the group with elements $F_{(a)}$ a **Lie group with r (essential) parameters** or in short an **r parameter Lie group**. By set of essential parameters, we mean that no parameter of this set can be expressed as a function of the remaining $r - 1$ parameters.

The set of maps $\hat{F}_{(a)}$ being a group, the product of two of its elements $F_{(a)}$ and $F_{(\delta a)}$, where δa is the infinitesimal r-vector

$$\delta a = \{\delta a^1 \dots \delta a^r\} \ , \tag{26.5a}$$

is another map; let us call it $F_{(a+da)}$, where the infinitesimal r-vector

$$da = \{da^1 \dots da^r\} \ , \tag{26.5b}$$

is defined by the relation

$$\hat{F}(\delta a) \circ \hat{F}_{(a)} = \hat{F}_{(a+da)} \ . \tag{26.6}$$

By a straightforward application of (26.2), rule III, there exists the relation

$$\Theta(a, \delta a) = a + da \ , \tag{26.7}$$

between $a, \delta a$ and da.

We now consider again the rule (26.2), III and put $a' = 0$: since $\hat{F}_{(0)} = 1$ (rule I),

$$\hat{F}_{(a)} = \hat{F}_{(a'')} \ , \tag{26.8a}$$

which means

$$a'' = a = \Theta(a, 0) \ . \tag{26.8b}$$

Using (26.7), one has

$$da = \Theta(a, \delta a) - a = \Theta(a, \delta a) - \Theta(a, 0) \ , \tag{26.9}$$

and this is a formula with a simple interpretation if one uses the formalism of the Jacobian matrix developed in Sec. 3, Level 1. The left-hand side of (26.9) can be thought of as a column of r elements da^λ; its right-hand side can be expressed, following (9.8), Level 1, as the product of the Jacobian matrix of the function $\Theta(a, b)$ of $2r$ variables (when the parameter b vanishes) with the column matrix $\delta\alpha$ with $\delta\alpha^\lambda$ as elements

$$da = \frac{\mathbf{D}(\Theta(a, b))}{\mathbf{D}(b)}\bigg|_{b=0} \delta a = \Delta(a)\delta a \ , \tag{26.10}$$

where $\Delta(a)$ is the previous r by r Jacobian matrix for $b = 0$. We now return to (26.3) and write

$$M_{(a+da)} = \hat{F}_{(a+da)}(M) = \hat{F}_{(\delta a)} \circ \hat{F}_{(a)}(M) = \hat{F}_{(\delta a)}(M_a) \ . \tag{26.11a}$$

Locally, such a result means

$$x_{(a+da)} = F(\delta a)(x_{(a)}) \tag{26.11b}$$

and also

$$x_{(a+da)} - x_{(a)} = \frac{\mathbf{D}(F_{(b)}(x_{(a)}))}{\mathbf{D}(b)}\bigg|_{b=0} da = U(x_{(a)})da \ , \tag{26.12}$$

where we have again used the formalism of the Jacobian matrix and defined clearly the matrix $U(x_{(a)})$ as an r by r matrix.

Explicitly, formula (26.12) takes the form

$$x^i_{(a+da)} = x^i_{(a)} + \delta a^\lambda U^l_\lambda(x_{((a)})\frac{\partial}{\partial x^l_{(a)}}x^i_{(a)} \quad \begin{matrix} i = 1 \ldots N \ , \\ \lambda = 1 \ldots r \ . \end{matrix} \tag{26.13}$$

We now introduce the r first-order differential operators called **generators**:

$$X_\lambda(x_{(a)}, \partial) = U^l_\lambda(x_{(a)})\frac{\partial}{\partial x^l_{(a)}} \ . \tag{26.14}$$

(26.13) can be written as

$$x^i_{(a+da)} = x^i_{(a)} + \delta a^\lambda X_\lambda(x_{(a)}, \partial)x^i_{(a)} \ , \tag{26.15a}$$

or

$$x_{(a+da)} = \{I + \delta a^\lambda X_\lambda(x_{(a)}, \partial)\}x_{(a)} \ . \tag{26.15b}$$

The number r of generators denotes the **dimension** of the Lie group under study*

Each generator X_λ is, following Sec. 4, a vector of $\mathcal{M}_N$ and their commutator is just another vector. The commutator $[X_\alpha, X_\beta]$ can be calculated following the method we have been using in formulae (4.19) to (4.21). It turns out that

$$
[X_\alpha, X_\beta] = \left[U_\alpha^m \frac{\partial}{\partial x_{(a)}^m}, U_\beta^n \frac{\partial}{\partial x^n{}_{(a)}} \right]
$$
$$
= \left(U_\alpha^m \frac{\partial U_\beta^i}{\partial x^m{}_{(a)}} - U_\beta^m \frac{\partial U_\alpha^i}{\partial x^m{}_{(a)}} \right) \frac{\partial}{\partial x^i{}_{(a)}} \; ;
$$

(26.16)

we want now to show that the bracket in the expression of the last term in (26.16) is a linear combination of the generators with coefficients depending on a^λ only. The calculation is indeed tedious but unavoidable: let us return to (26.10); one has

$$
\delta_\alpha = (\Delta(a))^{-1} da = \Delta^{-1}(a) da \;,
$$

(26.17)

an expression to be brought into formula (26.12). We obtain

$$
dx_{(a)} = U(x_{(a)} \Delta^{-1}(a) da \;,
$$

(26.18a)

or

$$
dx_{(a)}^m = U_\lambda^m \Delta^{-1\lambda}{}_\mu da^\mu : \begin{array}{l} m = 1 \ldots N \\ \mu, \lambda = 1 \ldots r \;, \end{array}
$$

(26.18b)

which is an expression of the total differential $dx^m{}_{(a)}$ as a function of the differential da^μ. The well-known integrability conditions

$$
\frac{\partial^2 x^m{}_{(a)}}{\partial a^\rho \partial a^\sigma} = \frac{\partial^2 x_{(a)}^m}{\partial a^\sigma \partial^\rho}
$$

(26.19a)

lead to

$$
\frac{\partial}{\partial a^\rho} \{ U_\lambda^i \Delta^{-1\lambda}{}_\sigma \} = \frac{\partial}{\partial a^\sigma} \{ U_\lambda^i \Delta^{-1\lambda}{}_\rho \} \;.
$$

(26.19b)

A simple algebraic manipulation shows that (26.19) can be brought into the form

$$
U_\lambda^i \left(\frac{\partial \Delta^{-1\lambda}{}_\sigma}{\partial a^\rho} - \frac{\partial \Delta^{-1\lambda}{}_\rho}{\partial a^\sigma} \right) + \Delta^{-1\lambda}{}_\sigma \frac{\partial U_\lambda^i}{\partial a^\rho} - \Delta^{-1\lambda}{}_\rho \frac{\partial U_\lambda^i}{\partial a^\sigma} = 0 \;.
$$

(26.20)

*For a finite group, its order is the number of its elements.

Now the $U_\lambda^i = U_\lambda^i(x_{(a)})$ are functions of $\{x^1 \ldots x^N\}$ and, through the variable $x_{(a)}$, of the parameters $\{a^1 \ldots a^r\}$. For a given point x, the U_λ^i are then functions of a^λ and

$$\frac{\partial U_\lambda^i}{\partial a^\rho} da^\rho = \frac{\partial U_\lambda^i}{\partial x_{(a)}^m} \frac{\partial x_{(a)}^m}{\partial a^\rho} da^\rho = \frac{\partial U_\lambda^i}{\partial x_{(a)}^m} dx^m{}_{(a)} \ . \tag{26.21}$$

We may express the $dx^m{}_{(a)}$ as function of a^λ using (26.18), then

$$\frac{\partial U_\lambda^i}{\partial a^\rho} da^\rho = \frac{\partial U_\lambda^i}{\partial x_{(a)}^m} U_\nu^m \Delta^{-1\nu}{}_\rho da^\rho \ , \tag{26.22}$$

and consequently we obtain $\partial U_\lambda^i / \partial a^\rho$, which we may introduce in the second and third terms of (26.20). After a few transformations, one gets

$$U_\lambda^i \left(\frac{\partial \Delta^{-1\lambda}{}_\sigma}{\partial a^\rho} - \frac{\partial \Delta^{-1\lambda}{}_\rho}{\partial a^\sigma} \right) + \Delta^{-1\lambda}{}_\sigma \Delta^{-1\tau}{}_\rho \left(U_\tau^m \frac{\partial U_\lambda^i}{\partial x_{(a)}^m} - U_\lambda^m \frac{\partial U_\tau^i}{\partial x_{(a)}^m} \right) = 0 \ . \tag{26.23}$$

We multiply this expression by $\Delta^\sigma{}_\alpha \Delta^\rho_\beta$, contract, and obtain

$$U_\alpha^m \frac{\partial U_\beta^i}{\partial x_{(a)}^m} - U_\beta^m \frac{\partial U_\alpha^i}{\partial x_{(a)}^m} = \Delta^\sigma_\alpha \Delta^\rho_\beta \left(\frac{\partial \Delta^{-1\lambda}{}_\sigma}{\partial a^\rho} - \frac{\partial \Delta^{-1\lambda}{}_\rho}{\partial x^\sigma} \right) U_\lambda^i \ . \tag{26.24}$$

The introduction of the **structure constants**

$$c_{\alpha\beta}^i = \Delta^\sigma_\alpha \Delta^\rho_\beta \left(\frac{\partial \Delta^{-1\lambda}{}_\sigma}{\partial a^\rho} - \frac{\partial \Delta^{-1\lambda}{}_\rho}{\partial a^\sigma} \right) \tag{26.25}$$

leads to the final formula

$$U_\alpha^m \frac{\partial U_\beta^i}{\partial x_{(a)}^m} - U_\beta^m \frac{\partial U_\alpha^i}{\partial x_{(a)}^m} = c_{\alpha\beta}^\lambda U_\lambda^i \ . \tag{26.26}$$

We now return to (26.16) and write the **Lie relations**:

$$[X_\alpha, X_\beta] = c_{\alpha\beta}^\lambda U_\lambda^i \partial_i = c_{\alpha\beta}^\lambda X_\lambda \ , \tag{26.27}$$

which express that the generators X^α build up an algebra, called the **Lie algebra**, of the Lie group under consideration.

It remains to show that the structure constants are independent of the parameters a^λ: we take the derivative of (26.26) with respect to a^ρ and let the operator

$$\frac{\partial x^k_{(a)}}{\partial a^\rho} \frac{\partial}{\partial x^k_{(a)}}$$

act on the result, then

$$U^i_\lambda \frac{\partial c^\lambda_{\alpha\beta}}{\partial a^\rho} = 0 \ . \tag{26.28}$$

Let the index ρ be given, then (26.28) describes a system of $N(i = 1 \ldots N)$ linear equations for the r unknowns $\partial c^\lambda_{\alpha\beta}/\partial a^\rho (\lambda = 1 \ldots r)$. Their elimination gives rise to a certain number of relations among the U^i_λ. As a consequence a certain number of U^i_λ are functions of the others, and as another consequence λ cannot take all the values ranging from 1 to r, in contradiction with the hypothesis that there must be r essential parameters a^λ. Then the only possible solution of the previous system is the value 0 for the unknowns $\partial c^\lambda_{\alpha\beta}/\partial a^\rho$. This is a property which means that the coefficients $c^\lambda{}_{\alpha\beta}$ are independent of the parameters a^ρ. We may then arbitrarily choose this set of parameters: their common value

$$a^1 = a^2 = \ldots a^r = 0$$

brings important simplifications. Indeed, (26.15) we may choose $a = 0$;

$$x_{(da)} = \{I + \delta a^\lambda \chi_\lambda\}x \ , \tag{26.29a}$$

and define the Lie algebra in the neighborhood of the identity map $(F_{(0)})$. Then, we may use an exponential representation of $x_{(da)}$:

$$x_{(da)} = e^{i\delta a^\lambda X_\lambda}x \ , \tag{26.29b}$$

which is a very useful formula in the study of Lie groups of matrices (see Problems 2.10 and 2.11).

One shows further that the structure constants have interesting properties:

(1) antisymmetry $c^\lambda_{\alpha\beta} = -c^\lambda_{\beta\alpha}$,
(2) they obey the Jacobi identity

$$c^\rho_{\alpha\beta}c^\sigma_{\rho\nu} + c^\rho{}_{\beta\nu}c^\sigma{}_{\rho\alpha} + c^\rho_{\nu\alpha}c^\sigma_{\rho\beta} = 0 \ ,$$

(3) they are components of a tensor of order 3, twice covariant and once contravariant,

(4) the knowledge of the structure constants determines completely a Lie group, the determination is unique (modulo an isomorphism) if the group is simply connected.

The proof of the three first properties is straightforward; the proof of the fourth is involved and will not be given (however, see M. Hammermesh, *loc. cit.* p. 304).

A last concept is useful because of its physical implications: the **rank** of a group gives the number of the generators which commute. The rank of SO(3) is, for instance, equal to 1 (only 1 of the generators J_1, J_2, J_3 is diagonalizable) while the rank of SU(3) is 2.

We may conclude this physicist's point of view by considering the special case $r = 1$: the corresponding group is then a **one-parameter Lie group**. Formulae become simpler than before, for instance the set of the U_λ^j matrices is replaced by the multiplet of functions $\{U^1(x_{(a)}) \ldots U^N(x_{(a)})\}$ and the set of generators is reduced to a single generator

$$X(x_{(a)}, \partial) = U^l(x_{(a)}) \frac{\partial}{\partial x^l_{(a)}} \ . \tag{26.30}$$

Eq. (26.18) is also simplified:

$$dx^m_{(a)} = U^m(x_{(a)}) \frac{da}{\Delta(a)} \ , \tag{26.31}$$

i.e.

$$\frac{dx^1(a)}{U^1(x_{(a)})} = \ldots = \frac{dx^N_{(a)}}{U^N(x_{(a)})} = \frac{da}{\Delta(a)} \ , \tag{26.32a}$$

i.e.

$$\frac{dx^1(a)}{da} = \frac{1}{\Delta(a)} U^1(x_{(a)}), \ldots \frac{dx^N_{(a)}}{da} \frac{1}{\Delta(a)} U^N(x_{(a)}) \ . \tag{23.32b}$$

We may introduce a new variable t:

$$dt = \frac{da}{\Delta(a)} \ : \ t = \int_0^a \frac{d\mu}{\Delta(\mu)} \ . \tag{26.33}$$

The last term of (26.32a) now becomes dt and, denoting by $x^k(t)$ the new unknown function, the system of differential Eqs. (26.32a) is replaced by

$$\frac{dx^1}{U^1(x(t))} = \ldots = \frac{dx^N}{U^N(x(t))} = dt \ , \tag{26.34}$$

which can be summarized by a single equation

$$\frac{dx(t)}{dt} = U(x(t)) \ . \tag{26.35}$$

The condition corresponding to $F_{(0)} = 1$ now reads $x(0) = \xi$. We may then calculate the successive derivatives of $x(t)$,

$$\frac{d^2 x^i}{dt^2} = \{U^j \partial_j\} U^i(x(t)) \ldots \frac{d^n x^i}{dt^n} = \{U^j \partial_j\}^{n-1} U^i(x(t)) \ldots \ , \tag{26.36}$$

and for $t = 0$ one has

$$\begin{aligned}
\frac{d^n x^i(t)}{dt^n} &= \left\{ U^j(\xi) \frac{\partial}{\partial \xi^j} \right\}^{n-1} U^i(\xi) = \left\{ U^j(\xi) \frac{\partial}{\partial \xi^j} \right\}^n \xi^i \\
&= \left\{ X \left(\xi, \frac{\partial}{\partial \xi} \right) \right\}^n \xi^i \ ,
\end{aligned} \tag{26.37}$$

with the corresponding definition of $X(\xi, \partial/\partial\xi)$. Finally we may give the following expansion of $x(t)$ in a neighborhood of $t = 0$:

$$x^i(t) = \left\{ 1 + tX \left(\xi, \frac{\partial}{\partial \xi} \right) + \frac{t^2}{2!} \left(X \left(\xi, \frac{\partial}{\partial \xi} \right) \right)^2 + \ldots \right\} \xi^i = e^{tX(\xi, \partial/\partial\xi)} \xi^i \ . \tag{26.38}$$

As we did in the previous paragraph (25.5), we may associate to (26.35) a first-order partial derivative equation

$$U^j(x) \partial_j F(x) = 0 \ , \tag{26.39}$$

and consider its $N - 1$ first integrals

$$\phi_1(x) = C_1 \ldots \quad \phi_{N-1}(x) = C_{N-1} \ . \tag{26.40}$$

In our previous considerations we used the formalism developed in Sec. 25: Eq. (26.35) is very similar to Eqs. (25.1) and (25.2). In both cases, we want to find the trajectories of vector fields. We can still use the evolution operator formalism as in (25.6) and (25.12): it is presently the exponential of the differential operator $X(\xi, \partial/\partial\xi)$ defined by (26.38):

$$x(t) = \mathcal{U}_X(t)\xi = \{\exp tX\}\xi = e^{tX(\xi, \partial/\partial\xi)}\xi \ , \tag{26.41}$$

with all corresponding group properties (as in Sec. 25) of the exponential. Examples of Lie groups and their uses in physics are extremely numerous. Indeed, during the 19th century, group theory was recognized as an independent discipline and became an important tool in science: in crystallography first of all and after a round of observations (the group pest period!) in atomic physics and elementary particles theory. Even classical mechanics uses some properties of SO(2) or SO(3), the group of rotations; electromagnetism needs properties of the Lorentz group SL(2, C), quantum mechanics the spin group SU(2) and theory of elementary particles such groups of different structures as SO(3), SO(5), SO(4),... (see also Problems 2.10 and 2.11).

It may be observed that we dealt in the present paragraph with point transformations of the space $\mathcal{E}_N$. But physics leads us to consider situations in which the objects to be transformed are not only points but fields of mechanical or electromagnetic forces. In such cases, we have to start again at the very beginning and define generators and structure coefficients. A generalization of the present formalism should be welcomed and we shall present it in Part (B).

(B) *The mathematician's point of view:* as an illustration of the conclusions we have developed in Part (A), let us consider a scalar field $A(\mathbf{x})$ in the affine Euclidean space $\mathcal{E}_3$ with metric $g_{ij} = \delta_{ij}$. Even if in such a space the position of the indices has no meaning, we still use them in the right positions. Consider an infinitesimal rotation defined by the infinitesimal skew-symmetric tensor

$$\delta\omega^i{}_j = -\delta\omega^j{}_i \ . \tag{26.42}$$

Such a rotation transforms the vector $\mathbf{x}$ with components x^i into a vector $\mathbf{x}$ with components x'^i:

$$x'^i = x^i + \delta\omega^i_j x^j \ . \tag{26.43}$$

Since $A(\mathbf{x})$ is a scalar, its transformation law reads

$$A(\mathbf{x}) = A'(\mathbf{x}') = A'(\mathbf{x}) + x^i \delta\omega^j_i \partial_j A'(\mathbf{x}) \ , \tag{26.44}$$

up to the first-order in $\delta\omega^i_j$. If one takes into account the antisymmetry of $\delta\omega^i_j$ one may write

$$\begin{aligned}
A'(\mathbf{x}') &= A'(\mathbf{x}) + \frac{1}{2}\delta\omega^i_j\{x^j\partial_i - x^i\partial_j\}A'(\mathbf{x}) \\
&= \left\{I + \frac{1}{2}\delta\omega^i_j J^j_i(\mathbf{x},\partial)\right\}A(\mathbf{x}) \ ,
\end{aligned} \tag{26.45a}$$

with the generators

$$J_i^i = x^j \partial_i - x^i \partial_j \ . \tag{26.46b}$$

The J_i^j are antisymmetric and allow the definition of a vector

$$J_k = J_j^i \ .$$

One verifies

$$[J_i, J_j] = J_k = \varepsilon_{ij}^m J_m \ , \tag{26.46}$$

where ε_{ijm} is the Ricci numerical tensor (see Problem 1.7). The structure constants are

$$c^k{}_{ij} = \varepsilon^k{}_{ij} \ .$$

The reader has certainly noticed that in quantum mechanics, the J_j^i are the components of the angular momentum and the complex number i should be included in their definition. A more complicated but fundamental example is the one of Lorentz group: see Problems 1.19(c) and 2.9.

It is by now clear that a more general framework for the Lie groups theory should be welcomed. We shall now study successively the Lie groups and the Lie groups of transformations.

(1) **Lie groups.** Let $G = \{g, g', \ldots h, h', \ldots\}$ be a set which can be endowed with the two following structures:

G is a group,

G is a differentiable manifold such that the map $G \times G \to G$, with elements $g'^{-1}g$, is a differentiable map.

Since G is a manifold, one may consider a homeomorphsim $\varphi : G \to \mathbb{R}^r$ such that

$$\overline{g} = \varphi(g) : , \overline{g} = \{g^1 \ldots g^r\} \ ; \tag{26.47}$$

the dimension of G is thus r. Furthermore, if G is a group, from $gg' = g''$ one gets

$$\varphi(g'') = \varphi(g, g') \ .$$

If $\overline{g''}, \overline{g}, \overline{g'}$ are the images of g'', g, g' by φ, we may locally write

$$\overline{g}'' = \Theta(\overline{g}, \overline{g'}) : \ g''^k = \Theta^k(g^1 \ldots g^r, g'^1 \ldots g'^r) \ , \tag{26.48}$$

a relation equivalent to the rule III of (26.2).

We now consider two diffeomorphisms $G \to G$ denoted by $L_{(g)}$ and $R_{(g)}$, where $g \in G$, and called **left** and **right translations** respectively. They are defined as follows: for any $h \in G$, one has

$$L_{(g)}h = gh , \quad R_{(g)}h = hg . \tag{26.49}$$

When g goes over G, these diffeomorphisms define respectively two groups. One has for instance

$$L_{(g)} \circ L_{(g')}h = L_{(g)}g'h = gg'h = L_{(gg')}h , \tag{26.50a}$$

$$L_{(g)} \circ L_{(g^{-1})}h = h = L_{(g^{-1})} \circ L_{(g)}h = L_{(e)}h = eh , \tag{26.50b}$$

where e is the identity in G. One derives therefore

$$L_{(g)} \circ L_{(g')} = L_{(gg')} , \quad L_{(g)}^{-1} = L_{(g^{-1})} , \tag{26.50c}$$

$$L_{(e)} = I , \tag{26.50d}$$

I being the identity of the groups generated by $L_{(g)}$ and $R_{(g)}$. It is also clear that $L_{(g)}$ and $R_{(g)}$ act on G: **transitively**, if g' and g'' are elements of G, then there exists an operator $L_{(g)}$ (and $R_{(g)}$) such that

$$g'' = L_{(g)g'} ,$$

since one has indeed $g = g''g'^{-1}$. **Effectively**: if for any $h \in G$, one has

$$L_{(g)}h = h ,$$

then $g = e$; the converse is also true.

Let us consider the local point of view: we remark first of all that formula (24.49) expresses the group multiplication law: $g'' = gg'$. Locally, one has

$$\varphi(L_{(g)}g') = \varphi(g'') : \quad g''^k = \Theta^k(\overline{g}, \overline{g}') \tag{26.51}$$

as a consequence of (26.47) and (26.48). With different notations, we find again formula (26.2).

We notice furthermore that if $g' = e$,

$$g = ge = eg : \quad g^k = \Theta^k(\overline{g}, \overline{e}) = \Theta^k(\overline{e}, \overline{g}) . \tag{26.52}$$

The differential $DL_{(g)}$ of $L_{(g)}$ is the map $D\phi_M^{M'} : T_{\mathcal{M}'}(M) \to T_{\mathcal{M}'}(\mathcal{M})$ defined in (6.9) with the following notations:

$$\phi = L_{(g)} , \quad M = g' , \quad M' = L_{(g)}g' = g'' ,$$

and $DL_{(g)}$ is a shorthand notation for the map

$$(DL_{(g)})_{g'}^{g''} : T_G(g') \to T_G(g'') ,$$

such that for two vectors v_M and w_M, denoted $v_{g'}, w_{L(g)g'}$, one has

$$T_G(g') \stackrel{DL_{(g)}}{\to} T_G(gg') : \ DL_{(g)}v_{g'} = w_{L(g)g'} = w_{gg'} = w_{g''} . \qquad (26.53)$$

If $g' = e$, then

$$DL_{(g)}v_e = w_g . \qquad (26.54)$$

To obtain the local expressions of the preceding formulae, we start from formula (6.4) which reads

$$v'_{M'}(\hat{F}(M')) = v_M(\hat{F} \circ \phi)(M) .$$

Translating this formula in our present notations, we have

$$w_{g''}(\hat{F}(g'')) = v_{g'}(\hat{F}(L_{(g)g'}) = v_{g'}(\hat{F}(g'')) , \qquad (26.55)$$

where $\hat{F}$ is any C^∞ function on the manifold G. The left-hand side of (26.55) reads

$$w^i(\overline{g}'')\frac{\partial F(\overline{g}'')}{\partial g''^i} , \qquad (26.56a)$$

while its right-hand side can be written as

$$v^i(\overline{g}')\frac{\partial F(\overline{g}'')}{\partial g''^i} = v^j(\overline{g}')\frac{\partial \Theta^i(\overline{g}, \overline{g}')}{\partial g'^j}\frac{\partial F(\overline{g}'')}{\partial g''^i} \qquad (26.56b)$$

and, by comparing these two last formulae, one derives

$$\begin{aligned} w^i(\overline{g}'') &= v^j(\overline{g}')\frac{\partial \Theta^i(\overline{g}, \overline{g}')}{\partial g'^j} \\ &= \Delta_j^i(\overline{g}, \overline{g}')v^j(g') \end{aligned} \qquad (26.56c)$$

where $\Delta_j^i(\overline{g}, \overline{g'})$ is a real function of the multiplets $\overline{g}$ and $\overline{g'}$. We may now define the **vector space E_{left}** at g of the left invariant vectors on the manifold G. The vector field v_M is said to be left invariant, if in formula (26.53), one has $w = v$, i.e.,

$$DL_{(g)}v_{g'} = v_{gg'} = v_{g''} \ . \tag{26.57}$$

In particular, formula (26.54) is replaced by

$$DL_{(g)}v_e = v_g \tag{26.58}$$

and any vector of E_{left} can be generated starting from v_e. We now consider the local form of (26.56),

$$v^i(\overline{g}) = \Delta_j^i(\overline{g}, \overline{e})v^j(\overline{e}) \ , \tag{26.59}$$

since $g'' = ge = g$. Given the vector v_e of components $\{v^j(\overline{e})\}$, one obtains the same vector at the point g with components $\{v^j(\overline{g})\}$: E_{left} and $T_G(e)$ are thus isomorphic and their dimensions are the same, equal to r. Notice also that all the notions we developed are also valid for $R_{(g)}$.

We may now introduce in our formalism the notion of Lie algebra of the Lie group G under study. But we need to modify certain formulae of Sec. 6; let us rewrite formulae (6.1), (6.3), (6.4):

$$P = \phi(M) \ , \quad w_P = D\phi_P v_M \ , \\ w_P(\hat{F}(P)) = (D\phi_M^P v_M)\hat{F}(P) = v_M(\hat{F}(P)) \tag{26.60}$$

and show that for two vectors u_M and v_M of $\mathcal{M}_N$ and an arbitrary C^∞ function $\hat{F}$ one has

$$(D\phi_M^P[u_M, v_M])\hat{F}(P) = [D\phi_M^P u_M, D\phi_M^P v_M]\hat{F}(P) \ . \tag{26.61}$$

Indeed, since the commutator $[u_M, v_M]$ is a vector of $\mathcal{M}_N$, we can use (26.60) and write the left-hand side of the previous formula as

$$(D\phi_M^P[u_M, v_M])\hat{F}(P) = [u_M, v_M]\hat{F}(P) \ . \tag{26.62}$$

For the right-hand side of the same formula, we consider the product

$$(D\phi_M^P u_M)(D\phi_M^P v_M)(\hat{F}(P)) = (D\phi_M^P u_M)(v_M(\hat{F}(P)) \tag{26.63}$$

and finally have

$$(D\phi_M^P u_M)(D\phi_M^P v_M)(\hat{F}(P)) = u_M v_M \hat{F}(P) \ . \tag{26.64}$$

We calculate in the same way

$$D\phi_M^P v_M D\Phi_M^P u_M$$

and, since $\hat{F}$ in (26.61) is arbitrary, we have

$$D\phi_M^P [u_M, v_M] = [D\phi_M^P u_M, D\phi_M^P v_M] \ . \tag{26.65}$$

We now return to our considerations on Lie groups: let u_g and v_g be two vectors of the manifold G at the point $g \in G$ and suppose that they are left invariant. Then following (26.58),

$$u_g = DL_{(g)} u_e \ , \quad v_g = DL_{(g)} v_e \ . \tag{26.66}$$

We want furthermore to show that their commutator

$$[u_g, v_g] = w_g \ , \tag{26.67}$$

is also left invariant: indeed we apply formula (26.65) to a situation where $M = g, \phi = L_{(g)}$, then

$$\begin{aligned} w_g = [u_g, v_g] &= [DL_{(g)} u_e, DL_{(g)} v_e] \\ &= DL_{(g)} [u_e, v_e] = DL_{(g)} w_e \ , \end{aligned} \tag{26.68}$$

when one takes (26.65) into account. As a consequence, we may state that w_g belongs to the same space E_{left} with u_g and v_g as elements. But E_{left}, endowed with the internal operation [,], may be considered as an algebra: the **Lie algebra G of the Lie group** G. The dimension – say r – of that algebra represents the dimension of the Lie group G. The space E_{left} at g is by definition the space of all left invariant vectors tangent to G at g. But, since it can be obtained from the space E_{left} at e (identity of the group), then the Lie algebra may be simply defined by the vector space $T_G(e)$ endowed with the commutator operation.

If $\{X_1 \ldots X_r\}$ is a basis for the algebra associated with the vector space $T_G(e)$, then

$$[X_\alpha, X_\beta] = c^\lambda{}_{\alpha\beta} X_\lambda : \ \alpha, \beta, \lambda = 1 \ldots r \ . \tag{26.69}$$

The X_α's are called the **generators** of **G**. The **structure coefficients** $c^\lambda{}_{\alpha\beta}$ have all the properties stated in (26.29) (properties 1,2,3). Again we will not prove property 4, i.e. they locally determine the Lie group (see however L. Pontrjagin, *Topological Groups*, Princeton). One may thus define a Lie group by giving the 3^r structure coefficients.

Group of matrices: SO(2), SO(3), SU(2), SU(3), SU(5)... used by physicists (and particularly by elementary particle physicists!) are Lie groups (see Problem 2.11). Let us look more closely at the group built up by linear regular maps (bijective linear maps) $\mathbb{R}^N \to \mathbb{R}^N$: such a group is isomorphic to the one built up by $N \times N$ regular matrices. Such a group, called GL(N, $\mathbb{R}$), can be considered as open set of $\mathbb{R}^{N^2}$ (N^2 is the number of elements of a $N \times N$ regular matrix).

The map $\mathbb{R}^{N^2} \to \mathbb{R}$ is a polynomial map; it is thus continuous. The set

$$(\det)^{-1}\{\mathbb{R} - \{0\}\} = \mathrm{GL}(N, \mathbb{R}) \ , \qquad (26.70)$$

is an open set, can be endowed by a C^∞ structure and may be considered as an open set of $\mathbb{R}^{N^2}$. (See also [Bibl. 9], p. 165 or [Bibl. 10], Vol. 1, p. 509). For one-parameter Lie groups and Lie subgroups, see below.

(2) **Lie groups of transformations**. The physicist's point of view as developed in Part (A) is an application of the present considerations. Let us consider (26.71):

(a) a Lie group G with elements $g \in G$,

(b) a differentiable manifold $\mathcal{M}_N$ with points $M \in \mathcal{M}_N$ and a set of differentiable map $\overline{F} = \{\mathcal{F}\}$ such that

$$M \to \mathcal{F}(M) \in \mathcal{M}_N \ ,$$

(c) a differentiable map $G \to \mathcal{F} \times G$ such that for any $g \in G$ one has

$$\forall g \in G : \ \mathcal{F}_{(g)} \in \overline{F} \times G \ ,$$

each $\mathcal{F} \in \overline{F}$ being parametrized by a $g \in G$,

$$\{\mathcal{F}, G, M\} \to \mathcal{F}_{(g)}(M) \in \mathcal{M}_N \ ,$$

and $\bar{\mathcal{F}}$ being endowed with a group structure with the laws,

(d) $\mathcal{F}_{(g)} \circ \mathcal{F}_{(g')} = \mathcal{F}_{(gg')}$,

(e) $\mathcal{F}_{(e)}$ the identity element of $\overline{\mathcal{F}}$,

(f) then

$$\mathcal{F}_{(g^{-1})} = \mathcal{F}_{(g)}^{-1} .$$

The statements we just made have as consequences the statements I, II, III of (26.2). We notice first of all that the differential manifold G is supposed to be homeomorphic to $\mathbb{R}^r$, then the product $g'' = gg'$ can be written locally as

$$\overline{g}'' = \Theta(\overline{g}, \overline{g}') : \; g''^k = \Theta^k(g^1 \ldots g^r; g'^s \ldots g'^r) ; \qquad (26.71)$$

a form which was exhibited in (26.2) III (we recall that $\overline{g} = \{g^1 \ldots g^r\}$). We now come to the statements (b) and (c) of (26.71); let φ and φ_1 be two diffeomorphisms mapping certain neighborhoods of G and $\mathcal{M}_N$ into corresponding neighborhoods of $\mathbb{R}^r$ and $\mathbb{R}^N$ and let the point $\mathcal{F}_{(g)}(M) \in \mathcal{M}_N$ be denoted by $M_{(g)}$. Then

$$M_{(g)} = \mathcal{F}_{(g)}(M) = \mathcal{F}_{(\varphi^{-1}\overline{g})}(M) = F_{(g)}(M) . \qquad (26.72a)$$

This is a result (similar to the one given in (26.3)) which reads locally

$$\varphi_1(M_{(g)}) = \varphi_1 \circ F_{(\overline{g})} \circ \varphi^{-1}(x) : \; x = \varphi_1(M), \; M \in \mathcal{M}_N \qquad (26.72b)$$

and will be written as in (26.4)

$$x_{(\overline{g})} = F_{(\overline{g})}(x) . \qquad (26.72c)$$

The statements (d), (e), (f) of (26.71) correspond to I, II, III of (26.2); one has indeed

$$\begin{aligned} F_{(\overline{g})} \circ F_{(\overline{g}')} = F_{(\overline{g}'')} &: \; \overline{g}'' = \Theta(\overline{g}, \overline{g}') \\ F_{(\overline{e})} = I \; , \; F_{(\overline{g})}^{-1} &= F_{(g^{-1})} . \end{aligned} \qquad (26.73)$$

The link between the physicist's point of view and the mathematician's is then clearly established, including the part played by the generators which, for a physicist, were defined by considering transformations in an infinitesimal neighborhood of the element $e \in G$. We may remark also that G acts effectively and transitively on $\mathcal{M}_N$. The map $G \to \overline{\mathcal{F}}$ is a homeomorphism and defines a **realization** of G. If the corresponding maps $\mathcal{F}_{(g)}$ are linear, the realization becomes a **representation**, a concept familiar to all physicists.

We now shift our attention first to the study of **1-parameter local pseudo groups** and connect them to our comments of Sec. 25 on vector fields and their trajectories. Our considerations will be local and we shall use the language of Part (A) of the present paragraph, expressing the physicist's point of view. As we saw in (26.30), such a group which was called – inaccurately – a one-parameter group is described by a single generator in the neighborhood of $x_{(e)}$:

$$X(x_{(a)}, \partial) = U^l(x_{(a)}) \frac{\partial}{\partial x^l_{(a)}} \ , \tag{26.74a}$$

and

$$x_{a+da} = \{I + \delta a X(x_{(a)}, \partial)\} x_{(a)} \ . \tag{26.74b}$$

We now can associate with the generator X a vector field

$$\hat{U}(M_{(a)}) = U^l(x_{(a)}) \varepsilon_l^{M(a)} \tag{26.75}$$

and a curve C, image of a curve $\Gamma \subset \mathcal{M}_N$ which is a trajectory of that vector field,

$$\frac{dx^k_{(a)}}{da} = U^k(x_{(a)}) \ . \tag{26.76}$$

We may conform our notation to that generally used by denoting by t the parameter $a \in R$ and $x^k_{(a)}$ by $x^k(t)$; then the former equation reads

$$\frac{dx^k(t)}{dt} = U^k(x(t)) \ .$$

At that point we may use the formalism developed in Part (B) of the same Sec. 25 and consider the curve $\Gamma \subset \mathcal{M}_N$ defined by the map $g_t : \mathbb{R} \to \mathcal{M}_N$,

$$M_t = \hat{\phi}(t) \in \mathcal{M}_N \ ,$$

and the intrinsic form (26.76) given by (25.15d),

$$\frac{d\hat{\phi}(t)}{dt} = \hat{U}(M(t)) \ . \tag{26.77}$$

Then the formalism developed in Sec. 25 is clearly valid (see (25.15)) and following).

But local pseudo groups can also be studied from another point of view: consider differentiable maps $\hat{\phi}_t : \mathcal{M}_N \to \mathcal{M}_{N'}$, such that for any $M \in \mathcal{M}_N$,

$$M_t = \hat{\phi}_t(M) \in \mathcal{M}_N \ . \tag{26.78}$$

We then proceed by transforming the set $\{\hat{\phi}_t, t \in \mathbb{R}\}$ into a group and state the following laws:

$$\hat{\phi}_0 = I \ , \quad \hat{\phi}_t \circ \hat{\phi}_s = \hat{\phi}_{t+s} \ , \tag{26.79a}$$

statements which imply that

$$\hat{\phi}_t^{-1} = \hat{\phi}_{-t} \ . \tag{26.79b}$$

We may connect again such a family of maps with the considerations developed in Sec. 25. To do so, the easiest way is the recourse to local considerations: let φ be a homemorphism of $\mathcal{M}_N$ and $x(t)$ the coordinates of M_t:

$$x_t = \varphi(M_t) \ , \quad x = \varphi(M_0) \ ,$$

then the composition law above leads to

$$\begin{aligned}
\varphi(\hat{\phi}_{t+s}(M)) &= \varphi \circ \hat{\phi}_t \circ \varphi^{-1} \circ \varphi \circ \hat{\phi}_s \circ \varphi^{-1}(x) \\
&= \phi_t \circ \phi_s(x) = \phi_t(\phi_s(x)) = \phi_t(x_s) \ ,
\end{aligned} \tag{26.80}$$

with an obvious definition of ϕ_t. For M fixed on $\mathcal{M}_N$,

$$x(t) = \phi_t(x)$$

is the local description of a curve on $\mathcal{M}_N$. All the previous considerations of Sec. 25 are once more valid: for instance, the first of the pseudo group laws (26.79) gives an initial condition

$$x(0) = \phi_0(x) = x \ .$$

We may then state that any vector field $\hat{U}_M$ on $\mathcal{M}_N$ generates a 1-parameter pseudo group obeying the equation

$$\frac{d\hat{\phi}_t}{dt} = \hat{U}_{M_t} \ . \tag{26.81}$$

Among all other properties we may mention the exponential representation of the solution of the above equation,

$$\hat{\phi}_t = \exp t\hat{U}_M \ , \tag{26.82}$$

which was given in (25.16).

Careful and subtle distinctions due to the mathematician's point of view open the way to new considerations in the domain of Lie groups. A few of them will be surveyed as a conclusion of the present paragraph.

One defines a **1-parameter Lie group** (distinct from the pseudo-group subject of the above considerations!) as a 1-dimensional Lie group G. By this we mean that the manifold G has 1 as dimension, in other words to each element of G there corresponds a real number. Intuitively speaking such a group isomorphic to $\mathbb{R}$ should be abelian (commutative): topological considerations show that it is indeed the case. We also pointed out that the denomination used by physicists who call a 1-parameter group what is really a 1-parameter pseudo group appears here as inaccurate.

One-parameter subgroup of a Lie group G is another important and interesting concept. Such a subgroup is defined as a curve Γ of the manifold G such that each point $g \in G$ is now considered as a function of the real variable t. In other words, we consider a map $\mathbb{R} \rightarrow G$:

$$g = g(t) \ ,$$

which is endowed with the following properties:

$$g(0) = e \ , \quad g(t)g(s) = g(t+s) \ . \tag{26.83}$$

Each element $g \in G$ is then associated with a curve Γ. Let the dimension of the manifold G be r, then to any element g there corresponds the r-uplet:

$$g \rightarrow \overline{g} = \{g^1 \ldots g^r\} \tag{26.84}$$

and the local expression of the curve Γ is the curve C:

$$g^\lambda = g^\lambda(t): \ \lambda - 1 \ldots \tau \ . \tag{26.85a}$$

The identity element e of G has for components

$$e \rightarrow \overline{e} = \{e^1 \ldots e^r\} \ . \tag{26.85b}$$

The tangents along the curve Γ build up a vector field $U(g(t))$. Its image

$$U(\overline{g}(t)) = \{U^1(\overline{g}(t)) \ldots U^r(\overline{g}(t))\} \tag{26.85c}$$

defines its components $U^\alpha(\overline{g}(t))$: the trajectories of such a field are given by the intrinsic equation

$$\frac{dg(t)}{dt} = \hat{U}(g(t)) \ . \tag{26.86a}$$

Its image is the system of differential equations

$$\frac{dg^\lambda(t)}{dt} = U^\lambda(\overline{g}(t)) : \ \lambda = 1 \ldots r \ , \tag{26.86b}$$

which is exactly the system (25.1).

The expansion (26.38) represents an expansion of $g^\lambda(t)$ around the point $g^\lambda(0) = e^\lambda$. It can be written using the present notation as

$$g^\lambda(t) = e^\lambda + \left\{ t + \frac{t^2}{2!} X_e + \ldots \right\} U^\lambda(\overline{e}) \ , \tag{26.87a}$$

where X_e is the differential operator

$$X_e = U^\lambda(\overline{e}) \frac{\partial}{\partial e^\lambda} \ . \tag{26.87b}$$

The expression (26.87) of $\overline{g}(t)$ has a meaning each time the expansion itself makes sense: this is clearly a very severe restriction; it can be avoided by using global methods. See for instance [Bibl. 9], Part (D) of Chap. 3. In spite of all these restrictions, the geometrical meaning of (26.87) is very clear: it maps an element of the tangent vector space $T_G(e)$ into G. But since G is a Lie group, this is a map of the Lie algebra into G and is called an exponential map for the group G. It may be represented as in (25.16) as follows:

$$\mathbf{G} \overset{\exp U}{\rightarrow} G \ .$$

The application of the previous remarks to Lie groups of transformations does not call for extensive comments: the set of transformations $\{\mathcal{F}_{g(t)}, t \in \mathbb{R}\}$ acts on $\mathcal{M}_N$ and describes the curve of $\mathcal{M}_N$. The vector field which generates this set is called the **Killing vector field** on $\mathcal{M}_N$ relative to the group G. In the special case of Riemannian space, the group G is the group of its isometries. See Level 1, Sec. 12.

Complements, Exercises and Applications to Physics

Contents

Intrinsic Approach

> **Problem 2.1.** *Examples of manifolds*

(a) Comments on topological spaces that are not Hausdorff spaces.

(b) Consider the space $\mathbb{R}^N$, endowed with the "usual" metric: for x and y elements of $\mathbb{R}^N$:

$$d(x,y) = \sqrt{\sum_1^N (x^i - y^i)^2} \ .$$

Show that it is a topological space and a manifold. Show that any open subset of $\mathbb{R}^N$ is homeomorphic to $\mathbb{R}^N$.

(c) By the sphere S^1, we mean the set of points of $\mathbb{R}^2$ (with its usual topology).

$$S^1 = \{x \in \mathbb{R}^2 : d(x,0) = 1\} \ .$$

Is S^1 homeomorphic to $\mathbb{R}^2$? Define an atlas of S^1. Same questions for the ellipse

$$\left\{ x \in \mathbb{R}^2 : \ \frac{(x^1)^2}{a^2} + \frac{(x^2)^2}{b^2} = 1 \right\} \ ,$$

and for the cylinder in $\mathbb{R}^3$

$$\{x \in \mathbb{R}^3 : (x^1)^2 + (x^2)^2 = 1\} .$$

Is the half cone defined in $\mathbb{R}^3$ (with its usual toplogy),

$$\{x \in \mathbb{R}^3 : (x^1)^2 - (x^2)^2 - (x^3)^2 = 0 , \quad x^1 > 0\} ,$$

a manifold?

(d) Extend these considerations to the sphere S^2:

$$S^2 = \{x \in \mathbb{R}^3 : d(x, 0) = 1\} .$$

(e) In the complex plane C^2, does the complex function $Z = z^2$ define a nonsingular map between the complex plane of z and that of Z? Does the transformation define a diffeomorphism? Discuss $Z = z^k$, k real number.

(f) The obvious generalization of the torus studied in Ex. 0.7 is the manifold $S^1 \times S^1$ since a direct product of manifolds is a manifold. One may define the n-holed torus by the manifold $S^1 \times S^1 \times \ldots S^1$, a direct product of n factors.

(g) **Projective space:** we remind the reader that for a given relation $\mathcal{R}$ on a set S, **the quotient set** $S/\mathcal{R}$ is the set of all equivalence classes $\overline{x}$ defined by $\mathcal{R}$ such that for any element x of S,

$$x \sim x' \text{ iff } x \mathcal{R} x' : \ x' \in \overline{x} .$$

We now specialize our considerations to the set $S = E_2$, a vector space with its usual metric and the equivalence relation $\mathcal{P}$ for $x, x' \in E_2$,

$$x \sim x' \text{ iff } x' = \lambda x , \quad \lambda \in \mathbb{R} ,$$

the elements of $E_2/\mathcal{P}$, called **directions** or **rays**, build up a projective space. Each direction may be characterized by its unit vector $\|\mathbf{u}\| = 1$, and the distance between two elements $\mathbf{u}$ and $\mathbf{v}$ can be defined by $d(\mathbf{u}, \mathbf{v}) = \|\mathbf{u} - \mathbf{v}\|$. This is a topological structure on $E_2/\mathcal{P}$ which allows us to define open sets, homeomorphisms between $E_2/\mathcal{P}$ and $\mathbb{R}^2$, and finally a manifold such that its points correspond to directions in E_2. The same remarks apply to E_N.*

---

*In contradistinction to the quotient vector space E/F, where E and F are vector spaces ($F \subset E$), $E_2/\mathcal{P}$ is not a vector space.

(h) The usual groups of matrices of Problems 2.10 and 2.11 constitute also differentiable manifolds: such is the case of $GL(N, \mathbb{R})$, the group of regular $N \times N$ matrices, since to any matrix $A \in GL(N, \mathbb{R})$ there corresponds a differentiable manifold of dimension $\mathbb{R}^{N^2}$ whose points are parameterized by the N^2 elements $\{a_1^1 \ldots a_N^N\}$ of A. For $SL(N, \mathbb{R})$, the corresponding manifold is of dimension $N^2 - 1$ since one has a supplementary condition $\det A = 1$. Several other manifolds defined by the usual groups will be studied in Problem 2.11.

(i) A **complex analytic** manifold is a C^∞ differentiable manifold on which one can define holomorphic maps $\varphi \circ \varphi'^{-1}$ (see (1.4), Level 2): the new coordinates z'^k represent a system of N holomorphic functions of N complex variables. For a real analytic differentiable C^∞ manifold, the previous system of functions is to be real analytic, i.e., one should be able to expand each function in a convergent Taylor series in the neighborhood of each of its points.

Hints: (a) Consider two curves C and C' in parametrized form. Their equations are $C{:}M = M(\lambda)$, $C'{:}$ $P(\mu)$: the respective portions of C and C' corresponding to $\lambda < 0$ and $\mu < 0$ (strict inequalities!) are clearly topological space. Identify now the points belonging to these portions and consider the space built up in this way. It is a topological space but not a Hausdorff space since the points O and O' corresponding to $\lambda = 0$ and $\mu = 0$ have no disjoint neighborhoods.

(b) See any textbook on topology of $\mathbb{R}^N$.

(c) S^1 is not homeomorphic to $\mathbb{R}$: we can indeed build an open interval (homeomorphic to $\mathbb{R}$) of $\mathbb{R}$. Such an interval can be displayed in polar-coordinates where S^1 is represented by $x^1 = \cos\varphi, x^2 = \sin\varphi$; then clearly S^1 is not homeomorphic to the open interval $0 < \varphi < 2\pi$ (to obtain S^1 we should have included either 0 or 2π). To recover S^1 we need at least two charts: we may then proceed as in the following figures.

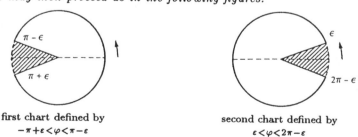

first chart defined by
$-\pi + \varepsilon < \varphi < \pi - \varepsilon$

second chart defined by
$\varepsilon < \varphi < 2\pi - \varepsilon$

Analogous considerations apply to the ellipse whose equation in parametric form reads $x^1 = a\cos\varphi, x^2 = b\sin\varphi$. The same is true for the cylinder. Notice that if Δ is a segment of a straight line in $\mathcal{E}_2$, a general cylinder is defined as the manifold $S^1 \times \Delta$. The half-cone is a manifold: it is homeomorphic to $\mathbb{R}^2$ if the origin is excluded. We notice that the double cone is not a manifold.

(d) Let $Z = X + iY$ and $z = x + iy$, then $X = x^2 - y^2, Y = 2xy$. The Jacobian

$$\frac{\partial(X,Y)}{\partial(x,y)} = 4(x^2 + y^2) \, ,$$

vanishes at the point $x = y = 0$ (and the Jacobian matrix $D(X,Y)/D(x,y)$ is of rank 1 only) and the map not being bijective at that point is not a diffeomorphism.

| Problem 2.2. | *Immersion, embedding of a manifold $\mathcal{M}_N$ in a manifold $\mathcal{M}'_{N'}$*

An approach of that problem was sketched in Sec. 6, Level 2: we now want to add some complements.

Let M be any point of $\mathcal{M}_N$, $U_{\mathcal{M}}(M)$ its neighborhood and a differentiable map $\phi : \mathcal{M}_N \to \mathcal{M}'_{N'}$. Consider the inverse map ϕ^{-1} restricted to the image $\phi(U_{\mathcal{M}}(\mathcal{M}))$ by ϕ; if ϕ^{-1} is differentiable for any $M \in \mathcal{M}_N$ and $N < N'$, one calls ϕ an **immersion**. One says that $\mathcal{M}_N$ is either immersed in $\mathcal{M}^v_{N'}$ or a submanifold of $\mathcal{M}'_{N'}$. If an immersion is injective, it is called an **embedding** and $\mathcal{M}_N$ is embedded in $\mathcal{M}'_{N'}$. If $N = N' - 1, \mathcal{M}_N$ is called a **hypersurface** in $\mathcal{M}'_{N'}$.

(a) Give a few examples of immersion which are not embeddings.

(b) Define a subset $\mathcal{M} \subset \mathcal{M}_N$ by p equations $\hat{f}_r(M) = 0 = f_r(x), r = 1 \ldots p$ (see Sec. 2 of Level 2), and suppose the functions to be differentiable. Consider then the map

$$\mathcal{M}_N \to \mathbb{R}^p : P \to (\hat{f}_1(P) \ldots \hat{f}_p(P)) \, .$$

If the Jacobian matrix $D(f_1 \ldots f_p)/D(x^1 \ldots x^N)$ is of rank p for any point of $\mathcal{M}$ then $\mathcal{M}$ is a differentiable submanifold of $\mathcal{M}_N$.

This is the way to give a precise meaning to the intuitive considerations of Ex. 1.6.

Hints: (a) Consider the following maps $\mathcal{E}_2 \to \mathbb{R}^2$:

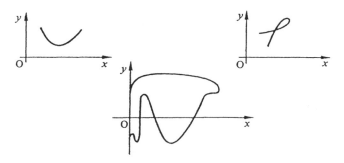

The first one is an embedding, the second one is an immersion. The third one is the curve $\sin(1/x)$ between $0 < x \le a$ (0 excluded) which has been smoothly joined to the $0y$ axis. It is an immersion of $\mathbb{R} \to \mathbb{R}^2$, but not an embedding.

(b) See [Bibl. 9], p. 228.

Problem 2.3.   *The inner product operation and the star operation*

Both operations will be defined for p-forms (on a $\mathcal{M}_N$ manifold) with components $\alpha_{i_1 \ldots i_p}(x)$; they will be supposed to be completely antisymmetrized and expressed as in (9.3b), Level 2:

$$\alpha_M = \frac{1}{p!}\alpha_{(i_1 \ldots i_p)}(x)\underline{dx}^{i_1} \wedge \ldots \underline{dx}^{i_p}$$

$$= \sum_{i_1 < i_2 < \ldots i_p} \alpha_{(i_1 \ldots i_p)}(x)\underline{dx}^{i_1} \wedge \ldots \underline{dx}^{i_p} \ .$$

A : i_v is an operation which maps the vector space of differentiable p-forms on $\mathcal{M}_N$ to the vector space of $(p-1)$-forms on the same manifold, namely for any vector v_M of $T_{\mathcal{M}}(M)$,

$$i_v \alpha_M = \frac{1}{(p-1)!}v^j(x)\alpha_{(ji_2 \ldots i_p)}\underline{dx}^{i_2} \wedge \ldots \underline{dx}^{i_p} \ ,$$

and represents in a certain sense a generalization of (9.1b) of Level 2. Then for any function $\hat{F}$,

$$i_v \hat{F}(M) = 0 \ ,$$

and for any dx^j,

$$i_v \underline{d}x^j = v^j(x) \ .$$

(a) Show that i_v is linear and is also an antiderivation:

$$i_v(\alpha_M + \beta_M) = i_v \alpha_M + i_v \beta_M \ ,$$
$$i_v\{\alpha_M \wedge \beta_M\} = (i_v \alpha_M) \wedge \beta_M + (-1)^p \alpha_M \wedge i_v \beta_M \ ,$$

if p is the degree of α_M.

(b) Show that i_v is a nilpotent operation on p-forms:

$$i_v^2 \alpha_M = 0 \ .$$

(c) For any α_M,

$$L_\xi \alpha_M = \underline{d} i_\xi \alpha_M + i_\xi \underline{d} \alpha_M \ .$$

Proceed by establishing this property for $N = 2$, then extending it to $N = p$.

(d) Prove that

$$[L_\xi, \underline{d}]\alpha_M = 0 \ , \quad [L_\xi, i_\xi] = 0$$
$$L_\xi i_{\xi'} \alpha_M - L_{\xi'} i_\xi \alpha_M = i_{[\xi,\xi']} \alpha_M \ .$$

B: Let $\mathcal{M}_N$ be an oriented $(\partial(x')/\partial(x) > 0)$ Riemannian manifold. We recall the following two formulae, the first one having been proved in (8.14b), Level 1,

$$\sqrt{|g'(x')|} = \frac{\partial(x)}{\partial(x')} \sqrt{|g(x)|} \ ,$$

and the second in (9.30), Level 2

$$\underline{d}x'^1 \wedge \dots \underline{d}x'^N = \frac{\partial(x')}{\partial(x)} \underline{d}x^1 \wedge \dots \underline{d}x^N \ .$$

We define the **volume element** η by

$$\eta(x) = \sqrt{|g(x)|} \underline{d}x^1 \wedge \dots \underline{d}x^N \ ,$$

a differential N-form in an N-dimensional manifold. It has only one strict (independent) component, the others being given by

$$\eta_{i_1 \dots i_N} = \sqrt{|g(x)|} \varepsilon_{i_1 \dots i_N}^{1 \dots N} \ ,$$

and the numerical tensor $\varepsilon^{1\ldots N}_{i_1\ldots i_N}$ has been defined in Problems 1.7b and 1.8a. We then consider a differential p-form α_M with $\alpha_{(i_1\ldots i_p)}$ as strict components (see formula (9.3c), Level 2) in an arbitrary basis $\{\omega^1 \ldots \omega^N\}$ of the cotangent vector space to $\mathcal{M}_N$:

$$\alpha_M = \frac{1}{p!}\alpha_{(i_1\ldots i_p)}\omega^{i_1} \wedge \ldots \omega^{i_p} .$$

We shall define α_M^* (or α_M^* or $*\,\alpha_M$) as a $(N-p)$-form with strict components

$$\alpha_M^* = \frac{1}{p!}\alpha^*_{(i_{p+1}\ldots i_N)}\omega^{i_{p+1}} \wedge \ldots \omega^{i_N}$$

$$= \sum_{i_{p+1}<i_{p+2}<\ldots i_N} \alpha^*_{i_{p+1}\ldots i_N}\omega^{i_{p+1}} \wedge \ldots \omega^{i_N} ,$$

such that

$$\alpha^*_{(i_{p+1}\ldots i_N)} = \frac{1}{(N-p)!}\eta_{i_1\ldots i_p i_{p+1}\ldots i_N}\alpha^{(i_1\ldots i_p)} .$$

(e) The numerical components $\varepsilon^{j_1\ldots j_p}_{i_1\ldots i_p}$ clearly generalize the Levi-Civita tensor components introduced in Ex. 1.3 and 1.7; consequently we shall define[*]

$$\varepsilon_{i_1\ldots i_N} = \varepsilon^{1\ldots N}_{i_1\ldots i_N} .$$

Calculate the contravariant components of the Levi-Civita tensor (of Problem 1.7) and those of the tensor η. Check for $N = 2, 3, 4$ that

$$\eta^{i_1\ldots i_p j_{p+1}\ldots j_N}\eta_{j_1\ldots j_p j_{p+1}\ldots j_N} = (N-p)!\varepsilon^{i_1\ldots i_p}_{j_1\ldots j_p} .$$

(f) Show that

$$\alpha_M^{**} = (-1)^{p(N-p)}\alpha_M$$

so that α_M^{**} is again a p-form. One defines for any differential p-form β, the operation $(*)^{-1}$, inverse of $*$, as follows:

$$(*)^{-1}\beta_M = (-1)^{p(N-p)}\beta_M^* ,$$

in accordance with the first formula of this same question (f).

---

[*]The definition given here differs slightly from the one used in Problem 1.7.

(g) Define the f-inner product of two p-forms (f denotes form) as follows:

$$\langle \underline{dx}^{i_1} \wedge \dots \underline{dx}^{i_p}, \underline{dx}^{j_1} \wedge \dots \underline{dx}^{j_p} \rangle_f$$
$$= \varepsilon^{i_1 \dots i_p}_{k_1 \dots k_p} g^{j_1 k_1}(x) \dots g^{j_p k_p}(x) ,$$

to show that for two differential p-forms α_M and β_M,

$$\langle \alpha_M, \beta_M \rangle_f = \frac{1}{p!} \alpha_{i_1 \dots i_p}(x) \beta^{i_1 \dots i_p}(x) .$$

This approach may be used to give another definition of the ∗-operation. See [Bibl. 9], p. 282.

Solution:

(a) Use the definition i_v.

(b) $i_v^2 \alpha_M = i_v i_v \left\{ \dfrac{1}{p!} \alpha_{(i_1 \dots i_p)} \underline{dx}^{i_1} \wedge \dots \underline{dx}^{i_p} \right\}$

$\qquad = \dfrac{1}{(p-1)!} i_v \{ v^j \alpha_{(ji_2 \dots i_p)} \underline{dx}^{i_2} \wedge \dots \underline{dx}^{i_p}$

$\qquad = \dfrac{1}{(p-2)!} v^j v^h \alpha_{(jhi_3 \dots i_p)} \underline{dx}^{i_3} \wedge \dots \underline{dx}^{ip}$

$\qquad = 0$

because of the antisymmetry in j, h of $\alpha_{(jhi_s \dots i_p)}$.

(c) Define

$$\alpha_M = \frac{1}{2!} \alpha_{(i_1 i_2)} \underline{dx}^{i_1} \wedge \underline{dx}^{i_2}$$

and calculate the first term of the right-hand side of (c). One has first

$$i_\xi \alpha_M = \frac{1}{1!} \xi^j \alpha_{(ji_2)} \underline{dx}^{i_2}$$

and

$$\underline{d} i_\xi \alpha_M = ((\partial_k \xi^j) \alpha_{(ji_2)} + \xi^k \partial_k \alpha_{(ji_2)} \underline{dx}^k \wedge \underline{dx}^{i_2} ,$$

all other terms vanishing since $\underline{d}^2 = 0$.

To calculate completely the second term of the right-hand side of (c), we note first

$$\underline{d} \alpha_M = \frac{1}{2!} \partial_j \alpha_{(i_1 i_2)} \underline{dx}^j \wedge \underline{dx}^{i_1} \wedge \underline{dx}^{i_2} ,$$

but we cannot use the definition of i_v, since this expression, although antisymmetric in i_1, i_2, is not antisymmetric with respect to the triplet i_1, i_2, j.

Let us antisymmetrize it by the exchange $j \leftrightarrow i_1, j \leftrightarrow i_2$, we then get two other expressions, collecting finally all terms,

$$i_\xi \underline{d}\alpha_M = \frac{1}{3!}(\partial_j \alpha_{(i_1 i_2)} - \partial_{i_1}\alpha_{(j i_2)} - \partial_{i_2}\alpha_{(i_1 j)})\underline{d}x^j \wedge \underline{d}x^{i_1} \wedge \underline{d}x^{i_2} \ ,$$

$$i_\xi \underline{d}\alpha_M = \frac{1}{2!}(\xi^j \partial_y \alpha_{(i_1 i_2)} - \xi^j \partial_{i_1}\alpha_{(j i_2)} - \xi^j \partial_{i_2}\alpha_{(i_1 j)})\underline{d}x^{i_1} \wedge \underline{d}x^{i_2} \ .$$

This last expression completes the right-hand side of (c).

We now calculate $\mathcal{L}_\xi \alpha_M$ using formula (10.21), Level 2:[*]

$$L_\xi\{\alpha_{(i_1 i_2)}\underline{d}x^{i_1} \wedge \underline{d}x^{i_2}\} = (\xi^j \partial_j \alpha_{(i_1 i_2)} + 2\alpha_{(j i_2)}\partial_{i_1}\xi^j)\underline{d}x^{i_1} \wedge \underline{d}x^{i_2}$$

and compare the two sides of (c) to obtain the required result.

The extension to any differential p-form is left to the reader.

(d) The proof of the first two relations is immediate; to prove the last one it is recommended to use local coordinates.

(e) $\varepsilon^{i_1 \ldots i_N} = g^{i_1 j_1}\ldots g^{i_N j_N}\varepsilon_{j_1 \ldots j_N} = \det\{g^{i_1 j_1}\ldots g^{i_N j_N}\}$

$$= \begin{vmatrix} g^{i_1 1} & g^{i_1 2} \ldots & g^{i_1 N} \\ \ldots & & \\ g^{i_N 1} \ldots & & g^{i_N N} \end{vmatrix} = \frac{1}{|g|}\varepsilon_{i_1 \ldots i_N} \ ,$$

$$\eta^{i_1 \ldots i_N} = \frac{1}{\sqrt{|g|}}\eta_{i_1 \ldots, i_N} \ .$$

Notice also

$$g = \varepsilon^{j_1 \ldots j_N}_{1 \ldots N} g_{j_1 1}\ldots g_{j_N N} \ .$$

Checking of the last formula of question (e) is left to the reader.

(f) From the definition of α_M^* given above,

$$\alpha^{*i_{p+1}\ldots i_N} = \frac{1}{p!}\eta^{i_1 \ldots i_p i_{p+1}\ldots i_N}\alpha_{i_1 \ldots i_p}$$

---

[*]Note that (10.22) gives for α_M

$$L_\xi\Big\{\frac{1}{p!}\alpha_{(i_1 \ldots i_p)}\underline{d}x^{i_1} \wedge \ldots \underline{d}x^{i_p}$$

$$= (\xi^j \partial_j \alpha_{(i_1 \ldots i_p)} + p\alpha_{(j i_1 \ldots i_p)}\partial_{i_1}\xi^j)\underline{d}x^{i_1} \wedge \ldots \underline{d}x^{i_p} \ .$$

is a $(N-P)$-form, then α^{**} will be a $N-(N-p)=p$ form and one has

$$\alpha^{**}_{j_1\ldots j_p} = \frac{1}{(N-p)!}\eta_{j_{p+1}\ldots j_N j_1\ldots j_p}\alpha^{*j_{p+1}\ldots j_N}$$

$$= \frac{1}{(N-p)!p!}\eta_{j_{p+1}\ldots j_N j_1\ldots j_p}\eta^{i_1\ldots i_p j_{p+1}\ldots j_N}\alpha_{i_1\ldots i_p} .$$

For the contracted product of the η's, we use the formula in (e) and notice that we must make $p(N-p)$ transpositions in the first η if we bring the indices in the order indicated in (e). We are then left with

$$\alpha^{**}_{j_1\ldots j_p} = (-1)^{p(N-p)}\frac{1}{p!}\varepsilon^{i_1\ldots i_p}_{j_1\ldots j_p}\alpha_{i_1\ldots i_p} ,$$

and finally

$$\alpha^{**}_M = \alpha^{**}_{j_1\ldots j_p}\underline{d}x^{j_1}\wedge\ldots\underline{d}x^{j_p}$$

$$= (-1)^{p(N-p)}\frac{1}{p!}\alpha_{\langle j_1\ldots j_p\rangle}\underline{d}x^{j_1}\wedge\ldots\underline{d}x^{j_p}$$

$$= (-1)^{p(N-p)}\alpha .$$

Problem 2.4. *Differential forms and Maxwell's equations*

There is an alternative way of presenting Maxwell's equations, differing from the more traditional one of Ex. 1.20. One uses the tensor $F_{\mu\nu}(x)$ (as given in Problem 1.20) to write down differential 2-forms.

(A) The Riemannian manifold we are going to use is that of special relativity, indices being raised and lowered with the metric $\eta_{\mu\nu}$. From the tensor $F_{\mu\nu}(x)$ of Ex. 1.20 we build up the differential 2-form*

$$\underline{F}_M = \frac{1}{2}F_{\mu\nu}(x)\underline{d}x^\mu\wedge\underline{d}x^\nu .$$

(a) Show that the equation

$$\underline{d}\,\underline{F}_M = 0$$

---

*Following the notations adopted in formulae (9.3), Level 2 and in the previous exercise, the field $F_{\mu\nu}(x)$ should have been denoted $F_{\langle\mu\nu\rangle}$. Do not confuse the differential form F_M with $\hat{F}(M)=F(x)$, a function.

recapitulates the two Maxwell's equations:

$$\operatorname{curl} \mathbf{E} = -\frac{\partial \mathbf{B}}{\partial t} \ , \quad \operatorname{div} \mathbf{B} = 0 \ .$$

(b) As seen in the previous Problem 2.3(d),

$$\eta_{\alpha\beta\gamma\delta} = \sqrt{|g|}\varepsilon_{\alpha\beta\gamma\delta}, \quad \eta^{\alpha\beta\gamma\delta} = \frac{1}{\sqrt{|g|}}\varepsilon^{\alpha\beta\gamma\delta} \ .$$

We then define $F^*_{\alpha\beta}$ as

$$F^*_{\alpha\beta} = \frac{1}{2!}\eta_{\mu\nu\alpha\beta}F^{\mu\nu} = \frac{1}{2\sqrt{|g|}}\varepsilon_{\mu\nu\alpha\beta}F^{\mu\nu} \ .$$

In the case of the metric $\eta : \sqrt{|g|} = 1$, we find the matrix $F^*_{\alpha\beta}$ in the form given in Problem 1.20b and there is a corresponding 2-form

$$\underline{F}^*_M = \frac{1}{2}F^*_{\mu\nu}\underline{d}x^\mu \wedge \underline{d}x^\nu \ .$$

Calculate the differential 3-form $dF^*{}_M$ and show that the other pair of Maxwell's equations

$$\boldsymbol{\nabla} \wedge \mathbf{B} = \frac{\partial \mathbf{E}}{\partial t} + \mathbf{J} \ , \quad \boldsymbol{\nabla} \cdot \mathbf{E} = \rho$$

can also be written as

$$\underline{d}\,\underline{F}^*_M + \gamma_M = 0 \ ,$$

γ_M being a differential 3-form to be determined.

(c) The 3-form γ_M may be used to define a 1-form using the $*$ operation, but in order to write an equation similar to the last one, we have to introduce a new operator denoted $\underline{\delta}$. Suppose α_M to be a differential p-form, we define the $\underline{\delta}$ operation:

$$\underline{\delta}\alpha_M = (-1)^p(*)^{-1}\underline{d}\alpha^*_M \ ,$$

where the inverse $(*)^{-1}$ operator of $*$ has been defined in the preceding problem: for any q-form β_M one has

$$(*)^{-1}\beta_M = (-1)^{q(N-q)}\beta^*_M \ .$$

Show that the pair of Maxwell equations defined in (c) can take the form

$$\underline{\delta}\,\underline{F}_M + \mathcal{I}_M = 0\ ,$$

where $\mathcal{I}_M$ is a 1-form deduced from γ_M.

(B) We now go over to any manifold $\mathcal{M}_4$. The starting point will still be the matrix $(F_{\alpha\beta})$ of Problem 1.20, but the resulting calculations will be more involved since the raising and lowering of the indices will be made through the use of the metric $g_{\alpha\beta}(x)$ (relative to $\mathcal{M}_4$) instead of the metric $\eta_{\alpha\beta}$ of section (A).

(d) Verify that the equation $\underline{d}\,\underline{F}_M = 0$ of question (a) is still valid.

(e) The problem concerning $\underline{\delta}\,\underline{F}_M + \mathcal{I}_M = 0$ of (c) is more involved because of the metric. For instance, in $\mathcal{M}_4$ one has

$$F^{\alpha\beta} = g^{\alpha\mu}g^{\beta\nu}F_{\mu\nu}\ .$$

We do not want to go into details, retaining only that the result of (c) is still valid. The interested reader may consult [Bibl. 9], p. 318.

(f) Consider an open set of $\mathcal{M}_4$ homeomorphic to $\mathbb{R}^4$. Show that there exists a 1-form A_M such that

$$\underline{F}_M = \underline{d}A_M\ ,$$

where $A_\mu(x)$ is the electromagnetic field such that $F_{\mu\nu} = \nabla_\mu A_\nu - \nabla_\nu A_\mu$ (see Problem 1.20).

The reader interested in the applications of the studied formalism should refer to [Bibl. 9], p. 318. He will find there, among other applications, an intrinsic definition of a pure radiation field.

On the other hand, the operator $\underline{\delta}$ is of use in the definition of differential operators of differential forms on a manifold $\mathcal{M}_4$ (see [Bibl. 9], p. 305). It has furthermore interesting properties, for instance, its property of nilpotency ($\underline{\delta}^2 = 0$) leads to a conservation law for the 1-form current: $\underline{\delta}\mathcal{I}_M = 0$.

Solution:

(a) One has

$$\begin{aligned}
\underline{d}\,\underline{F}_M &= \frac{1}{2}\partial_\rho F_{\mu\nu}\underline{d}x^\rho \wedge \underline{d}x^\mu \wedge \underline{d}x^\nu \\
&= \frac{1}{3!}[\partial_\rho F_{\mu\nu} + \partial_\mu F_{\nu\rho} + \partial_\nu F_{\rho\mu}]\underline{d}x^\rho \wedge \underline{d}x^\mu \wedge \underline{d}x^\nu
\end{aligned}$$

and the bracket is now antisymmetrized in its three indices. We may then introduce the strict (independent) components which are linearly independent and write

$$\underline{d}\,\underline{F}_M = \sum_{\rho < \mu < \nu} [\partial_\rho F_{\mu\nu} + \partial_\mu F_{\nu\rho} + \partial_\nu F_{\rho\mu}]\underline{d}x^\rho \wedge \underline{d}x^\mu \wedge \underline{d}x^\nu \ .$$

$\underline{d}\,\underline{F}_M = 0$ *leads then to the four equations*

$$\partial_\rho F_{\mu\nu} + \partial_\mu F_{\nu\rho} + \partial_\nu F_{\rho\mu} = 0$$

and, following Problem 1.20(c) and the beginning of question (d), they summarize indeed the given pair of Maxwell's equations.

(b) Following the methods used in question (a),

$$\underline{d}\,\underline{F}_M^* = \sum_{\rho < \mu < \nu} (\partial_\rho F_{\mu\nu}^* + \partial_\mu F_{\nu\rho}^* + \partial_\nu F_{\rho\mu}^*)\underline{d}x^\rho \wedge \underline{d}x^\mu \wedge \underline{d}x^\nu \ .$$

We pick up the terms which are linearly independent:

$$\begin{aligned}
\underline{d}\,\underline{F}_M = {}& (\partial_0 F_{12}^* + \partial_1 F_{20}^* + \partial_2 F_{01}^*)\underline{d}x^0 \wedge \underline{d}x^1 \wedge \underline{d}x^2 \\
& + (\partial_0 F_{13}^* + \partial_1 F_{30}^* + \partial_3 F_{01}^*)\underline{d}x^0 \wedge \underline{d}x^1 \wedge \underline{d}x^3 \\
& + (\partial_0 F_{23}^* + \partial_2 F_{30}^* + \partial_3 F_{02}^*)\underline{d}x^0 \wedge \underline{d}x^2 \wedge \underline{d}x^3 \\
& + (\partial_1 F_{23}^* + \partial_2 F_{31}^* + \partial_3 F_{12}^*)\underline{d}x^1 \wedge \underline{d}x^2 \wedge \underline{d}x^3 \ .
\end{aligned}$$

We now bring the matrix $F_{\alpha\beta}^*$ as given in Ex. 1.20(b) into the previous formula and obtain

$$\begin{aligned}
\underline{d}\,\underline{F}_M^* = {}& (-\partial_0 E_z + (\mathrm{curl}\,\mathbf{B})_z\underline{d}x^0 \wedge \underline{d}x^1 \wedge \underline{d}x^2 \\
& + (\partial_0 E_y - (\mathrm{curl}\,\mathbf{B})_y)\underline{d}x^0 \wedge \underline{d}x^1 \wedge \underline{d}x^3 \\
& + (-\partial_0 E_x + (\mathrm{curl}\,\mathbf{B})_x)\underline{d}x^0 \wedge \underline{d}x^2 \wedge \underline{d}x^3 - \boldsymbol{\nabla} \cdot \mathbf{E}\underline{d}x^1 \wedge \underline{d}x^2 \wedge \underline{d}x^3 \ .
\end{aligned}$$

We then use the Maxwell equations

$$\boldsymbol{\nabla} \wedge \mathbf{B} = \mathbf{J} + \frac{\partial \mathbf{E}}{\partial t}$$

to write the relation above as a differential 3-form γ_M:

$$\begin{aligned}
\underline{d}\,\underline{F}_M^* = {}& J_z(x)\underline{d}x^0 \wedge \underline{d}x^1 \wedge \underline{d}x^2 \\
& - J_y(x)\underline{d}x^0 \wedge \underline{d}x^1 \wedge \underline{d}x^3 \\
& + J_x(x)\underline{d}x^0 \wedge \underline{d}x^2 \wedge \underline{d}x^3 \\
& - \rho(x)\underline{d}x^1 \wedge \underline{d}x^2 \wedge \underline{d}x^3 = -\gamma_M \ .
\end{aligned}$$

(c) With $\alpha_M = \underline{F}_M$, one has

$$\underline{\delta}\,\underline{F}_M^* = (-1)^2(*)^{-1}\underline{d}\,\underline{F}_M^* = (-1)^2(-1)^3(\underline{d}\,\underline{F}_M^*)^* = \gamma_M^* \ ,$$

where it is clear that $(\underline{d}\,\underline{F}_M^*)^*$ is a 1-form. In order to simplify the calculations, we write γ_M as

$$\gamma_M = \gamma_{(012)}(x)\underline{d}x^0 \wedge \underline{d}x^1 \wedge \underline{d}x^2 + \gamma_{(013)}(x)\underline{d}x^0 \wedge \underline{d}x^1 \wedge \underline{d}x^3$$
$$+ \gamma_{(023)}(x)\underline{d}x^0 \wedge \underline{d}x^2 \wedge \underline{d}x^3 + \gamma_{(123)}(x)\underline{d}x^1 \wedge \underline{d}x^2 \wedge \underline{d}x^3 \ ,$$

then

$$\gamma_\alpha^* = \frac{1}{3!}\varepsilon_{\lambda\mu\nu\alpha}\gamma^{(\lambda\mu\nu)} \ .$$

But since $\gamma^{(\lambda\mu\nu)}$ is antisymmetric, the factor 1/3! can be omitted and

$$\gamma_3^* = \varepsilon_{0123}\gamma^{(012)} = -\gamma_{(012)} \ ,$$
$$\gamma_2^* = \varepsilon_{0132}\gamma^{(013)} = \gamma_{(013)} \ ,$$
$$\gamma_1^* = \varepsilon_{0231}\gamma^{(023)} = -\gamma_{(023)} \ ,$$
$$\gamma_0^* = \varepsilon_{1230}\gamma^{(123)} = \gamma_{(123)} \ .$$

We now use the equation $\underline{\delta}F_M = \gamma_M^*$ (first formula of (c)) to write

$$\begin{aligned}
\underline{\delta}\,F_M &= \gamma_M^* = \gamma_0^* dx^0 + \gamma_k^* \underline{d}x^k \\
&= \gamma_{(123)}\underline{d}x^0 - \gamma_{(023)}\underline{d}x^1 + \gamma_{(013)}\underline{d}x^2 - \gamma_{(012)}\underline{d}x^3 \\
&= -\rho(x)\underline{d}x^0 - J_x\underline{d}x^1 - J_y\underline{d}x^2 - J_z\underline{d}x^3 \\
&= -\mathcal{I}_M \ ,
\end{aligned}$$

where $\mathcal{I}_M$ is the differential 1-form current. Question (c) is then solved.
   (d) No comment is necessary.
   (e) See the quoted reference.
   (f) An application of Poincaré's lemma (Sec. 11, Level 2).

| **Problem 2.5.** | *Dynamical systems and symplectic manifolds* |

We begin first with a few classical reminders on dynamical systems. A dynamical system with N degrees of freedom is described by an N-uplet of C^∞ functions: $q(t) = \{q^1(t), \ldots q^N(t)\}$ where the real parameter t is the time. One also associates with $q(t)$ another function $L(q(t), \dot{q}(t), t)$ where

$\dot{q}(t)$ is the time derivative of $q(t)$; L is called the **Lagrangian** (and is for a mechanical system the difference between its kinetic and potential energies) of the system. Finally, the time evolution of $q(t)$ is obtained by requiring that the **action integral**

$$A = \int_{t_1}^{t_{11}} L(q, \dot{q}, t)dt \ ,$$

be stationary. Indications on this problem were given in Sec. 4, Level 0, Sec. 13, Level 1 and Sec. 24, Level 2; as a consequence, the $q(t)$ should obey the Euler-Lagrange equations

$$\frac{\partial L}{\partial q^k} - \frac{d}{dt}\frac{\partial L}{\partial \dot{q}^k} = 0 \quad k = 1 \ldots N \ ,$$

which are a system of N differential equations of $2°$ order.

It is very convenient for theoretical as well as practical purposes to replace the $2°$ order system of the Euler-Lagrange equations by $2N$ equations of first order, called **Hamilton equations**. One may proceed as follows: define the conjugate N-uplet $p(t) = \{p_1(t) \ldots p_N(t)\}$:

$$p_k(t) = \frac{\partial L}{\partial \dot{q}^k} \quad k = 1 \ldots N$$

and suppose thast these N equations allow us to express q_k as a function of p_j, q^j, then the **Hamiltonian**

$$H(q^1 \ldots q^N; p_1 \ldots p_N; t) = p_k \dot{q}^k - L(q, \dot{q}, t)$$

can be truly considered as a function of the independent variables q and p. In all textbooks on mechanics, it is shown that the Euler-Lagrange equations can be brought into Hamilton's form:

$$\dot{q}^k = \frac{\partial H}{\partial p_k} \ , \quad p_k = -\frac{\partial H}{\partial q^k} \quad k = 1 \ldots N \ .$$

One may observe at this point that the function H is a function of a contravariant vector q and a covariant one p: such an object has to be carefully defined. Our next task then is one of clarification and abstraction, and in doing so we shall come across many concepts of importance like **Poisson brackets, canonical transformations**...

We now come to the main questions: $\mathcal{M}_4$ is a differentiable manifold, $T_{\mathcal{M}}(M)$ its tangent vector space at M with v_M as elements, $T^*_{\mathcal{M}}(M)$ its cotangent vector space at M with elements v^*_M (instead of the notation α_M previously used) which are differential 1-forms. In Sec. 3, Level 2 we defined the tangent bundle, $T_{\mathcal{M}}$; we shall define the cotangent bundle $T^*_{\mathcal{M}}$ as the union of all the cotangent vector spaces to $\mathcal{M}$. On the other hand, if $M \in \mathcal{M}_N$ with local coordinates $\{q^1 \ldots q^N\} = q$ and $v^*_M \in T^*_{\mathcal{M}}(M)$ with local coordinates $\{p_1 \ldots p_N\} = \underline{p}$, the $2N$-uplet $\{q, \underline{p}\}$ are coordinates of a certain point of $T^*_{\mathcal{M}}$.

A symplectic manifold is the couple $\{\mathcal{M}, \sigma_M\}$, where σ_M is a differentiable closed 2-form and $\mathcal{M}$ is of even dimension (see question (e)). Such a manifold is somewhat similar to a Riemmanian manifold with the only difference that instead of a symmetric 2-form G we have an antisymmetric one, the 2-form σ_M.

We shall proceed in two steps.

(I) Let us write explicitly σ_M in its natural basis:

$$\sigma_M = \sigma_{ij}(x)\underline{d}x^i \wedge \underline{d}x^j \ ,$$

where the matrix σ_{ij} is skew-symmetric ($\sigma_{ij} = -\sigma_{ji}$) and will play the part of the matrix g_{ij} of a Riemannian manifold in the raising or lowering rule for indices, and furthermore

$$\sigma^{il}\sigma_{lj} = \delta^i_j \ .$$

On the symplectic manifold $\mathcal{M}_{2N}$ we consider a function $\hat{F}(M) = F(x)$ and the 1-form

$$\underline{d}\hat{F}(M) = \partial_i F(x)\underline{d}x^i \ ,$$

which defines the covariant vector $\partial_i F(x) = \nabla_i F(x)$.

Let Γ be a curve on $\mathcal{M}_{2N} : M = M(t)$ or in local coordinates $x = x(t)$; the field lines are then

$$\frac{dx_k}{dt} = \nabla_k F(x) \ ,$$
$$\frac{dx^k}{dt} = \sigma^{kj}\nabla_j F(x) \qquad k = 1 \ldots N$$

and one says that these equations describe a **gradient flow**.

These preliminaries having been laid down, we look for a certain diffeomorphism $\varphi : \{M, v_M^*\} \overset{\varphi}{\to} \hat{H} \in C^\infty$. If $\{q^1 \ldots q^N\}$ are the coordinates of M and $\{p_1 \ldots p_N\}$ the components of v_M^*, the corresponding parametrization of $\hat{H}$ is by no means intrinsic, it is strictly local.

(a) Choose first the diffeomorphism φ such that $\hat{H}$ is parametrized as $H(q^1, p_1; q^2, p_2, \ldots q^N, p_N)$, then seek the field lines $\underline{d}H$ and determine the matrix σ^{ij} such that the system of differential equations for these field lines will be Hamilton's equation (which describe a **Hamiltonian flow**). Same question for the parametrization $H(q^1 \ldots q^N; p_1 \ldots p_N)$.

(b) Consider a Hamiltonian flow

$$\frac{dx^k}{dt} = \sigma^{kj} \nabla_j H(x)$$

and let $\hat{F}(M(t))$ be any C^∞ function on the field lines of this flow. Show that

$$\frac{d}{dt} \hat{F}(M(t)) = \nabla_k F(x) \sigma^{kj} \nabla_j H(x) .$$

Write down this last equation using a matrix notation.

(c) A **Poisson bracket** is an internal operation of the vector space of C^∞ functions (transforming such a space in algebra). For two C^∞ functions $\hat{F}$ and $\hat{G}$, it is locally defined as

$$[F, G]_{\text{cl}} = \partial F^T \sigma \partial G .$$

The Poisson brackets have the following properties:

$$[F, G]_{\text{cl}} = -[G, F]_{\text{cl}}$$
$$[\alpha F_1 + \beta F_2, G]_{\text{cl}} = \alpha [F_1, G]_{\text{cl}} + \beta [F_2, G]_{\text{cl}}$$
$$\alpha, \beta \in \mathbb{R} .$$

They satisfy the Jacobi identity

$$[F, [G, H]_{\text{cl}}]_{\text{cl}} + [G, [H, F]_{\text{cl}}]_{\text{cl}} + [H, [F, G]_{\text{cl}}]_{\text{cl}} = 0 .$$

A proof of this last property can be based on the following formula:

$$\nabla^i [F(x), G(x)]_{\text{cl}} = -L_{\partial F}(\partial G(x))^i ,$$

where, using the notations of formula (11.7), Level 1,

$$L_\xi X^i(x) = [\xi(x), X(x)]_{\text{cl}}^i = \xi^k \partial_k X^i - X^k \partial_k \xi^i .$$

(d) Show that Hamilton's equations can also be written as

$$\dot{q}^k = [q^k, H]_{\text{cl}} \, , \quad \dot{p}_k = [p_k, H]_{\text{cl}}$$

and conclude that for any C^∞ function $F(q, p)$

$$\dot{F} = [F, H]_{\text{cl}} \, .$$

(e) Under the assumption that the coefficients σ_{ij} (of 2-form σ_M) are constant (as in the parametrizations in (a)), one can conclude that $\underline{d}\sigma_M = 0$ and also that the C^∞ functions build up a Lie algebra, the internal operation being defined by the Poisson bracket. We now want to examine the case where the σ^{ij} coefficients are functions of x; the C^∞ functions will build up a Lie algebra if they still obey the Jacobi identity, the other two properties being clearly fulfilled. Show that such functions build up indeed a Lie algebra, if and only if σ_M is a closed form, i.e. $\underline{d}\sigma_M = 0$.

(f) Show that there exists a parametrization of the Hamiltonian $\hat{H}$ such that σ_M can be reduced to the canonical form

$$\sigma_M = \underline{d}q^i \wedge \underline{d}p_i$$

and that in general

$$\underbrace{\sigma_M \wedge \sigma_M \wedge \ldots \sigma_M}_{N} = N! \sqrt{\det \sigma_M} \, \underline{d}q^1 \wedge \ldots \underline{d}q^N \wedge \underline{d}p_1 \wedge \ldots \underline{d}p_N \, ,$$

where $\det \sigma_M$ is the determinant of the matrix with σ_{ij} as elements. Furthermore, from its canonical form, one can conclude that $\underline{d}\sigma_M = 0$. But the following general statement also holds: if σ_M is a closed differential 2-form and the exterior product $\sigma_M \wedge \ldots \sigma_M$ (with N factors) does not vanish, there exists a system of local coordinates $\{q^1 \ldots q^N; p_1 \ldots p_N\}$ in which σ_M can be brought into its canonical form. This property is known as Darboux's theorem, we do not intend to prove it (see however, [Bibl. 9], p. 269).

(g) On the field lines of a Hamiltonian flow, σ_M becomes a function of $t : \sigma_M = \sigma_{M(t)}$, where $M = M(t)$ is the equation of any field line. Show that

$$\frac{d}{dt} \sigma_{M(t)} = 0$$

and deduce Liouville's theorem.

Some of the calculations we have used in the solution of this problem are true enough, lengthy. We followed the methods of [Bibl. 12], Chap. 6 Sec. 31. The reader will find in [Bibl. 15] proofs whose spirit is more in line with Level 2.

Solution:

(a) For the parametrization $\{q^k, p_k\}$ the components of the 1-form $\underline{d}H$ are

$$\left\{ \frac{\partial H}{\partial q^1}, \frac{\partial H}{\partial p_1} ; \frac{\partial H}{\partial p^2}, \frac{\partial H}{\partial q_2} ; \dots \right\} .$$

We build up then a $2N \times 2N$ matrix called A which is a direct sum of N matrices $\begin{pmatrix} 0 & 1 \\ -1 & 0 \end{pmatrix}$. From its very definition, the matrix elements a^{ij} of A should vanish except that $a^{i,i+1} = 1$ and $a^{i+1,i} = -1$. Then the system of equations $dx^k/dt = a^{kj} \partial_j H$ reads

$$\begin{pmatrix} \dot{q}^1 \\ \dot{p}_1 \\ \dot{q}^2 \\ \dot{p}_2 \\ \vdots \end{pmatrix} = \begin{pmatrix} 0 & 1 & & & & \\ & & & 0 & & \cdots \\ -1 & 0 & & & & \\ & & & 0 & 1 & \\ & 0 & & & & \cdots \\ & & & -1 & 0 & \\ & \cdots & & \cdots & & \end{pmatrix} \begin{pmatrix} \partial H/\partial q^1 \\ \partial H/\partial p_1 \\ \partial H/\partial q^2 \\ \partial H/\partial p_2 \\ \vdots \end{pmatrix} ,$$

and is the system of Hamilton's equations.

For the second parametrization, the components of $\underline{d}H$ are

$$\left\{ \frac{\partial H}{\partial q^1} \cdots \frac{\partial H}{\partial q^N} ; \frac{\partial H}{\partial p_1} \cdots \frac{\partial H}{\partial p_N} \right\} ,$$

and, using a matrix σ^{ij} still antisymmetric but different from the previous one, we obtain the following system of equations:

$$\begin{pmatrix} \dot{q}^1 \\ \vdots \\ \dot{q}^N \\ \dot{p}_1 \\ \vdots \\ \dot{p}_N \end{pmatrix} \begin{pmatrix} \overbrace{0 \quad 0 \dots 0}^{N} & & \\ \cdots & N & I_N \\ 0 \dots \quad 0 & & \\ & 0 \dots 0 & \\ -I_N & \cdots & \\ & \underbrace{0 \dots \quad 0}_{N} & \end{pmatrix} \begin{pmatrix} \partial H/\partial q^1 \\ \vdots \\ \partial H/\partial q^N \\ \partial H/\partial p_1 \\ \vdots \\ \partial H/\partial p_N \end{pmatrix}$$

where I_N is the $N \times N$ unit matrix. One obtains again Hamilton's system of equations.

(b) One has indeed

$$\frac{d}{dt}\hat{F}(M(t)) = \frac{\partial F(x)}{\partial x^k}\frac{dx^k}{dt} = \frac{\partial F(x)}{\partial x^k}\sigma^{kj}\frac{\partial H}{\partial x^j}$$
$$= \nabla_k F(x)\nabla^k H \ .$$

If F depends explicitly on t: $\hat{F}_t(M(t))$, one should add to the right-hand side the term $\partial F(x,t)/\partial t$.

Let us be more specific and suppose that $\hat{F}$ is parametrized as in the last formula of question (a): $H(q^1 \ldots q^N, p_1 \ldots p_N)$. We consider the matrix σ with elements σ^{ij}

$$\sigma = \begin{pmatrix} 0 & I_N \\ -I_N & 0 \end{pmatrix} \ ,$$

and also

$$\partial F = \begin{pmatrix} \partial F/\partial q^1 \\ \vdots \\ \partial F/\partial q^N \\ \partial F/\partial p_1 \\ \vdots \\ \partial F/\partial p_N \end{pmatrix}$$

and the transposed matrix ∂F^T. We also remark that

$$\sigma\partial F = \begin{pmatrix} \partial F/\partial p_1 \\ \vdots \\ \partial F/\partial p_N \\ -\partial F/\partial q^1 \\ \vdots \\ -\partial F/\partial q^N \end{pmatrix} \ .$$

We may bring the equation giving the time change of $F(x(t))$ the matrix form

$$\frac{\partial F(x(t))}{\partial t} = \partial F^T \sigma \partial H$$
$$= \left\{ \frac{\partial F}{\partial q^l}\frac{\partial}{\partial p_l} - \frac{\partial F}{\partial p_l}\frac{\partial}{\partial q^l} \right\} H(q,p) = \mathcal{D}_F H(q,p) \ ,$$

with an obvious definition of the differential operator $\mathcal{D}_F$.

(c) We first notice that the expression of a Poisson bracket can be brought into the following form through the introduction of the differential operator $\mathcal{D}_F$:

$$[F(x), G(x)]_{\mathrm{cl}} = \nabla_k F \nabla^k G = \partial F^T \sigma \partial G = \mathcal{D}_F G .$$

The proof of the first and second property of the Poisson bracket is straightforward, the proof of the Jacobi identity is a direct consequence of the last property. We want indeed to prove

$$\nabla^i [F(x), G(x)]_{\mathrm{cl}} = -L_{\partial F}(\partial G(x))^i .$$

From the above definition of the Poisson bracket, the left-hand side can be brought into the form

$$\nabla^i [F(x), G(x)]_{\mathrm{cl}} = \sigma^{ij} \partial_j \mathcal{D}_F G(x) ,$$

which is the i-th component of the following vector:

$$\begin{pmatrix} \vdots \\ \partial/\partial p_k \\ \vdots \\ -\partial/\partial q^k \\ \vdots \end{pmatrix} \left(\frac{\partial F}{\partial q^l} \frac{\partial G}{\partial p_l} - \frac{\partial F}{\partial p_l} \frac{\partial G}{\partial q^l} \right) \quad k = 1 \ldots N .$$

We now turn our attention to the right-hand side; from the above definition of $L_{\partial F}(\partial G)^i$, one has

$$\begin{aligned} L_{\partial F}(\partial G)^i &= \sigma^{kj}(\partial_j F)\partial_k \{\sigma^{im} \partial_m G\} - \sigma^{kj}(\partial_j G)\partial_k \{\sigma^{im} \partial_m G\} \\ &= -(\partial_j F)\sigma^{jk}\partial_k \{\sigma^{im} \partial_m G\} + (\partial_j G)\sigma^{jk}\partial_k \{\sigma^{im} \partial_m F\} \\ &= -(\partial F)^T \sigma \partial \{\sigma^{im} \partial_m G\} + (\partial G)^T \sigma \partial \{\sigma^{im} \partial_m F\} . \end{aligned}$$

But

$$(\partial F)^T \sigma \partial = \mathcal{D}_F , \quad (\partial G)^T \sigma \partial = \mathcal{D}_G ,$$

then

$$L_{\partial F}(\partial G)^i = -\mathcal{D}_F \sigma^{im} \partial_m G + \mathcal{D}_G \sigma^{im} \partial_m F$$

and $L_{\partial F}(\partial G)^i$ appears as the i-th component of the vector

$$
-\left\{ \frac{\partial F}{\partial q^l}\frac{\partial}{\partial p_l} - \frac{\partial F}{\partial p_l}\frac{\partial}{\partial q^l} \right\}
\begin{pmatrix} \partial G/\partial p_1 \\ \vdots \\ -\partial G/\partial q^1 \\ \vdots \end{pmatrix}
$$

$$
+\left\{ \frac{\partial G}{\partial q^l}\frac{\partial}{\partial p_l} - \frac{\partial G}{\partial p_l}\frac{\partial}{\partial q^l} \right\}
\begin{pmatrix} \partial F/\partial p_1 \\ \vdots \\ \partial F/\partial q^1 \\ \vdots \end{pmatrix} \ .
$$

We compare finally both sides to obtain the required relation.

(d) The three equalities can be easily verified using the last formula of (b) and the first formula of (c).

(e) The calculations are to be performed along the same lines as in question (c). We have only to keep the $\sigma^{ij}(x)$ in all the formulae. The use of the x^k variables will appear more convenient than the use of $\{p_k, q^k\}$ variables. Using the notation in (c), one has

$$
[F,G]_{\rm cl} = \mathcal{D}_F G : \ \mathcal{D}_F = (\partial F)^T \sigma \partial \ .
$$

Suppose now that Jacobi's identity is indeed fulfilled for σ^{ij}, which are functions of x^k, and seek its consequences. Its left-hand side $\nabla^i[F,G]_{\rm cl}$ can be written as

$$
\nabla^i[F,G]_{\rm cl} = \sigma^{ik}\partial_k\{\mathcal{D}_F G\} = \sigma^{ik}\partial_k\{\partial_l F \sigma^{lp}\partial_p G\}
$$
$$
= \sigma^{ik}\frac{\partial \sigma^{lp}}{\partial x^k}\frac{\partial F}{\partial x^l}\frac{\partial G}{\partial x^p}
$$
$$
+ \sigma^{ik}\sigma^{lp}\left(\frac{\partial^2 F}{\partial x^k \partial x^l}\frac{\partial G}{\partial x^p} + \frac{\partial^2 G}{\partial x^k \partial x^l}\frac{\partial F}{\partial x^l} \right) \ .
$$

On the other hand, from the results in question (c) one can express the right-hand side of Jacobi's identity as

$$
L_{\mathcal{D}F}(\partial G)^i = \mathcal{D}_F\{\sigma^{im}\partial_m G\} - \mathcal{D}_G\{\sigma^{im}\partial_m F\}
$$
$$
= \sigma^{lp}\left(\frac{\partial F}{\partial x^l}\frac{\partial \sigma^{im}}{\partial x^p}\frac{\partial G}{\partial x^m} - \frac{\partial G}{\partial x^l}\frac{\partial \sigma^{im}}{\partial x^p}\frac{\partial F}{\partial x^m} \right)
$$
$$
= \sigma^{im}\sigma^{lp}\left(\frac{\partial F}{\partial x^l}\frac{\partial^2 G}{\partial x^m \partial x^p} - \frac{\partial G}{\partial x^l}\frac{\partial^2 F}{\partial x^m \partial x^p} \right) \ .
$$

We then compare the two sides: the two last brackets $\sigma^{ij}\sigma^{lp}(\dots)$ cancel and we are left with

$$\left(\sigma^{ik}\frac{\partial\sigma^{lm}}{\partial x^k}+\sigma^{pl}\frac{\partial\sigma^{im}}{\partial x^p}+\sigma^{mp}\frac{\partial\sigma^{im}}{\partial x^p}\right)\frac{\partial F}{\partial x^l}\frac{\partial G}{\partial x^m}=0$$

and, since $\partial_l F$ and $\partial_m G$ can be arbitrarily chosen, it turns out that the parenthesis should vanish.

Multiply then this bracket by $\sigma_{ri}\sigma_{ls}\sigma_{tm}$ to obtain

$$* \qquad \sigma_{ls}\sigma_{tm}\frac{\partial\sigma^{lm}}{\partial x^r}+\sigma_{ri}\sigma_{tm}\frac{\partial\sigma^{im}}{\partial x^s}+\sigma_{ri}\sigma_{ls}\frac{\partial\sigma^{il}}{\partial x^t}=0$$

and observe that from

$$\sigma^{lm}\sigma_{ls}=-\delta^m_s\ ,\quad \sigma^{im}\sigma_{ri}=\delta^m_r\ ,\quad \sigma^{il}\sigma_{ri}=\delta^l_r\ ,$$

one can derive

$$\frac{\partial\sigma^{lm}}{\partial x^r}\sigma_{ls}=-\sigma^{lm}\frac{\partial\sigma_{ls}}{\partial x^r}\ ,\quad \frac{\partial\sigma^{im}}{\partial x^s}\sigma_{ri}=-\sigma^{im}\frac{\partial\sigma_{rl}}{\partial x^s}\ ,$$
$$\frac{\partial\sigma^{il}}{\partial x^t}\sigma_{ri}=-\sigma^{il}\frac{\partial\sigma_{ri}}{\partial x^t}\ ,$$

and finally equation (*) can be written as

$$\frac{\partial\sigma_{ts}}{\partial x^r}+\frac{\partial\sigma^{rt}}{\partial x^s}+\frac{\partial\sigma^{rs}}{\partial x^t}=0\ .$$

But one also has

$$\underline{d}\sigma_M=\underline{d}\{\sigma_{rs},\underline{d}x^r\wedge\underline{d}x^s\}=\frac{\partial\sigma_{rs}}{\partial x^t}\underline{d}x^t\wedge\underline{d}x^r\wedge\underline{d}x^s$$
$$=\frac{\partial\sigma_{rt}}{\partial x^s}\underline{d}x^s\wedge\underline{d}x^r\wedge\underline{d}x^t=\frac{\partial\sigma_{ts}}{\partial x^r}\underline{d}x^r\wedge\underline{d}x^t\wedge\underline{d}x^s$$
$$=\frac{1}{3}\left(\frac{\partial\sigma_{rs}}{\partial x^t}+\frac{\partial\sigma_{rt}}{\partial x^s}+\frac{\partial\sigma_{ts}}{\partial x^r}\right)\underline{d}x^t\wedge\underline{d}x^r\wedge\underline{d}x^s\ .$$

Comparing these two last equations, we conclude that if Jacobi's identity is fulfilled then the σ_M 2-form is closed. The converse proposition could also be proved along the same lines and the theorem we just established explains why we included in the definition of σ_M the condition that it ought to be closed.

(f) If there exists a system of coordinates $\{x^k\}$ denoted by $\{q^k, p_k\}$ such that the σ matrix is the direct sum of N matrices $\begin{pmatrix} 0 & 1 \\ -1 & 0 \end{pmatrix}$, it is straightforward to see that σ_M is canonical,

$$\sigma_M = \underline{d}q^i \wedge \underline{d}p_i \ ,$$

and then to verify the second formula of question (f). This is valid for any system of coordinates, since this second formula of (f) is form invariant. It may be recalled that $\sqrt{\det \sigma_M}$, which is polynomial into σ^{ij}, is called a **pfaffian.**

(g) One has for the canonical form of σ_M,

$$\frac{d}{dt}\sigma_{M(t)} = \frac{d}{dt}\{\underline{d}q^i \wedge \underline{d}p_i\}$$

$$= \frac{d}{dt}\{\underline{d}q^i\} \wedge \underline{d}p_i + \underline{d}q^i \wedge \frac{d}{dt}\underline{d}p_i \ .$$

But

$$\frac{d}{dt}\underline{d}q^i = \underline{d}\frac{dq^i}{\partial t} = \underline{d}\frac{\partial H}{\partial p_i}$$

$$= \frac{\partial^2 H}{\partial p_i dq^j}\underline{d}q^j + \frac{\partial^2 H}{\partial p_i \partial p_j}\underline{d}p_j \ ,$$

$$\frac{d}{dt}\underline{d}p_i = \underline{d}\frac{dp_i}{dt} = -\underline{d}\frac{\partial H}{\partial q^i}$$

$$= -\frac{\partial^2 H}{\partial q^i \partial q^j}\underline{d}q^j - \frac{\partial^2 H}{\partial q^i \partial p_j}\underline{d}p_j \ ;$$

a simple calculation of $\underline{d}\sigma_{M(t)}$ will show that it vanishes. For the proof of Liouville's theorem, since $\underline{d}\sigma_{M(t)} = 0$,

$$\underline{d}\{\sigma_{M(t)} \wedge \sigma_{M(t)} \wedge \ldots\} = 0$$

and

$$\underline{d}\{\sigma_{M(t)} \wedge \sigma_{M(t)} \wedge \ldots\} = \underline{d}\{\underline{d}q^{i_1} \wedge \underline{d}p_{i_1} \wedge \underline{d}q^{i_2} \wedge \underline{d}p_{i_2} \wedge \ldots\} \ .$$

Thus the elementary volume $dq^{i_1} \wedge \underline{d}p_{i_1} \wedge \underline{d}q^{i_N} \wedge dp_{i_N}$ of the phase space $\{q^i, p_i\}$ is conserved along a field line of a Hamiltonian flow. Such a transformation is a typical canonical transformation.

$\boxed{\textit{Problem 2.6.}}$ *Pfaff systems, thermodynamics, first-order partial differential equations and the Hamilton-Jacobi equation*

We shall make a sketchy study of a very extensive subject. All our results will be strictly local, valid on a neighborhood of a point $M \in \mathcal{M}_N$: we thus confine ourselves to $\mathbb{R}^N$ space. We shall also leave aside all algebraic considerations which are so valuable in a deep study of first-order partial differential systems of equations; the interested reader may consult [Bibl. 9], Chap. 4, Part (C).

(a) Let ω_M be a differential 1-form and $\hat{\mu}(M) = \mu(x)$ be a C^∞ function on $\mathcal{M}_N$. Under which conditions is $\hat{\mu}(M)\omega_M$ an exact differential 1-form? $\hat{\mu}(M)$ is called an **integrating factor**. Consider the special case of $\mathcal{M}_2$: application to phenomenological thermodynamics.

(b) Consider r differential 1-form $\omega_M^{\{\alpha\}} : \alpha = 1 \ldots r$, defined on a differentiable manifold $\mathcal{M}_N$, and the system of equations

$$\omega_M^{(\alpha)} = 0 \ , \quad \alpha = 1 \ldots r \ .$$

Locally it can be written as

$$\omega_M^{(\alpha)} = \omega_i^\alpha(x)\underline{d}x^i \quad \begin{array}{l} \alpha = 1 \ldots r \\ \varepsilon = 1 \ldots N \end{array} \ .$$

Such a system will be called a **Pfaff system** of rank r if the matrix with $\omega_i^\alpha(x)$ as elements is of rank r. A Pfaff system will be well adapted for discussion if it is equivalent to a system of exact differential 1-forms: in such a case, it will be called **completely integrable**. Let us be more specific: suppose that locally, one can find a regular $r \times r$ matrix $S(x)$ (the **integrating matrix**) such that

$$\omega_M^{(\alpha)} = S_\lambda^\alpha(x)\underline{d}z^\lambda : \ S_\lambda^\alpha(x) : \ C^\infty \text{ functions} \ ,$$

where α and λ run from 1 to r and $z^\lambda(x)$ are new coordinates. Then

$$\underline{d}z^\alpha(x) = S^{-1\alpha}{}_\rho(x)\omega_M^{(\rho)} = S^{-1\alpha}{}_\rho(x)\omega_i^\rho(x)\underline{d}x^i$$

and the Pfaff system $\omega_M^{(\alpha)} = 0$ is now equivalent to a system of r exact differential equation $\underline{d}z^\alpha(x) = 0, \alpha = 1 \ldots r$. There is a theorem due to Frobenius which states a necessary and sufficient condition for a Pfaff system to be completely integrable; it reads

$$\underline{d}\omega^{(\alpha)} \wedge \omega^{(1)} \wedge \omega^{(2)} \wedge \ldots \omega^r = 0 : \ \alpha = 1 \ldots r \ .$$

We shall omit a general proof of this theorem (however, the interested reader may consult [Bibl. 9], p. 239) and we will check it for $\mathcal{M}_3, r = 1$.

We now suppose that the original Pfaff system $\omega^{(\alpha)} = 0, \alpha = 1 \ldots r$ is completely integrable; then this system is equivalent to

$$\underline{d}z^1(x) = 0 \ldots \underline{d}z^r(x) = 0 \ .$$

Consider the αth-equation,

$$\underline{d}z^\alpha(x^1 \ldots x^N) = \partial_k z^\alpha(x)\underline{d}x^k = 0 \ .$$

It implies successively $\partial_k z^\alpha(x) = 0, k = 1 \ldots N$, and

$$z^\alpha(x^1 \ldots x^N) = c^\alpha : \ c^\alpha \in \mathbb{R} \ ,$$

the $z^\alpha(x)$ build up a set of r **first integrals** of the Pfaff system. We have r constant first integrals which depend on N variables x^k : $N - r$ of the coordinates are independent variables and can be chosen *ad libitum*. Then the point $M \in \mathcal{M}_N$ has for coordinates

$$x^1 = u^1, \quad x^2 = u^2, \ldots x^{N-r} = u^{N-r}$$
$$x^{N-r+1} = x^{N-r+1}(u^1 \ldots u^{N-r}, c^1 \ldots c^r) \ ,$$

$$\ldots$$

$$x^N = x^N(u^1 \ldots u^{N-r}; c^1 \ldots c^r) \ ,$$

where $u^1 \ldots u^{N-r}$ is just another name for $x^1 \ldots x^{N-r}$. The preceding system defines a submanifold of $\mathcal{M}_N$ of dimension $N - r$: such a submanifold is called **an integral manifold**. We now turn our attention to the determination of the C^α constants: suppose $P \in \mathcal{M}_N$ to be a point of a convenient neighborhood of $M \in \mathcal{M}_N$ and ask for P to belong also to the integral manifold in question. If $\{\xi^1 \ldots \xi^N\}$ are the coordinates of P, and the C^α area given by the relations

$$c^\alpha = z^\alpha(\xi^1 \ldots \xi^N) \ ,$$

our problem is solved. We may state then that given a point P in the neighborhood of M there exists an integral manifold going through P.

Check the Frobenius theorem for a single Pfaff equation in $\mathcal{M}_2$ and $\mathcal{M}_3$, proving it in the last case. Consider its consequences for thermodynamics. Show that the forms of ranks N and $N-1$ on $\mathcal{M}_N$ are completely integrable.

(c) We begin with a few reminders: we saw that the field lines of the vector field X_M are given by Eq. (25.15), Level 2,

$$\frac{d\tilde{\mathcal{F}}_t}{dt} = X_{\hat{F}_t} \,,$$

where $\tilde{\mathcal{F}}$ is a map of a real interval into a neighborhood of $M \in \mathcal{M}_N$ such that $\tilde{\mathcal{F}}_t = M_t$. Locally this equation is equivalent, by (25.2) to the system of differential equations

$$* \qquad\qquad \frac{dx^1(t)}{X^1(x(t))} = \ldots = \frac{dx^N(t)}{X^N(x(t))} = dt$$

and any first integral $\hat{F}(M)$ of that system fulfills Eq. (25.18),

$$i_X \overline{d} \hat{F}(M) = 0 \,,$$

equivalent to (25.2). We also observe that

$$i_X \hat{F}(M) = 0 \,,$$

as in Problem 2.3(a). These results can be generalized for any differential form ω_M invariant under the system $*$:

$$i_X \omega_M = 0 \,, \quad i_x \underline{d} \omega_M = 0 \,.$$

Another important problem related to these considerations is the study of absolute and relative integral invariants. We find again in the same paragraph 25 that, using Eq. (25.17), $\exp\{tX_M\}$ can be looked upon as an element of a one-parameter pseudo group which sends any point M_0 in a certain neighborhood of $M \in \mathcal{M}_N$ into a point M_t. Locally, if x_0 and $x(t)$ are the coordinates of M_0 and M_t, one states that a point $x_0 \in D_0 \subset \mathbb{R}^N$ is transformed into a point $x(t)$ of $x_t \in D_t \subset \mathbb{R}^N$, D_0 and D_t being compacts of $\mathbb{R}^N$ and D_t the transformed D_0. Both these compacts may be vector sub-spaces of $\mathbb{R}^N$; we shall explicitly consider only the case where D_0 and D_t have N as dimensions. We finally consider the integral

$$J(t) = \int_{D_t} f(x^1 \ldots x^N) \underline{d}x^1 \wedge \ldots \underline{d}x^N \,,$$

where $f(x)$ is a C^∞ function. $J(t)$ will be an **absolute invariant integral** if $J(t)$ is independent of t when one performs any of the transformations

$\exp\{tX_M\}$. Show that the conditions under which $J(t)$ is an absolute integral reads

$$\partial_k\{f(x)X^k(x)\} = 0 \ .$$

If ω_M is any differential p-form, we have to consider a p-chain denoted by $D_t^{(p)}$, following the notations introduced in Sec. 14, Level 2, and the following integral

$$J(t) = \int_{D_t^{(p)}} \omega$$

will be an absolute invariant integral if it is again independent of t. One shows that the previous condition on $f(x)$ should be replaced by the following:

$$i_X\omega_M = 0 \ , \quad i_X\underline{d}\omega_M = 0 \ .$$

Besides the absolute invariant integral, we shall define a **relative** one,

$$J^{(\mathrm{rel})}(t) = \int_{\partial D_t^{(p)}} \omega \ ,$$

(where $\partial D_t^{(p)}$ is the boundary of $D_t^{(p)}$) as being again independent of t. The conditions given above are reduced to one only:

$$i_X\underline{d}\omega_M = 0 \ .$$

Prove that the 3-form

$$\omega_M = \frac{1}{3!}f(x)\underline{d}x^1 \wedge \underline{d}x^2 \wedge \underline{d}x^3$$

is an absolute invariant integral in $\mathbb{R}^3$ with $(+,+,+)$ as signature if the two conditions given above are fulfilled.

The reader will find the complete proofs of the preceding assertions in [Bibl. 9], p. 249.

(d) Another important field of application for the Pfaff system is the theory of partial differential first-order equations. Again we will recall just a few classical properties.

Consider the relation

$$F(x^1 \ldots x^N; z; p_1 \ldots p_N) = 0 \ ,$$

when F is C^∞ function, $z = z(x^1 \ldots x^N)$ the unknown function, and

$$p_k = \frac{\partial z}{\partial x^k} ,$$

(not to be confused with the momenta in Hamilton's theory).

We shall use a slightly different notation by defining

$$x^{N+1}(x^1 \ldots x^N) = z$$

and looking for a solution of the form

$$\varphi(x^1 \ldots x^{N+1}) = C \qquad C \text{ real constant} ,$$

such that the equation $F(\ldots) = 0$ can be brought into the form

$$p + \phi(x^1 \ldots x^N, x; p_1 \ldots p_N) = 0 ,$$

where x is one of the variables x^k and $p = \partial x^{N+1}/\partial x$. Such an equation is called a **Hamilton-Jacobi equation**.

For completeness we should have added to the preceding remarks the study of characteristic curves and strips which lead to the solution of any Cauchy problem. On these points, the reader should consult either a classical analysis textbook such as, for instance, [Bibl. 2], Vol 4, Chap. 3 or [Bibl. 9], Chap. 4: the problem is studied there within the framework of Pfaff systems.

(e) We come back to Hamilton's system of differential equations. Show that Hamilton's equations can be deduced from the **eiconal** W using a variational principle:

$$W(p, q) = \int_{t_I}^{t_{II}} (p_k \dot{q}^k - H(p, q)) dt ,$$

where $\dot{q}^k$ is to be considered as a function of p, q and $\delta p_k(t)$ and $\delta q^k(t)$ vanish on the boundaries t_I, t_{II} of the domain of integration.

Show also that Hamilton's equations are equivalent to the system of differential equations describing the field lines of the vector field X with components

$$** \qquad X : \left\{ \frac{\partial H}{\partial p_1} \ldots \frac{\partial H}{\partial p_N}; \; 1; \; -\frac{\partial H}{\partial q^1} \ldots -\frac{\partial H}{\partial q^N} \right\} .$$

We add to these considerations some comments in the language of exterior analysis. We notice first of all that there are three different differentiable manifolds to be considered in the theory of Hamiltonian systems: first the **configuration space** of dimension N with points $\{q^1 \ldots q^N\}$, then the **phase space** of dimension $2N$ with points $\{q^1 \ldots q^N, p_1 \ldots p_N\}$, and finally the **state space** of dimension $2N + 1 : \{q^1 \ldots q^N; t; p_1 \ldots p_N\}$.

The state space will be the object of our main interest: the differential forms corresponding to the coordinates $\{q, t, p\}$ will be $\underline{d}q^k, \underline{d}t, \underline{d}p_k$ given as functions of the field X_M:

$$* * * \qquad \underline{d}q^k\{X_M\} = \frac{\partial H}{\partial p_k} \ , \quad \underline{d}t\{X_M\} = 1 \ ,$$

$$\underline{d}p_k\{X_M\} = -\frac{\partial H}{\partial q^k} \ .$$

One considers the differential 1-form

$$w = p_k \underline{d}q^k - H\underline{d}t \ .$$

Comment. Show that with respect to Hamilton's equation: w is a relative integral invariant, $\underline{d}w$ is an absolute one and also an invariant of the Hamiltonian system. Comment.

(f) **Canonical transformations.** Application to the Hamilton-Jacobi equation.

Solution:

(a) If $\hat{\mu}(M)\omega_M$ is an exact differential, then

$$\underline{d}\{\hat{\mu}(M)\omega_M\} = \underline{d}\hat{\mu}(M) \wedge \omega_M + \hat{\mu}(M)\underline{d}\omega_M = 0 \ ,$$

or

$$\frac{\underline{d}\hat{\mu}(M)}{\hat{\mu}(M)} \wedge \omega_M = -\underline{d}\omega_M \ , \tag{I}$$

which is a sufficient condition but also clearly a necessary one. If that condition is fulfilled, so is

$$\underline{d}\omega^M \wedge \omega_M = 0 \ , \tag{II}$$

which is a sufficient but not necessary condition.

Consider the case $\mathcal{M}_2$. Eq. (I) reads

$$\frac{\partial_k \mu(x)}{\mu(x)}\omega_i(x)\underline{d}x^k \wedge \underline{d}x^i = -\partial_k \omega^i(x)\underline{d}x^k \wedge \underline{d}x^i \ ,$$

and since $\underline{dx}^k \wedge \underline{dx}^i$ is defined in $\mathcal{M}_2$ (formula (8.16), Level 2),

$$\varepsilon^{ki} \partial_k \mu(x) \omega_i(x) \underline{dx}^1 \wedge \underline{dx}^2 = -\varepsilon^{ki} \mu(x) \partial_k \omega_i(x) \underline{dx}^1 \wedge \underline{dx}^2 \ ,$$

which implies

$$\omega_2 \frac{\partial \mu}{\partial x^1} - \omega_1 \frac{\partial \mu}{\partial x^2} = \mu \left(\frac{\partial \omega_1}{\partial x^2} - \frac{\partial \omega_2}{\partial x^1} \right) \ . \tag{III}$$

In thermodynamics, the absolute temperature can be introduced as an integrating factor: consider e.g. a divariant thermodynamical system (a perfect gas for instance) where states are described by the pressure p and the volume v. During an infinitesimal transformation the quantity of heat exchanged by the system with the surrounding medium, say $\delta Q(p, v)$ is a differential 1-form in $\underline{dp}, \underline{dv}$, but not an exact one. The principle of entropy states that there exists an integrating factor $T(p, v)$ such that

$$\frac{\delta Q(p, v)}{T(p, v)} = \underline{dS}(p, v) \ ,$$

where T is the absolute temperature and S the entropy of the system.

(b) For a single Pfaff equation ($r = 1$) in $\mathcal{M}_N$, the complete integrability condition reads

$$\underline{d\omega}_M \wedge \omega_M = \partial_k \omega_j(x) \omega_i(x) \underline{dx}^k \wedge \underline{dx}^j \wedge \underline{dx}^i = 0 \ .$$

For $\mathcal{M}_2$, this condition is clearly fulfilled since all 3-forms on $\mathcal{M}_2$ do vanish.

  Consider now $\mathcal{M}_3$ and the differential 1-form

$$\omega_M = \omega_i(x) \underline{dx}^i \qquad i = 1, 2, 3 \ .$$

As in question (a) let $\hat{\mu}(M) = \mu(x)$ be an integrating factor, i.e.,

$$\hat{\mu}(M) \omega_M = \underline{dz}(x) \ ,$$

or

$$\mu(x) \omega_i(x) \underline{dx}^i = \partial_i z(x) \underline{dx}^i \ .$$

Identification of the two sides and subsequent partial derivations lead to the compatibility conditions

$$\frac{\partial}{\partial x^3} \{\mu \omega_2\} = \frac{\partial}{\partial x^2} \{\mu \omega_3\} \ ,$$

$$\frac{\partial}{\partial x^1} \{\mu \omega_3\} = \frac{\partial}{\partial x^3} \{\mu \omega_1\} \ ,$$

$$\frac{\partial}{\partial x^2} \{\mu \omega_1\} = \frac{\partial}{\partial x^1} \{\mu \omega_2\} \ .$$

If one adds the three equations after respective multiplications by $\omega_1, \omega_2, \omega_3$ one easily finds

$$\omega_1 \left(\frac{\partial \omega_2}{\partial x^3} - \frac{\partial \omega_3}{\partial x^2} \right) + \omega_2 \left(\frac{\partial \omega_3}{\partial x^1} - \frac{\partial \omega_1}{\partial x^3} \right)$$
$$+ \omega_3 \left(\frac{\partial \omega_1}{\partial x^2} - \frac{\partial \omega_2}{\partial x^3} \right) = 0 \ .$$

In a rectangular frame, such an equation can be interpreted as meaning that

$$\boldsymbol{\omega} \cdot \operatorname{curl} \boldsymbol{\omega} = 0 \ ,$$

where $\boldsymbol{\omega}$ has $\{\omega_1, \omega_2, \omega_3\}$ as components.

We now go back to the Frobenius condition (as stated at the beginning of (b)) in $\mathcal{M}_3$. One has

$$\underline{dx}^k \wedge \underline{dx}^j \wedge \underline{dx}^i = \varepsilon^{kji} \underline{dx}^1 \wedge \underline{dx}^2 \wedge \underline{dx}^3 \ ,$$

following formula (8.16), Level 2. Then the Frobenius condition reads

$$\varepsilon^{kji} (\partial_k \omega_j) \omega_i \underline{dx}^1 \wedge \underline{dx}^2 \wedge \underline{dx}^3 = 0 \ ,$$

which means that

$$\boldsymbol{\omega} \cdot \operatorname{curl} \boldsymbol{\omega} = 0$$

as before. There is then a complete equivalence between the Frobenius condition and the existence of an integrating factor in the case $\mathcal{M}_3$ and $r = 1$. As far as thermodynamical considerations are concerned, let us introduce as in (a) an infinitesimal transformation of the same divariant system. The exchanged quantity of heat $\delta Q(p, v)$ is given by

$$\delta Q = dU - p dv \ ,$$

where $U(p, v)$ is the internal energy of the system; δQ is indeed a differential 1-form but not an exact differential. Suppose that the infinitesimal transformation is an adiabatic one, then it is described by the Pfaff equation

$$\delta Q(p, v) = 0 \ .$$

But as noticed above such a Pfaff equation admits an integrating factor $\mu(p, v) = 1/T(p, v)$ and we obtain the entropy $S(p, v)$:

$$\delta Q(p, v) = 0 \Leftrightarrow \underline{d}S(p, v) = 0 \ .$$

The entropy is a first integral and an adiabatic (reversible) transformation is an isentropic one.

 Consider finally the Frobenius condition for a Pfaff system of rank r; it reads

$$d\omega^{(\alpha)} \wedge \omega^{(1)} \wedge \ldots \omega^{(r)}$$
$$= (\partial_k \omega_i^\alpha)\omega_{j_1}^1 \ldots \omega_{j_r}^r \underline{dx}^k \wedge \underline{dx}^i \wedge \underline{dx}^{j_1} \wedge \ldots \underline{dx}^{j_r}$$
$$= 0 ,$$

i.e. a differential $(r+2)$-form. If we take $r = N$ or $r = N-1$ we have to define respectively a differential $N+2$ or $(N+1)$-form on $\mathcal{M}_N$; they necessarily vanish and Pfaff systems of rank N and $N-1$ are always completely integrable.

 (c) $J(t)$ will be invariant under a solution of the differential system of this same question (c) provided that $dJ(t)/dt = 0$, a derivative which will be calculated in a most straightforward way. When t becomes $t+h$, then $x(t)$ becomes

$$x^i(t+h) = x^i(t) + h\frac{dx^i(t)}{dt} = x^i(t) = hX^i(x(t))$$

up to the first order of h. The integral $J(t)$ is then

$$J_{t+h} = \int_{D_{t+h}} f(x'^1 \ldots x'^N)\underline{dx}'^1 \wedge \ldots \underline{dx}'^N ,$$

where since x' is a point of D_{t+h} it is related to $x(t)$ by the last but one relation.

 On the other hand, since D_{t+h} is the transformed of D_t, by a change of variables

$$x' = x'(x,t)$$

the integration domain D_{t+h} can be brought back to D_t and one may write

$$J(t+h) = \int_{D_t} f(x')\frac{\partial(x')}{\partial(x)}\underline{dx}^1 \wedge \ldots \underline{dx}^N .$$

But

$$f(x') = f(x(t+h)) = f(x(t)) + h\frac{df(x(t))}{dt}$$
$$= f(x(t)) + h\partial_k f(x)\frac{dx^k(t)}{dt} = f(x(t)) + hX^k(x(t))\partial_k f(x) ,$$

and one also has up to the first order in h

$$\frac{\partial(x')}{\partial(x)} = \begin{vmatrix} \frac{\partial x'^1}{\partial x^1} & \cdots & \frac{\partial x'^1}{\partial x^N} \\ \cdots & & \\ \frac{\partial x'^N}{\partial x^1} & \cdots & \frac{\partial x'^N}{\partial x^N} \end{vmatrix} = \begin{vmatrix} 1 + h\frac{\partial X^1}{\partial x^1} & \cdots & h\frac{\partial X^1}{\partial x^N} \\ \cdots & & \\ h\frac{\partial X^N}{\partial x^1} & \cdots & 1 + h\frac{\partial X^N}{\partial x^N} \end{vmatrix}.$$

$$= 1 + h\partial_i X^i(x) \ .$$

Finally $J(t+h)$ can be brought into the following form:

$$J(t+h) = \int_{D_t} f(x')\frac{\partial(x')}{\partial(x)}\underline{d}x^1 \wedge \ldots \underline{d}x^N$$

$$= \int_{D_t} (f + hX^k\partial_k f)(1 + h\partial_i X^i)\underline{d}x^1 \wedge \ldots \underline{d}x^N$$

$$= \int_{D_t} (f(x) + h\partial_i\{f(x)X^i\})\underline{d}x^1 \wedge \ldots \underline{d}x^N \ ,$$

so that

$$\frac{dJ(t)}{dt} = \lim_{h=0} \frac{J(t+h) - J(t)}{h} = \int_{D_t} \partial_i\{f(x)X^i(x)\}\underline{d}x^1 \wedge \ldots \underline{d}x^N \ .$$

Since the choice of D_t is arbitrary, one obtains

$$\partial_k\{f(x)X^k(x)\} = 0$$

*as a necessary and sufficient condition for $J(t)$ to be an invariant under transformations obeying the differential system $*$ of question (c): $f(x)$ is said to be a **Jacobi's factor** of this differential system.*

We consider next the differential 3-form

$$\omega_M = \frac{1}{3!}f(x)\underline{d}x^1 \wedge \underline{d}x^2 \wedge \underline{d}x^3 : \ x \in \mathbb{R}^3$$

and the integral

$$J(t) = \int_{D_t} \omega_M = \frac{1}{3!}\int_{D_t} f(x)\underline{d}x^1 \wedge \underline{d}x^2 \wedge \underline{d}x^3 \ .$$

The condition $i_x\underline{d}\omega_M = 0$ is clearly fulfilled since in $\mathbb{R}^3$, $\underline{d}\omega_M = 0$. The second condition reads

$$i_X\omega_M = \frac{1}{2!}X^i(x)\varepsilon_{ijk}\underline{d}x^j \wedge \underline{d}x^k f(x)$$

and in $\mathbb{R}^3$ *with* $(+,+,+)$ *as signature*

$$i_X \omega_M = X^k(x) N_k(x) dS(x) f(x) \;,$$

which represents the elementary flux of the vector field $\mathbf{X}(x)$ *across the surface* S *with normal* $\mathbf{N}$.

But, the condition $\partial_k\{f X^k\} = 0$ *implies that*

$$\int_{D_t} \partial_k\{f(x) X^k(x)\}\underline{d}x^1 \wedge \underline{d}x^2 \wedge \underline{d}x^3 = 0 \;,$$

which can be transformed by Stokes' theorem in

$$\oint_{\partial D_t} f(x)\mathbf{X}(x) \cdot \mathbf{N}(x) dS = 0 \;,$$

and since ∂D_t *is arbitrary, the vanishing of the integral means* $i_x \omega_M = 0$. *Both conditions are thus proved for this special case.*

(d) Let us take the partial derivative of φ *with respect of one of the variables* $\{x^1 \dots x^N\}$, *say* x^k:

$$\partial_k \varphi + \frac{\partial \varphi}{\partial x^{N+1}} \frac{\partial x^{N+1}}{\partial x^k} = \partial_k \varphi + p_k \frac{\partial \varphi}{\partial x^{N+1}} = 0 \;,$$

from which one deduces

$$p_k = -\frac{\partial \varphi / \partial x^k}{\partial \varphi / \partial x^{N+1}} \;,$$

and equation $F(\dots) = 0$ *takes the form*

$$F\left(x^1 \dots x^N; x^{N+1}; -\frac{\partial \varphi / \partial x^1}{\partial \varphi / \partial x^{N+1}} \dots -\frac{\partial \varphi / \partial x^N}{\partial \varphi / \partial x^{N+1}}\right) = 0 \;.$$

If this equation can be solved with respect of one of the variables $\partial \varphi / \partial x^i$, *we then put* $x^i = x$ *and* $\partial \varphi / \partial x = p$. *Then all the 2N remaining variables are denoted* $x^1 \dots x^N$ *and* $p^1 \dots p^N$ *and the previous equation can be brought into the Hamilton-Jacobi form:*

$$p + \phi(x^1 \dots x^N; x, p_1 \dots p_N) = 0 \;.$$

(e) It is easier to calculate directly $\delta W(p,q)$ (instead of going through the Euler-Lagrange equations). One has

$$\delta W = \int_{t_1}^{t_{11}} \left(p_k d\dot{q}^k + \delta p_k \dot{q}^k - \frac{\partial H}{\partial p_k} \delta p_k - \frac{\partial H}{\partial q^k} \delta q^k \right) dt$$

$$= \int_{t_1}^{t_{11}} \left[\left(\dot{q}^k - \frac{\partial H}{\partial p_k} \right) \delta p_k + \left(p_k \frac{d}{dt} \delta q^k - \frac{\partial H}{\partial q^k} \delta q^k \right) \right] dt \ .$$

The first term of the second bracket can be integrated by parts and taking the boundary conditions for $\delta p_k, \delta q^k$, one obtains

$$\delta W = \int_{t_1}^{t_{11}} \left[\left(\dot{q}^k - \frac{\partial H}{\partial p_k} \right) \delta p_k - \left(\dot{p}_k + \frac{\partial H}{\partial q^k} \right) \delta q^k \right] dt \ .$$

The condition $\delta W = 0$ leads directly to Hamilton's equations. We now observe that these equations can be brought into the following form:

$$\frac{dq^1}{\partial H/\partial p_1} = \cdots \frac{dq^N}{\partial H/\partial p_N} = dt$$

$$= -\frac{dp_1}{\partial H/\partial q^1} = \cdots - \frac{dp_N}{\partial H/\partial q^N} \ ,$$

which is exactly of the form $*$ in question (c) with X_M having as components those given by $**$ in question (e).

We consider now the 1-form

$$w = p_k \underline{dq}^k - H \underline{dt}$$

with no terms in $\underline{dp}_i$. The exterior differential of w is

$$dw = \underline{dp}_k \wedge \underline{dq}^k - \underline{dH} \wedge \underline{dt} \ .$$

We then calculate $i_x \underline{dw}$ with the help of relations $***$. We have

$$i_X\{\underline{dp}_i \wedge \underline{dq}^i\} = \underline{dp}_i(X)\underline{dq}^i - \underline{dp}_i \underline{dq}^i(X)$$

$$= -\frac{\partial H}{\partial p_i}\underline{dq}^i - \frac{\partial H}{\partial p_i}\underline{dp}_i$$

$$i_X\{\underline{dH} \wedge \underline{dt}\} = i_X \left\{ \frac{\partial H}{\partial p_i}\underline{dp}_i \wedge \underline{dt} + \frac{\partial H}{\partial q^i}\underline{dq}^i \wedge \underline{dt} \right\}$$

$$= \left(\frac{\partial H}{\partial p_i}\frac{\partial H}{\partial q^i}\underline{dt} - \frac{\partial H}{\partial p_i}\underline{dp}_i - \frac{\partial H}{\partial q^i}\frac{\partial H}{\partial p_i}\underline{dt} - \frac{\partial H}{\partial q^i}\underline{dq}^i \right)$$

where we have used the following relation of Problem 2.3(a):

$$i_X\{\alpha \wedge \beta\} = (i_X\alpha) \wedge \beta + (-1)^p \alpha \wedge i_X\beta \; .$$

Collecting all the terms, one gets

$$i_X \underline{d}w = 0$$

and the statements about w and $\underline{d}w$ are thus proved. As a straightforward application, we could obtain Liouville's theorems proved above.

(f) Consider the two sets

$$\{p_k, q^k; \; H(q,p) \in C^\infty\} \; , \quad \{p'_k, q'^k; \; H'(q',p') \in C^\infty\} \; ,$$

where k runs from 1 to N, and the two 1-forms

$$w(p,q,t,H) = p_k\underline{d}q^k - H\underline{d}t = w \; ,$$
$$w(p',q',t,H') = p'_k\underline{d}q'^k - H'\underline{d}t = w' \; .$$

Let us determine the $2N$ variables q', p' such that

$$w - w' = \underline{d}F(q,p; \; q',p'; \; t) \; , \tag{IV}$$

where F is any arbitrary C^∞ function. We now write explicitly the previous relation as

$$p_k\underline{d}q^k - H\underline{d}t - (p'_k\underline{d}q'^k - H'\underline{d}t)$$
$$= \frac{\partial F}{\partial q^k}\underline{d}q^k + \frac{\partial F}{\partial p_k}\underline{d}p_k + \frac{\partial F}{\partial q'^k}\underline{d}q'^k + \frac{\partial F}{\partial p'_k}\underline{d}p'_k + \frac{\partial F}{\partial t}\underline{d}t \; .$$

Suppose that all functions are functions of t, then the preceding equation becomes one between partial derivatives

$$p_k\dot{q}^k - H - (p'_k\dot{q}'^k - H') = \frac{d}{dt}F(q,p; \; q',p',t) \; .$$

Taking the integral of both sides between t_{I} and t_{II}, one gets

$$W(q,p) - W(q',p') = F_{\mathrm{II}} - F_{\mathrm{I}} \; ,$$

where F_{II} and F_{I} are the values taken by F at the times t_{II} and t_{I}.

Suppose furthermore that p, q, H describe a Hamiltonian flow: we take the variations of both sides, requiring the variations $\delta q, \delta p, \delta q', \delta p'$ to vanish when t takes the values t_I and t_II. The variation of the right-hand side vanishes because of the boundary conditions on $\delta q, \delta p, \delta q', \delta p'$; we are then left with $\delta W(q', p') = 0$ which signifies that q', p', H' describe indeed a Hamiltonian flow. Such a transformation is called a **canonical transformation** *which means that the new primed variables are also solutions of a Hamiltonian system.*

As an example take

$$F(q, p; q', p', t) = f(q, p', t) - p'_k q'^k .$$

The identification of both sides of (IV) leads to the relations

$$p_k = \frac{\partial f(q, p', t)}{\partial q^k} ,$$

$$\frac{\partial f(q, p', t)}{\partial t} = H'(q', p') - H(q, p) ,$$

$$q'^k = \frac{\partial f(q, p', t)}{\partial p'_k} .$$

We look for a possible transformation such that

$$H'(q', p') = 0 ,$$

which means $p'_k = 0, q'^k = 0; p'_k$ and q'_k are then independent of t. The first equation allows the calculation of p' as a function of q, p and this value is reported in f which as a function of q, p is denoted $S(q, p, t)$; then the second equation is a partial derivative equation for S and reads

$$\frac{\partial S(q, p, t)}{\partial t} + H\left(q^1, \ldots q^N; \frac{\partial S}{\partial q^1} \ldots \frac{\partial S}{\partial q^N}\right) = 0 .$$

This is the **Hamilton-Jacobi** *equation, an important equation for its theoretical and practical applications.*

For a more complete study, see [Bibl. 15], Chaps. 7, 8 and 9; [Bibl. 9], p. 25, and any classical textbook on analytical dynamics.

Problem 2.7.

We now come to some important questions which deal with the expressions of Riemann and Ricci tensors using intrinsic methods. The taxonomic method of Level 1 is a consistent and efficient way for the calculation of these entities since it can be programmed for the use of computers. But, it is also clear that it leads to tedious and cumbersome calculations by hand. The use of quicker methods is then to be welcomed even if they appear less methodical than the methods quoted above.

We will study two of these methods which both require some intuition: the first one will use the theory of differential forms while the second one will use a variational approach.

Differential forms and curvature. Robertson-Walker metric

We start by collecting some formulae which will be important for further use: all the quoted formulae will refer to Level 2. In formula (22.1) we introduced the 2-symmetric form $\underline{ds}^2$ which we denoted by $\mathbf{G}$ and we shall write it simply, dropping the sign $\otimes$ indicating tensor multiplication, as

$$\mathbf{G}_M = g_{\mu\nu}(x)\omega^\mu \otimes \omega^\nu = g_{\mu\nu}\omega^\mu\omega^\nu \ . \tag{22.1}$$

The curvature form $\mathcal{R}^\alpha_\beta$ has been introduced through formulae (21.1):

$$\mathcal{R}^\alpha_\beta = \frac{1}{2} R^\alpha{}_{\beta\mu\nu}\omega^\mu \wedge \omega^\nu = \sum_{\mu<\nu} R^\alpha{}_{\beta\mu\nu}\omega^\mu \wedge \omega^\nu \ . \tag{21.1}$$

The connection forms Ω^α_β were defined by formula (21.3):

$$\Omega^\alpha_\beta = \gamma^\alpha{}_{\rho\beta}(x)\omega^\rho \ . \tag{21.3}$$

Between the curvature form and the connection forms, there exists the relation

$$\mathcal{R}^\alpha_\beta = \underline{d}\Omega^\alpha_\beta + \Omega^\alpha_\mu \wedge \Omega^\mu{}_\beta \ . \tag{21.2}$$

Since a Riemannian manifold has a vanishing torsion form, (21.7) leads to the relation

$$\underline{d}\omega^\alpha + \Omega^\alpha_\mu \wedge \omega^\mu = 0 \ . \tag{21.7}$$

With $\Omega_{\alpha\beta} = g_{\alpha\rho}\Omega^\rho{}_\beta$, formula (23.9) can be written as follows:

$$\underline{d}g_{\alpha\beta} = \Omega_{\alpha\beta} + \Omega_{\beta\alpha} \ , \tag{23.9}$$

which represents a restriction on $\Omega_{\alpha\beta}$. If $g_{\alpha\beta}(x)$ is constant,

$$\Omega_{\alpha\beta} + \Omega_{\beta\alpha} = 0 \tag{23.9'}$$

and $\Omega_{\alpha\beta}$ becomes antisymmetric.

The rules of the game are very simple indeed: guess the expression of Ω^{α}_{β} from the previous formulae, calculate the 2-form $\mathcal{R}^{\alpha}_{\beta}$ from which one can deduce $R^{\alpha}{}_{\beta\gamma\delta}, R_{\alpha\beta}$ and finally Einstein's equations. Once a solution is found, it is the sole correct solution since (23.12) shows that the problem has a unique solution. If the guessing is too hard, one can use the same formula (23.12) to express $\Omega_{\alpha\beta}$ as a linear combination of ω^{μ}; but it often happens, however, that the calculations become as involved as with the taxonomic method.

Before applying the method just sketched, we solve some simple exercises on structure coefficients.

(a) Let M be a point of a manifold $\mathcal{M}_3$ with coordinates $x = (x^1, x^2, x^3)$ and $\{\varepsilon_1, \varepsilon_2, \varepsilon_3\}$ the natural basis of $T_{\mathcal{M}}(M)$ with $\varepsilon_i = \partial_i$. Choose a new coordinate basis of $T_{\mathcal{M}}(M)$, say (e_1, e_2, e_3),

$$e_1 = \varepsilon_1 \ , \quad e_2 = \frac{1}{x^1}\varepsilon_1 \ , \quad e_3 = \varepsilon_3$$

and the reader has certainly noted that for an Euclidean space $(+, +, +)$ e_1, e_2, e_3 are the components

$$\partial\rho \ , \frac{1}{\rho}\partial\rho \ , \partial z \ ,$$

of the gradient in cylindrical coordinates. Determine the dual basis $(\omega^1, \omega^2, \omega^3)$, the commutators $[e_i, e_j]$, the structure coefficients, the Cartan-Maurer equations and their solutions.

(b) Same questions for

$$e_1 = \varepsilon_1 \ , \quad e_2 = \frac{1}{x^1}\varepsilon_2 \ , \quad e_3 = \frac{1}{x^1 \sin x^2}\varepsilon_3 \ ,$$

which, in an Euclidean space are components of the gradient in spherical coordinates.

The aim of the two previous exercises is to familiarize the reader with the manipulation of differential forms. We now go to the main problem:

how to obtain the Riemann and Ricci tensors for a significant case. We shall consider the Robertson-Walker case already studied in Problem 1.24.

(c) As a differential form, the Robinson-Walker metric reads

$$\underline{ds}^2 = \mathbf{G}_M = (\underline{dx}^0)^2 - \frac{R^2}{1 - kr^2}\underline{dr}^2 - R^2 r^2 \underline{d\theta}^2 - R^2 r^2 \sin^2\theta \underline{d\varphi}^2 \ ,$$

where $(\underline{dx}^0)^2, \underline{dr}^2, \underline{d\theta}^2, \underline{d\varphi}^2$ mean $\underline{dx}^0 \otimes \underline{dx}^0, \underline{dr} \otimes \underline{dr}, \underline{d\theta} \otimes \underline{d\theta}$ and $\underline{d\varphi} \otimes \underline{d\varphi}$ and R is a function of x^0 only (not the total curvature!).

The metric coefficient are clearly

$$g_{00} = 1 \ , \quad g_{11} = -\frac{R^2}{1 - kr^2} \ , \quad g_{22} = -R^2 r^2 \ , \ g_{33} = -R^2 r^2 \sin^2\theta \ .$$

Perform the changes

$$\omega^0 = \underline{dx}^0 \ , \quad \omega^1 = \frac{R}{\sqrt{1 - kr^2}}\underline{dr} \ , \quad \omega^e = Rr\underline{d\theta} \ ,$$
$$\omega^3 = Rr\sin\theta\underline{d\varphi}$$

and the metric 2-form is now

$$\underline{ds}^2 = (\omega^0)^2 - (\omega^1)^2 - (\omega^2)^2 - (\omega^3)^2$$

with metric coefficients

$$\eta_{00} = \eta^{00} = 1 \ , \quad \eta_{11} = \eta^{11} = -1 \ , \quad \eta_{22} = \eta^{22} = -1 \ ,$$
$$\eta_{33} = \eta^{33} = -1$$

and all other $\eta_{\mu\nu}$ vanishing.

Calculate $\underline{d\omega}^0, \underline{d\omega}^1, \underline{d\omega}^2, \underline{d\omega}^3$.

(d) Consider the connection form Ω^α_β. Since $\underline{d\eta}_{\alpha\beta} = 0, \Omega_{\alpha\beta}$ is antisymmetric in α and β. What are symmetries of Ω^α_β?

(e) Use Eqs. (21.7) for $\alpha = 0, 1, 2, 3$ to show that the torsion of the considered Riemannian manifold vanishes.

(f) The system of corresponding relations builds up a system of equations for the unknown connection forms Ω^α_β. Guess a system of solutions.

(g) Use Eqs. (21.2) to express the curvature forms $\mathcal{R}^\alpha_\beta$ by the differential 1-form ω^α.

(h) Derive from the curvature forms the components of the Riemann and Ricci tensors and compare the results obtained with the ones in Problem 1.20.

The problem of the Robertson-Walker metric has been solved by two very different methods: the first belongs to the taxonomic approach of differential geometry, the second to the intrinsic approach. The reader will choose following his own taste and his own style of calculation.

Another example will be found in [Bibl. 9], p. 323 where the Schwarzschild metric is presented with a complete solution and [Bibl. 13], p. 365 gives a detailed solution in the case of the Friedmann metric. Other problems (with "hints" on their solutions) can be found on pp. 360-362.

There is also in [Bibl. 13] another method (p. 346) based on variational techniques; one starts from (24.7) (Level 2)

$$J(\lambda_{\mathrm{I}}, \lambda_{\mathrm{II}}) = \int_{\lambda_{\mathrm{I}}}^{\lambda_{\mathrm{II}}} g_{ij}(x(\lambda)) \frac{dx^i}{d\lambda} \frac{dx^j}{d\lambda} d\lambda \tag{24.7}$$

and applies standard variational techniques to obtain the geodesics in canonical form:

$$\frac{d^2 x^k}{d\lambda^2} + \Gamma^k{}_{ij}(x(\lambda)) \frac{dx^i}{d\lambda} \frac{dx^j}{d\lambda} = 0 , \tag{24.8}$$

from which one tries to derive the $\Gamma^k{}_{ij}$ in the second term of the right-hand side.

Solution:

(a) The dual basis $(\omega^1, \omega^2, \omega^3)$ is such that

$$\omega^1(e_1) = \omega^1(\varepsilon_1) = 1 , \quad \omega^2(e_2) = \frac{1}{x^1}\omega^2(\varepsilon_2) = 1 ,$$

$$\omega(e_3) = \omega^3(\varepsilon_3) = 1 ,$$

then

$$\omega^1 = \underline{d}x^1 , \quad \omega^2 = x^1\underline{d}x^2 , \quad \omega^3 = \underline{d}x^3 .$$

The commutator $[e_1, e_2]$ reads for any C^∞ function

$$[e_1, e_2]\hat{F}(M) = \left[\partial_1, \frac{1}{x^1}\partial_2\right] F(x) = -\frac{1}{x^1}e_2\hat{F}(M) ,$$

all the other commutators being 0. Then all the structure coefficients vanish except

$$c_{12}^2 = -c_{21}^2 = -\frac{1}{x^1} .$$

The Cartan-Maurer equations

$$\underline{d}\omega^k = -\frac{1}{2}c^k{}_{ij}\omega^i \wedge \omega^j = -\sum_{i<j}c^k{}_{ij}\omega^i \wedge \omega^j \ ,$$

have the solutions: $\omega^1 = \underline{d}x^1, \omega^3 = \underline{d}x^3$, *which are evident (Poincaré's lemma) and* ω^2 *is given by*

$$\underline{d}\omega^2 = -c^2{}_{12}\underline{d}x^1 \wedge \omega^2 = -\frac{1}{x^1}\underline{d}x^1 \wedge \omega^2 \ .$$

One looks for a solution $\omega^2 = Cx^1\underline{d}x^2$ *(C constant) to be determined by the conditions* $\omega^2(e_2) = 1$.

(b) One has

$$\omega^1(e_1) = \omega^1(\varepsilon_1) = 1 \ , \quad \omega^2\left(\frac{1}{x^1}\varepsilon_2\right) = 1 \ ,$$

$$\omega^3\left(\frac{1}{x^1\sin x^2}\varepsilon_3\right) = 1$$

and, as a result,

$$\omega^1 = \underline{d}x^1 \ , \quad \omega^2 = x^1\underline{d}x^2 \ , \quad \omega^3 = x^1\sin x^2\underline{d}x^3 \ .$$

(c) One has successively

$$\underline{d}\omega^0 = 0 \ ,$$

$$\underline{d}\omega^1 = \left[(\partial_0 R)\frac{1}{(1-kr^2)^{1/2}}\underline{d}x^0 + R\frac{\partial}{\partial r}\{(1-kr^2)^{-1/2}\}\underline{d}r\right] \wedge dr$$

$$= \frac{\overset{\circ}{R}}{R}\omega^0 \wedge \omega^1 \ ,$$

$$\underline{d}\omega^2 = \frac{\overset{\circ}{R}}{R}\omega^0 \wedge \omega^2 + \frac{(1-kr^2)^{1/2}}{Rr}\omega' \wedge \omega^2 \ ,$$

$$\underline{d}\omega^3 = \frac{\overset{\circ}{R}}{R}\omega^0 \wedge \omega^3 + \frac{(1-kr^2)^{1/2}}{Rr}\omega' \wedge \omega^3 + \frac{\cot\theta}{Rr}\omega^2 \wedge \omega^3 \ ,$$

where $\overset{\circ}{R} = dR/dx^0$.

(d) One goes from $\Omega_{\alpha\beta} = -\Omega_{\beta\alpha}$ to Ω_β^α by using the formula

$$\Omega_\beta^\alpha = \eta^{\alpha\mu}\Omega_{\mu\beta} \begin{cases} \Omega_i^0 = \Omega_0^i = \Omega_{01} \ , \\ \Omega_j^i = -\Omega_i^j = -\Omega_{ij} \ , \end{cases}$$

then Ω_β^α can be written as the following matrix:

$$(\Omega_\beta^\alpha) = \begin{pmatrix} 0 & \Omega_1^0 & \Omega_2^0 & \Omega_3^0 \\ \Omega_1^0 & 0 & \Omega_2^1 & \Omega_3^1 \\ \Omega_2^0 & -\Omega_2^1 & 0 & \Omega_3^2 \\ \Omega_3^0 & -\Omega_3^1 & -\Omega_3^2 & 0 \end{pmatrix} \ .$$

(e) We differentiate the expressions of ω^α given in (a) to obtain

$$\alpha = 0 : \underline{d}\omega^0 + \Omega_\mu^0 \wedge \omega^\mu = \Omega_\mu^0 \wedge \omega^\mu = 0 \ ,$$

$$\alpha = 1 : \underline{d}\omega^1 + \Omega_\mu^1 \wedge \omega^\mu = \frac{\overset{\circ}{R}}{R}\omega^0 \wedge \omega^1$$
$$+ \Omega_0^1 \wedge \omega^0 + \Omega_2^1 \wedge \omega^2 + \Omega_3^1 \wedge \omega^3 = 0 \ ,$$

$$\alpha = 2 : \underline{d}\omega^2 + \Omega_\mu^2 \wedge \omega^\mu = \frac{\overset{\circ}{R}}{R}\omega^0 \wedge \omega^2 + \frac{(1-kr^2)^{1/2}}{Rr}\omega^1 \wedge \omega^2$$
$$+ \Omega_0^2 \wedge \omega^0 + \Omega_1^2 \wedge \omega^1 + \Omega_3^2\omega^3 = 0 \ ,$$

$$\alpha = 3 : \underline{d}\omega^3 + \Omega_\mu^3 \wedge \omega^\mu = \frac{\overset{\circ}{R}}{R}\omega^0 \wedge \omega^3 + \frac{(1-kr^2)^{1/2}}{Rr}\omega^1 \wedge \omega^2$$
$$+ \frac{cotg\,\theta}{Rr}\omega^2 \wedge \omega^3 + \Omega_0^3 \wedge \omega^0$$
$$+ \Omega_1^3 \wedge \omega^1 + \Omega_2^3\omega^2 = 0 \ .$$

(f) The inspection of the $\alpha = 0$ case shows that a possible solution would be Ω_μ^0, proportional to ω^μ; the inspection of $\alpha = 3$ leads by identification to expressions for $\Omega^3{}_0, \Omega^3{}_1, \Omega^3{}_2$ and so for the $\alpha = 1, \alpha = 2$ cases. Taking into account the symmetries of Ω_β^α as shown in (d) we obtain the following tableau:

$$\Omega_1^0 = \Omega_0^1 = \frac{\overset{\circ}{R}}{R}\omega^1 \quad \Omega_2^1 = -\Omega_1^2 = -\frac{(1-kr^2)^{1/2}}{Rr}\omega^2 \ ,$$

$$\Omega_2^0 = \Omega_0^2 = \frac{\overset{\circ}{R}}{R}\omega^2 \quad \Omega_3^1 = -\Omega_1^3 = -\frac{(1-kr^2)^{1/2}}{Rr}\omega^3 \ ,$$

$$\Omega_3^0 = \Omega_0^3 = \frac{\overset{\circ}{R}}{R}\omega^3 \quad \Omega_3^2 = -\Omega_2^3 = -\frac{cotg\,\theta}{Rr}\omega^3 \ .$$

It has also been noticed that this is the unique solution.

(g) We use Eq. (21.2) to obtain the curvature forms. Let us calculate explicitly some of the $\mathcal{R}^\alpha_\beta$, for instance,

$$\mathcal{R}^0_1 = \underline{d}\Omega^0_1 + \Omega^0_\mu \wedge \Omega^\mu_1 \ ,$$

$$\underline{d}\Omega^0_1 = \frac{\overset{\circ\circ}{R}R - \overset{\circ}{R}^2}{R^2}\underline{d}x^0 \wedge \omega^1 + \frac{\overset{\circ}{R}}{R}\underline{d}\omega^1$$

$$= \frac{\overset{\circ\circ}{R}R - \overset{\circ}{R}^2}{R^2}\omega^0 \wedge \omega^1 + \frac{\overset{\circ}{R}^2}{R^2}\omega^0 \wedge \omega^1$$

$$= \frac{\overset{\circ\circ}{R}}{R}\omega^0 \wedge \omega^1 \ .$$

One also has

$$\Omega^0_\mu \wedge \Omega^\mu_1 = 0 \ ,$$

and finally

$$\mathcal{R}^0_1 = \frac{\overset{\circ\circ}{R}}{R}\omega^0 \wedge \omega^1 \ .$$

It is also straightforward to see that

$$\mathcal{R}^1_0 = \mathcal{R}^0_1 \ .$$

The same method shows that

$$\mathcal{R}^0_2 = \frac{\overset{\circ\circ}{R}}{R}\omega^0 \wedge \omega^2 = \mathcal{R}^2_0 \ ,$$

$$\mathcal{R}^0_3 = \frac{\overset{\circ\circ}{R}}{R}\omega^0 \wedge \omega^3 = \mathcal{R}^3_0 \ .$$

The calculations are more involved for the other 3 curvature forms, for instance,

$$\mathcal{R}^1_2 = \underline{d}\Omega^1_2 + \Omega^1_\mu \wedge \Omega^\mu_2 \ ,$$

with

$$\underline{d}\Omega^1_2 = -(1 - kr^2)^{1/2}\frac{\overset{\circ}{R}}{R^2 r}\omega^0 \wedge \omega^2 + \frac{1}{R^2 r^2}\omega^1 \wedge \omega^2$$

$$= \frac{1 - kr^2}{R^2 r^2}\omega^1 \wedge \omega^2 \ ,$$

while

$$\Omega^1{}_\mu \wedge \Omega^\mu{}_2 = \frac{\overset{\circ}{R}{}^2}{R^2} \omega^1 \wedge \omega^2$$

and finally

$$\mathcal{R}^1_2 = -(1 - kr^2)^{1/2} \frac{\overset{\circ}{R}}{R^2 r} \omega^0 \wedge \omega^2 + \frac{1}{R^2} \overset{\circ}{R}{}^2 + k) \omega' \wedge \omega^2 .$$

One can also show that

$$\mathcal{R}^2{}_1 = \mathcal{R}^1{}_2 .$$

Similarly one gets

$$\mathcal{R}^1_3 = -(1 - kr^2)^{1/2} \frac{\overset{\circ}{R}}{R^2 r} \omega^0 \wedge \omega^3 + \frac{1}{R^2} (\overset{\circ}{R}{}^2 + k) \omega^1 \wedge \omega^3$$

and also

$$\mathcal{R}^3{}_1 = \mathcal{R}^1{}_3 .$$

The same method applies to $\mathcal{R}^2{}_3$ although certain cancellations are more tricky:

$$\mathcal{R}^2_3 = \frac{1}{R^2} (\overset{\circ}{R}{}^2 + k) \omega^2 \wedge \omega^3 - \frac{\cot g\, \theta}{R^2 r} \omega^0 \wedge \omega^3 = -\mathcal{R}^3_2 .$$

One finally remarks that

$$\mathcal{R}^\alpha{}_\alpha = 0 .$$

(h) From now on, the steps are straightforward. The calculation of the components of the Riemann tensor uses formula (21.1). We just quote the results for a single example:

$$\mathcal{R}^0{}_1 = \sum_{\mu < \nu} R^0{}_{1\mu\nu} \omega^\mu \wedge \omega^\nu = \frac{\overset{\circ\circ}{R}}{R} \omega^0 \wedge \omega^1 ,$$

with

$$R^0{}_{101} = R^1{}_{001} = \frac{\overset{\circ\circ}{R}}{R} .$$

We also have

$$R^0{}_{202} = R^2{}_{002} = R^0{}_{303} = R^3{}_{003} = \frac{\overset{\circ\circ}{R}}{R} .$$

The calculations of the other components follow the same lines:

$$R^1{}_{212} = -R^2{}_{112} = \frac{\overset{\circ}{R}{}^2 + k}{R^2} \ ,$$

$$R^1{}_{202} = -R^2{}_{102} = -(1 - kr^2)^{1/2}\frac{\overset{\circ}{R}}{R^2 r} \ ,$$

$$R^1{}_{303} = -R^3{}_{103} = -(1 - kr^2)^{1/2}\frac{\overset{\circ}{R}}{R^2 r} \ ,$$

$$R^1{}_{313} = -R^3{}_{113} = \frac{\overset{\circ}{R}{}^2 + k}{R^2} \ ,$$

$$R^2{}_{323} = -R^3{}_{123} = \frac{\overset{\circ}{R}{}^2 + k}{R^2} \ ,$$

$$R^2{}_{303} = -R^2{}_{203} = -\cot\theta / R^2 r \ .$$

All other non-quoted components vanish. The evaluation of the Ricci tensor presents no difficulty; one has

$$R_{\alpha\beta} = R^\mu{}_{\alpha\mu\beta} \ .$$

Only its diagonal components survive and one verifies

$$R_{00} = -3\frac{\overset{\circ\circ}{R}}{R} \ ,$$

$$R_{11} = R_{22} = R_{33} = \frac{\overset{\circ\circ}{R}R + 2\overset{\circ}{R}{}^2 + 2k}{R^2} \ .$$

It remains now to go from the basics $(\omega^0, \omega^1, \omega^2, \omega^3)$ of $T^*_{\mathcal{M}}(M)$ to $(\underline{d}x^0, \underline{d}x^1 = \underline{d}r, \underline{d}x^2 = \underline{d}\theta, \underline{d}x^3 = \underline{d}\varphi)$. If $\overline{R}_{\alpha\beta}$ are the components of the Ricci tensor in this new basis, then

$$\sum_{\mu<\nu} R_{\mu\nu}\omega^\mu \wedge \omega^\nu = \sum_{\mu<\nu} \overline{R}_{\mu\nu}\underline{d}x^\mu \wedge \underline{d}x^\nu \ ,$$

i.e.,

$$\overline{R}_{00} = -3\frac{\overset{\circ\circ}{R}}{R} \ , \qquad\qquad \overline{R}_{11} = \frac{\overset{\circ\circ}{R}R + 2\overset{\circ}{R}{}^2 + 2k}{1 - kr^2} \ ,$$

$$\overline{R}_{22} = (\overset{\circ\circ}{R}R + 2\overset{\circ}{R}{}^2 + 2k)r^2 \ , \quad \overline{R}_{33} = (\overset{\circ\circ}{R}R + 2\overset{\circ}{R}{}^2 + 2k)r^2\sin^2\theta \ .$$

These results are to be compared to those of Problem 1.24, question (b). Noting that the present $\overline{R}_{\alpha\beta}$ were denoted there by $R_{\alpha\beta}$, the reader will verify that both sets of expressions are identical.

Problem 2.8. *Clifford algebra. Pauli and Dirac spinors — Quaternions*

Let E_N be a real vector space* with the $\{u, v, w \ldots\}$ as elements and $\{e_1 \ldots e_N\}$ as basis. We shall endow E_N with a real algebra structure C_N by defining an internal operation denoted simply uv and called **Clifford product** with the following properties:

1. associativity and distributivity with respect to the operation $+$ of E_N,
2. $uv = vu$ implies $u = \lambda v, \lambda \in \mathbb{R}$ and conversely.

As for any algebraic product, we have

$$uv = \frac{1}{2}(uv + vu) + \frac{1}{2}(uv - vu)$$
$$= \frac{1}{2}[u, v]_+ + \frac{1}{2}[u, v]_- \ .$$

We denote

$$\frac{1}{2}[u, v]_+ = u \circ v \ , \quad \frac{1}{2}[u, v]_- = u \wedge v$$

and deduce

$$uv = u \circ v + u \wedge v \ ,$$

with

$$u \circ v = v \circ u \ , \quad u \wedge v = -v \wedge u \ .$$

(a) Show that $uv = 0$ implies $u \wedge v = uv$. The basis $\{e_1 \ldots e_N\}$ when endowed with the Clifford product builds up a corresponding basis for the real Clifford algebra C_N; this new set is then rightly called the set of generators of C_N to which corresponds an infinite basis $\{I; e_1 \ldots e_N; e_1 e_2, \ldots e_i e_j \ldots e_1 e_2 e_3, \ldots e_i e_j e_k, \ldots\}$. But it would be interesting to look for an added condition such that the basis becomes finite. This new condition, called the reduction postulate, reads as follows:

$$e_i \circ e_j = g(e_i, e_j)e \ , \tag{III}$$

---

*i.e., only linear combinations $\alpha u + \beta v + \gamma w$ with $\alpha, \beta, \gamma \in \mathbb{R}$ are to be considered.

where $g_{ij} = g(e_i, e_j)$ is real; e being the unit element of C_N with respect to the Clifford product ($ue = u$).

(b) Define the reduction mechanism and give an example.

(c) Show that the dimension of C_N is then 2^N.

(d) **Pauli spinors** — Comment.

(e) **Dirac spinors** — Comment.

(f) **Quaternions** — Comment.

Solution:

(a) *If $u \circ v = 0$, then $uv = -vu$ and*

$$u \wedge v = \frac{1}{2}[u, v]_- = uv .$$

(b) *We use (III) and notice that $e_i \circ e_i = e^2{}_i = g_{ii}$, furthermore,*

$$e_i e_j = 2g_{ij}e - e_j e_i .$$

Consider as an example the triple Clifford product

$$e_i e_j e_i = (2g_{ij}e - e_j e_i)e_i = 2g_{ij}e_i - g_{ii}e_j .$$

It is then clear that any element of the algebraic basis with more than N factors contains necessarily at least two identical e_j: it may be automatically reduced to a linear combination of elements containing at most N factors and the last term of the basis is necessarily $e_1 e_2 \ldots e_N$.

(c) *Let us now count the number of elements of the C_N basis. By*

$$\binom{N}{p} = \frac{N!}{p!(N-p)!} ,$$

we denote the number of arrays we can build with p different objects chosen among N different objects. Then the different numbers of factors of the elements $e, e_k, e_j, e_k, \ldots$ of the basis can be found in the following tableau:

$$e \qquad\qquad : \binom{N}{0} = 1 ,$$

$$\{e_k\} \qquad\qquad : \binom{N}{1} = N ,$$

$$\{e_j e_k\}, j \neq k \quad : \binom{N}{2} = \frac{N!}{2(N-2)!} ,$$

$$\ldots$$

$$e_1 e_2 \ldots e_N \qquad \binom{N}{N} = 1 .$$

Thus, the number of elements of the basis is 2^N, following a well-known result of combinatorial analysis. It results that the dimension of C_N is 2^N.

(d) To any vector space E_3 with $\{e_1, e_2, e_3\}$ as a basis, we may associate a Clifford algebra C_3 with $\{e_1, e_2, e_3\}$ as generators if we define an internal product

$$e_i \circ e_j \begin{cases} 0 & \text{if } i \neq j \\ e & \text{if } i = j \end{cases} = \delta_{ij} e \ ,$$

e being the unit element of C_3. Such an algebra can be realized using the well-known Pauli matrices: $e_1 = \sigma_1, e_2 = \sigma_2, e_3 = \sigma_3, e = 1$. (See also Problem 2.10).

Any Pauli spinor u is defined as an element of the C_3 algebra corresponding to the vector space E_3.

The dimension of C_3 can be read in the following tableau:

$$e \qquad\qquad \binom{3}{0} = 1 \ ,$$

$$\{e_i\} \qquad\qquad \binom{3}{1} = 3 \ ,$$

$$\{e_i e_j\} \ , \ i \neq j \ \binom{3}{2} = 3 \ ,$$

$$e_1 e_2 e_3 \qquad\qquad \binom{3}{3} = 1 \ ,$$

which gives $2^3 = 8$ as the dimension. Clearly the elements $\{e, e_1, e_2, e_3\}$ are linearly independent (see question (e) for instance) and so are the 8 elements of the present basis. We may remark that $\sigma_1 \sigma_2 \sigma_3 = ie : ie$ is therefore an independent element of the present real Clifford algebra.

We first consider special elements of C_3, also called pure quaternions, with components $A^i \in \mathbb{R}$

$$A = A^i e_i \ .$$

They may be considered as elements of a 3-vector space and have some remarkable properties, for instance,

$$AA = A \circ A = A^2 = (A^1)^2 + (A^2)^2 + (A^3)^2 \ ,$$

and if $B = B^i e_i$,

$$A \circ B = A^i B^j e_i \circ e_j = \sum A^i B^i \ .$$

As another application, we decompose any Pauli spinor with respect to a given unit spinor s into $u_\|$ and $u_\perp$ such that

$$u = u_\| + u_\perp$$

and

$$u_\| \wedge s = 0 , \quad u_\perp \circ s = 0 .$$

Then with respect to the Clifford product $u_\|$ commutes with s while $u_\perp$ anticommutes with s, i.e.,

$$u_\| s = s u_\| , \quad u_\perp s = -s u_\perp .$$

One also remarks that $u_\| = \lambda s, \lambda = u_\| \circ s$ and

$$u_\| = (u_\| \circ s)s .$$

One also notices that $u_\|$ and $u_\perp$ are linearly independent: the proof is straightforward. The vector space E_3 is a direct sum of a space denoted by $\perp$ and another denoted by $\|$.

One may also define even and odd Pauli spinors: consider an idempotent element $\Sigma_+ = \frac{e+e_3}{2}$ of C_3 with

$$(\Sigma_+)^2 = \left(\frac{e + e_3}{2} \right)^2 = \frac{e + e_3}{2} ,$$

and a second $\Sigma_- = \frac{e-e_3}{2}$; then an even or odd spinor is u is defined by $u_\pm = \Sigma_\pm u$. One verifies also that

$$e_3 u_+ = e_3 \Sigma_+ u = \Sigma_+ u = u_+$$

and that $e_3 u_- = -u_-$. It is clear that $u_+ + u_- = u$ and that

$$\Sigma_+ \Sigma_- = 0 , \quad \Sigma_+ u_+ = u_+ , \quad \Sigma_- u_+ = 0 ,$$
$$\Sigma_- u_- = u_- , \quad \Sigma_- u_+ = 0 .$$

*These last properties can also be readily expressed in the language of abstract algebra: let A be a **ring** (i.e. an algebra without the external operation: multiplication of its elements by real numbers); a subring $\mathfrak{I} \subset A$ is an **ideal** if i_1, i_2 are elements of $\mathfrak{I}$, with $i_1 - i_2 \in \mathfrak{I}$ and if $i \in A$ then $ia \in \mathfrak{I}$*

(right ideal) and ai $\in$ $\mathfrak{I}$ (left ideal). Following these definitions, the sets $\{u_+\}$ $\{u_-\}$ are both ideals; furthermore, they are linearly independent and

$$C_3 = \{u_+\} \oplus \{u_-\} \ .$$

We may look for a basis for the set of even spinors. Starting from $u = c^0 e + c^i e_i$ one gets

$$u_+ = \Sigma_+ u = c_0 \Sigma_+ e + c^i \Sigma_+ e_i \ .$$

A possible basis is thus $\{\Sigma_+, \Sigma_+ e_1, \Sigma_+ e_2\}$; for the odd Pauli spinors a possible basis is $\{\Sigma_-, \Sigma_- e_1, \Sigma_- e_2\}$, and these two bases are orthogonal.

(e) We consider the real Clifford algebra C_4 with $\{e, e_0, e_1, e_2, e_3\}$ as a basis. Although, as far as the preceding results are concerned, we rarely used the whole Clifford product, we may, besides $e_i \circ e_j$, also define $e_i \wedge e_j$ in accordance with the well-known formula valid for Pauli matrices,

$$[\sigma_i, \sigma_j]_- = 2i\varepsilon_{ij}{}^k \sigma_k \ ,$$

and write

$$e_i \wedge e_j = i\varepsilon_{ij}{}^k e_k \ .$$

Then the Clifford product becomes

$$e_i e_j = e_i \circ e_j + e_i \wedge e_j \ .$$

We also note a connection with the product $\mathbf{a} \wedge \mathbf{b}$ of elementary vector calculus:

$$AB = A^i B^j e_i \wedge e_j = iA^i B^j \varepsilon_{ij}{}^k e_k$$
$$= i(A^2 B^3 - A^3 B^2)e_1 + i(A^3 B^1 - A^1 B^3)e_2 + i(A^1 B^2 - A^2 B^1)e_3 \ .$$

To any real vector space E_4 with $\{e, e_0, e_1, e_2, e_3\}$ as basis, we may associate a Clifford algebra C_4 by defining a product

$$e_\alpha \circ e_\beta = \eta_{\alpha\beta} e \ ,$$

where e is the unit element of C_4 and $\eta_{\alpha\beta}$ is the Lorentz metric already defined. Such a basis can obviously be realized by the well-known Dirac matrices.

The elements $\{e, e_\alpha : \alpha = 0, 1, 2, 3\}$ are linearly independent. Suppose indeed that $\lambda e_\alpha + \mu e_\beta = 0$ with $\lambda, \mu \in \mathbb{R}$. Then $(\lambda e_\alpha + \mu e_\beta) \circ e_\alpha = 0$, i.e., $\lambda e + \mu \eta_{\alpha\beta} e = 0$, and since $\alpha \neq \beta$, one concludes that $\lambda = 0$. We prove in the same way that $\mu = 0$ and that e is independent of the elements e_α.

The number of dimensions of C_4 is $2^4 = 16$: in the following tableau

$$e \, , \{e_\alpha\}$$
$$\{e_\alpha e_\beta : \alpha \neq \beta\}$$
$$\{e_\alpha e_\beta e_\gamma : \alpha \neq \beta \neq \gamma\}$$
$$e_0 e_1 e_2 e_3 \, ,$$

the sets have respectively 1 element, 4, 6, 4, elements, and again 1 element, a total of 16 elements which also are linearly independent. Let us show this last property for $e_0 e_1$ and $e_2 e_3$, for instance. Since they are all elements of C_4 we may form $e_0 e_1 \wedge e_2$ and $e_2 e_3 \wedge e_2$. The last expression vanishes, so that $\lambda e_0 e_1 + \mu e_2 e_3 = 0$ has obviously $\lambda = 0 = \mu$ as a consequence. Let us consider a special class of elements of C_4 which are also vectors of C_4. Then for two such elements ψ and φ,

$$\psi \circ \varphi = \psi^\mu \varphi^\nu e_\mu \circ e_\nu = \psi^\mu \varphi^\nu \eta_{\mu\nu} = (\psi^0 \varphi^0 - \Sigma \psi^i \varphi^i) e \, ;$$

the product is the Lorentz product of special relativity. Several of the comments on Pauli spinors apply equally well to Dirac spinors; the comments are left to the reader. For the use of Dirac spinors in physics, the interested reader should consult any textbook on quantum mechanics.

(f) Consider again the C_3 algebra as developed in (d). Let s, α, n^i be real numbers and $\hat{n} = n^i e_i$ such that

$$\hat{n}^2 = \hat{n}\hat{n} = \hat{n} \circ \hat{n} = n^i n^j e_i \circ e_j = \Sigma (n^i)^2 e \, .$$

We shall suppose that $\Sigma (n^i)^2 = 1$, then

$$\hat{n}^2 = e \, , \quad \hat{n}^3 = \hat{n}\hat{n}^2 = \hat{n} \, , \dots$$

A quaternion is defined as an element of C_3,

$$q = se + i\alpha\hat{n} \, .$$

We shall assign to any quaternion a norm and an inverse:

$$\|q\|^2 = s^2 + \alpha^2 \, , \quad q^{-1} = \frac{se - i\alpha\hat{n}}{\|q\|^2} \, .$$

We also define the unit quaternion

$$Q = \cos \omega\, e + i\hat{n} \sin \omega \ ,$$

with $Q^*Q = 1$. If we develop $\cos\theta$ and $\sin\theta$,

$$Q = \left(1 - \frac{\omega^2}{2} + \frac{\omega^4}{4!} - \dots\right) + i\hat{n}\left(\omega - \frac{\omega^3}{3!} + \frac{\omega^5}{5!} - \dots\right) \ .$$

But as seen above

$$\hat{n}\omega^3 = \hat{n}^3\omega^3 = (\hat{n}\omega)^3 \ , \quad \hat{n}\omega^5 = (\hat{n}\omega)^5, \dots \ ,$$

so

$$\begin{aligned}
Q =&(\cos\omega + i\hat{n}\sin\omega)e = 1 - \frac{\omega^2}{2!} + \frac{\omega^4}{4!} - \dots \\
&+ i\left(\hat{n}\omega - \frac{(\hat{n}\omega)^3}{3!} + \dots\right) = e^{i\hat{n}\omega} \ .
\end{aligned}$$

One interesting application of the quaternions concerns the representation of the rotation in E_3 (see Problem 2.9).

Remarks: In the representation where $e_i = \sigma_i$, a quaternion becomes a 2×2 matrix:

$$q = \alpha I + i\beta n^i \sigma_i \ ,$$

with α, β real. One defines the scalar product $\langle q, q' \rangle$ of two quaternions as

$$\langle q, q' \rangle = \tilde{q}q' \ ,$$

where $\tilde{q}$ is the adjoint matrix of q (complex, transposed). The norm $\|q\|$ can be derived from the scalar product

$$\|q\|^2 = \langle q, q \rangle = \alpha^2 + \beta^2 \ .$$

One often avoids the introduction of the complex i by introducing three generators i, j, k (do not confuse this i with the complex i!) and requiring that

$$i^2 = j^2 = k^2 = -1$$
$$ij = -ji = k \ , \quad jk = -kj = i \ , \quad ki = -ik = j \ .$$

Denoting by e the unit element, a quaternion takes the form

$$q = ae + bi + cj + dk \; .$$

Problem 2.9.

One cannot speak about rotations and the group of rotations without being more precise. In Euclid's geometry, it is readily shown that the product of two rotations around the same center in a plane $\mathbf{P} \subset \mathcal{E}_3$ is a planar rotation and so is the product of two rotations in $\mathcal{E}_3$ around the same axis. Such rotations build up, indeed, a commutative (abelian) group. But the product of two plane rotations around two different centers is a rotation provided the half-sum of the rotation angles is different from $k\pi(k$ integer) and the set of such transformations is a non-abelian group. If this half-sum is equal to $k\pi$, the resulting transformation is a translation. Similar remarks apply to rotations in $\mathcal{E}_3$: if the two rotation axes Δ and Δ' are in the same plane $P \subset \mathcal{E}_3$ and if they go through the same point, the resulting transformation is a rotation. But if the two axes Δ and Δ' are parallel, the resulting transformation is a translation. Finally, if Δ and Δ' are not in the same plane, the resulting transformation is a displacement in $\mathcal{E}_3$ and one may also prove that the displacements in $\mathcal{E}_3$ build up a group. Such considerations can be found in any textbook on the kinematics of the solid body. Note also that linear isometries in $\mathcal{E}_3$ (i.e., linear transformations which leave invariant the length of any vector of $\mathcal{E}_3$) also build up a group.

Rotations in $\mathcal{E}_3$: a geometric and algebraic introduction
Let E_2, E_3 be real vector spaces with $\{+, +\}$ and $\{+, +, +\}$ as the respective signatures and $\mathcal{E}_2, \mathcal{E}_3$ their associated affine spaces.

   (a) A rotation in $\mathcal{E}_3$ is a linear isometry in $\mathcal{E}_3$ with a fixed axis Δ. Let M be any point of $\mathcal{E}_3$ and M' its image through such a transformation. Consider a plane $\mathbf{P} \subset \mathcal{E}_3$ orthogonal to Δ and going through M, M' and let the point $\mathrm{O} = \mathbf{P} \cap \Delta$ be the intersection of $\mathbf{P}$ with Δ. To any point A of Δ (A being invariant under the rotation) correspond the vectors (bipoints) $\mathbf{AM}$ and $\mathbf{AM'}$; it is clear that $AM = AM', OM = OM'$. The angle $\mathbf{AM}, \mathbf{AM'} = \omega$ characterizes, jointly with Δ, the given rotation. Consequently, three parameters are sufficient for its definition: 2 for the Δ axis (its unit vector) and the angle ω. These 3 parameters can be considered as forming the triplet $\boldsymbol{\omega} = \{\omega^1, \omega^2, \omega^3\}$ vector of E_3 colinear to Δ. Any rotation may then be symbolized by $R(\boldsymbol{\omega})$. In a matrix notation $\mathbf{x} = A\mathbf{M}, \mathbf{x}' = A\mathbf{M}'$, one has

$$\mathbf{x}' = R(\boldsymbol{\omega})\mathbf{x} .$$

Decompose $\mathbf{x}'$ into its components along $\mathbf{x}, \boldsymbol{\omega}$ and $\boldsymbol{\omega} \wedge \mathbf{x}$. Show that

$$R(\boldsymbol{\omega}) = e^{\Omega} = e^{\boldsymbol{\omega} \cdot \mathbf{J}} ,$$

where $\mathbf{J}$ represents a triplet of 3×3 matrices $\{J_1, J_2, J_3\}$, called the **generators** of the rotation group.

(b) Special case $\boldsymbol{\omega} = \{0, 0, \omega^3\}$. Rotations in $\mathcal{E}_2$. Comment.

(c) **Infinitesimal rotation**, i.e. a rotation in an infinitesimal neighborhood of the identical rotation. Generators.

(d) Physical implications of the rotation theory. **Particles of spin 1**.

(e) Quaternionic representation of rotations in $\mathcal{E}_3$, application to the product of two rotations whose axes go through the same point $0 \in \mathcal{E}_3$.

Solution:

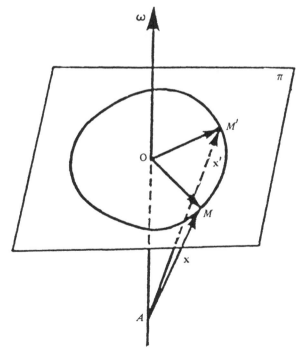

(a) Decompose the vector $\mathbf{x}$ *into two components,*

$$\mathbf{x} = \mathbf{x}_{\parallel} + \mathbf{x}_{\perp} ,$$

being elements of two supplementary vectors spaces: the axis Δ *and the vectorial plane* $\mathbf{P}$ *orthogonal to* Δ. *Then*

$$\mathbf{x}_{\parallel} = \left(\mathbf{x} \cdot \frac{\boldsymbol{\omega}}{\omega}\right) \frac{\boldsymbol{\omega}}{\omega} = \frac{\mathbf{x}\boldsymbol{\omega}}{\omega^2}\boldsymbol{\omega} ,$$
$$R(\boldsymbol{\omega})\mathbf{x}_{\parallel} = \mathbf{x}_{\parallel} .$$

We now seek the expression of $\mathbf{x}'_{\perp} = R(\boldsymbol{\omega})\mathbf{x}_1$. *We notice that* $\mathbf{x}'_{\perp}$ *makes the angle* ω *with* $\mathbf{x}_{\perp}$ *and that* $\boldsymbol{\omega} \wedge \mathbf{x}_{\perp}$ *is orthogonal to* $\mathbf{x}_{\perp}$. *The component of* $\mathbf{x}'_{\perp}$ *along* $\mathbf{x}_{\perp}$ *is then*

$$\|\mathbf{x}'_{\perp}\| \cos\omega \frac{\mathbf{x}_{\perp}}{\|\mathbf{x}_{\perp}\|} = \mathbf{x}_{\perp} \cos\omega ,$$

since $\|\mathbf{x}'_{\perp}\| = \|\mathbf{x}_{\perp}\|$; *its component along* $\boldsymbol{\omega} \wedge \mathbf{x}_{\perp}$ *can be written as*

$$\|\mathbf{x}'_{\perp}\| \sin\omega \frac{\boldsymbol{\omega} \wedge \mathbf{x}_{\perp}}{\|\boldsymbol{\omega} \wedge \mathbf{x}_{\perp}\|} = \boldsymbol{\omega} \wedge \mathbf{x}_{\perp} \frac{\sin\omega}{\omega} ,$$

and finally

$$\mathbf{x}'_{\perp} = \mathbf{x}_{\perp} \cos\omega + \boldsymbol{\omega} \wedge \mathbf{x}_{\perp} \frac{\sin\omega}{\omega}$$
$$= (\mathbf{x} - \mathbf{x}_{\parallel}) \cos\omega + \boldsymbol{\omega} \wedge (\mathbf{x} - \mathbf{x}_{\parallel}) \frac{\sin\omega}{\omega} .$$

Making use of the second formula at the beginning of this question, one sees that

$$\mathbf{x} = R(\boldsymbol{\omega})\mathbf{x} = R(\boldsymbol{\omega})\mathbf{x}_{\parallel} + R(\boldsymbol{\omega})\mathbf{x}_{\perp}$$
$$= \mathbf{x}_{\parallel} + (\mathbf{x} - \mathbf{x}_{\parallel}) \cos\omega + \boldsymbol{\omega} \wedge (\mathbf{x} - \mathbf{x}_{\parallel}) \frac{\sin\omega}{\omega}$$
$$= (1 - \cos\omega)\mathbf{x}_{\parallel} + \mathbf{x}_{\parallel} \cos\omega + \boldsymbol{\omega} \wedge \mathbf{x} \frac{\sin\omega}{\omega} ,$$

and, finally,

$$\mathbf{x}' = (1 - \cos\omega)\frac{\mathbf{x} \cdot \boldsymbol{\omega}}{\omega^2}\boldsymbol{\omega} + \mathbf{x} \cos\omega + \boldsymbol{\omega} \wedge \mathbf{x} \frac{\sin\omega}{\omega} .$$

We now want to bring the map $R(\boldsymbol{\omega})$ (a 3×3 matrix) in an exponential form. We introduce the three matrices

$$J_1 = \begin{pmatrix} 0 & 0 & 0 \\ 0 & 0 & 1 \\ 0 & -1 & 0 \end{pmatrix} \; , \quad J_2 = \begin{pmatrix} 0 & 0 & -1 \\ 0 & 0 & 0 \\ 1 & 0 & 0 \end{pmatrix} \; , \quad J_3 = \begin{pmatrix} 0 & -1 & 0 \\ 1 & 0 & 0 \\ 0 & 0 & 0 \end{pmatrix} \; ,$$

and, by a simple calculation, we verify that they satisfy the following commutation relations:

$$[J_i, J_j] = -\varepsilon_{ij}{}^k J_k \; .$$

We then calculate the matrix $\Omega = \omega^k J_k$ and verify first that

$$\Omega = \begin{pmatrix} 0 & -\omega^3 & -\omega^2 \\ \omega^3 & 0 & \omega^1 \\ \omega^2 & -\omega^1 & 0 \end{pmatrix} \; ,$$

and also that

$$\forall \mathbf{x} : \Omega \mathbf{x} = -\boldsymbol{\omega} \wedge \mathbf{x} \; .$$

Furthermore, one has

$$\Omega^2 \mathbf{x} = -\Omega(\boldsymbol{\omega} \wedge \mathbf{x}) = \boldsymbol{\omega} \wedge \boldsymbol{\omega} \wedge \mathbf{x} = -\mathbf{x}\omega^2 + (\boldsymbol{\omega} \cdot \mathbf{x})\boldsymbol{\omega} \; ,$$
$$\Omega^3 \mathbf{x} = -(\Omega \mathbf{x})\omega^2 + (\Omega \boldsymbol{\omega})(\boldsymbol{\omega}\mathbf{x}) = -\omega^2 \Omega \mathbf{x} \; ,$$
$$\ldots$$

and, since $\mathbf{x}$ can be arbitrarily chosen, one has

$$\Omega^3 = -\omega^2 \Omega \; .$$

Multiplying both sides by Ω, one obtains $\Omega^4, \Omega^5, \ldots$ and, more generally,

$$\Omega^4 = -\omega^2 \Omega^2 \; , \qquad\qquad \Omega^5 = -\omega^2 \Omega^3 = \omega^4 \Omega \; ,$$
$$\Omega^6 = \omega^4 \Omega^2 \; , \qquad\qquad \Omega^7 = \omega^4 \Omega^3 = -\omega^6 \Omega \; ,$$
$$\ldots$$
$$\Omega^{2n+1} = (-1)^n \omega^{2n} \Omega \; , \quad \Omega^{2n+2} = (-1)^{n+2} \omega^{3n} \Omega^2 \; .$$

We now consider the matrix e^Ω; using the two last formulae, we may write

$$e^\Omega = I + \Omega + \frac{\Omega^2}{2!} - \frac{\omega^2 \Omega}{3!} - \frac{\omega^2 \Omega^2}{4!} + \ldots$$
$$= I + \Omega \left(1 - \frac{\omega^2}{3!} + \frac{\omega^4}{4!} - \frac{\omega^6}{7!} + \ldots \right)$$
$$+ \Omega^2 \left(\frac{1}{2} - \frac{\omega^2}{4!} + \frac{\omega^4}{6!} - \ldots \right) \; .$$

The last bracket appears as an expansion of $\frac{1-\cos\omega}{\omega^2}$ and e^{Ω} takes the form

$$e^{\Omega} = I + \Omega\frac{\sin\omega}{\omega} + \Omega^2\frac{1-\cos\omega}{\omega^2} \ .$$

We compare $R(\boldsymbol{\omega})\mathbf{x}$ and $e^{\Omega}\mathbf{x}$ and conclude

$$R(\boldsymbol{\omega}) = e^{\Omega} = e^{\omega^k J_k} \ .$$

(b) In the special case $\mathcal{E}_2, \boldsymbol{\omega} = \{0, 0, \omega^3\}, \omega^3 = \omega(\omega = \pm\|\boldsymbol{\omega}\|)$,

$$\Omega = \omega J_3 \ , \quad \Omega^2 = \omega^2 (J_3)^2 \ ,$$

then

$$\Omega^2 = -\omega^2 \begin{pmatrix} 1 & 0 & 0 \\ 0 & 1 & 0 \\ 0 & 0 & 0 \end{pmatrix} \ ,$$

$$e^{\Omega} = \begin{pmatrix} 1 & 0 & 0 \\ 0 & 1 & 0 \\ 0 & 0 & 1 \end{pmatrix} + \sin\omega \begin{pmatrix} 0 & -1 & 0 \\ 1 & 0 & 0 \\ 0 & 0 & 0 \end{pmatrix}$$

$$- (1 - \cos\omega) \begin{pmatrix} 1 & 0 & 0 \\ 0 & 1 & 0 \\ 0 & 0 & 0 \end{pmatrix}$$

$$= \begin{pmatrix} \cos\omega & -\sin\omega & 0 \\ \sin\omega & \cos\omega & 0 \\ 0 & 0 & 1 \end{pmatrix} \ ,$$

a rather familiar formula. Any rotation in $\mathcal{E}_2$ has a matrix

$$e^{\Omega} = \begin{pmatrix} \cos\omega & -\sin\omega \\ \sin\omega & \cos\omega \end{pmatrix} \ .$$

One may notice that the eigenvalues of e^{Ω}, given by

$$\lambda^2 - 2\lambda\cos\omega + 1 = 0 \ ,$$

are complex and so is the diagonal form of e^{Ω}. As a matter of fact, to the components of $R(\boldsymbol{\omega})\mathbf{x}$,

$$x'^1 = x^1 \cos\omega - x^2 \sin\omega \ , \quad x'^2 = x^1 \sin\omega + x^2 \cos\omega \ ,$$

we may associate $x'_\pm = x'^1 \pm ix'^2$ and verify that

$$x'_+ = e^{i\omega}x_+ \ , \quad x'_- = e^{-i\omega}x_-$$

to obtain the complex diagonal form of e^Ω

$$\begin{pmatrix} e^{i\omega} & 0 \\ 0 & e^{-i\omega} \end{pmatrix} \ .$$

(c) The identical rotation corresponds clearly to $\omega = 0$. Let $\delta\omega$ be $\{\delta\omega^1, \delta\omega^2, \delta\omega^3\}$, the corresponding rotation $e^{\delta\Omega}$ being defined in an infinitesimal neighborhood of the identity. Then

$$R(\delta\omega) = e^{\delta\Omega} = I + \delta\Omega = I + \delta\omega^k J_k$$

and again

$$\delta\Omega\mathbf{x} = -\delta\omega \wedge \mathbf{x} \ .$$

One also has

$$\mathbf{x}' = \mathbf{x} + \begin{pmatrix} 0 & -\delta\omega^3 & -\delta\omega^2 \\ \delta\omega^3 & 0 & \delta\omega^1 \\ \delta\omega^2 & -\delta\omega^1 & 0 \end{pmatrix}\mathbf{x}$$

$$= \mathbf{x} + \delta\Omega\mathbf{x}$$

with components

$$x'^i = x^i + \delta\omega^i_j x^j \ .$$

The matrix $\delta\Omega$ is then defined by $\delta\omega^i_j = \delta\omega^k$, where i, j, k are in a cyclic permutation of 1 2 3, $\delta\Omega$ is a skew-symmetric matrix $\delta\Omega^T = -\delta\Omega$, and J_1, J_2, J_3 are the generators of infinitesimal rotations. We shall come back later to this question using the Lie groups approach.

(d) Generators similar to J_1, J_2, J_3 occur in the theory of particles of spin 1. But since they are components of the observable spin, they should be Hermitian (not real skew-symmetric matrices). Then instead of J_1, J_2, J_3 as defined in question (a), one introduces

$$J_1 = \begin{pmatrix} 0 & 0 & 0 \\ 0 & 0 & -i \\ 0 & i & 0 \end{pmatrix} \ , \quad J_2 = \begin{pmatrix} 0 & 0 & i \\ 0 & 0 & 0 \\ -i & 0 & 0 \end{pmatrix} \ , \quad J_3 = \begin{pmatrix} 0 & i & 0 \\ -i & 0 & 0 \\ 0 & 0 & 0 \end{pmatrix} \ ,$$

three Hermitian matrices $(J_k = J_k = J_k^{*T})$.   Their commutator is now complex:

$$[J_i, J_j] = i\varepsilon_{ij}{}^k J_k$$

and so is the corresponding rotation operator

$$e^{i\Omega} = e^{i\omega^k J_k} \ .$$

We may add a few remarks about the range of variation of the rotation angle ω. It is clear that, given a rotation axis Δ, to any value of ω there corresponds a rotation. But, if we want a one-to-one correspondence between the rotations and the domain of variation of ω, we have to restrict their domain to $0, 2\pi$ and to exclude one of the end points (periodicity of 2π), limiting the variation to the semi-open interval $0, 2\pi$. Such a remark occurred before in Problem 2.2 where we dealt particularly with the concepts of manifolds, charts and atlases.

(e) The vocabulary we are going to use is that of Problem 2.8, the generators $\{e_1, e_2, e_3\}$ of C^3 will be taken equal to the Pauli matrices $\{\sigma_1, \sigma_2, \sigma_3\}$. We shall consider two vectors of E_3 (the associated vector space to $\mathcal{E}_3$): the vector $\mathbf{x} = \{x^1, x^2, x^3\}$ and a unit vector $\mathbf{a} = \{a^1, a^2, a^3\}$ such that $\|\mathbf{a}\| = 1$ (metric $+, +, +$). To these vectors there will correspond two pure quaternions denoted by $\hat{x}$ and $\hat{a}$,

$$\mathbf{x} \to \hat{x} \to x^i \sigma_i \ , \quad \mathbf{a} \to \hat{a} = a^i \sigma_i \ ;$$

we note that (we shall write indifferently l or e)

$$a^2 = \hat{a}\hat{a} = \hat{a} \circ \hat{a} = a^i a^j \sigma_i \sigma_j$$
$$= \frac{1}{2} a^i a^j [\sigma_i, \sigma_j]_+ = (\hat{a} \circ \hat{a})I = \|\mathbf{a}\|^2 I \ .$$

It is also clear that one may consider $\hat{x}$ and $\hat{a}$ as elements of C^3 and write

$$\hat{a} = \tilde{\hat{a}}\hat{a} = \|\mathbf{a}\|^2 I \ .$$

We now consider the following map $\hat{x} \to \hat{x}$,

$$\hat{x}' = e^{-\frac{1}{2}\omega\hat{a}} \hat{x} e^{\frac{1}{2}\omega\hat{a}} \ .$$

It has the following properties: the length of any vector $\mathbf{x}$ is invariant. Indeed, on the one hand,

$$\tilde{\hat{x}}'\hat{x}' = \|\hat{x}'\|^2 I = \|\hat{x}\|^2 I \ ,$$

and on the other

$$\tilde{x}'\,\hat{x}' = (e^{-\frac{1}{2}\omega\hat{a}}\hat{x}e^{\frac{1}{2}\omega\hat{a}})^\sim (e^{-\frac{1}{2}\omega\hat{a}}\hat{x}e^{\frac{1}{2}\omega\hat{a}})$$

$$= (e^{-\frac{1}{2}\omega\hat{a}}\hat{x}e^{\frac{1}{2}\omega\hat{a}})(e^{-\frac{1}{2}\omega\hat{a}}\hat{x}e^{\frac{1}{2}\omega\hat{a}})$$

$$= e^{-\frac{1}{2}\omega\hat{a}}\|\hat{x}\|^2 e^{\frac{1}{2}\omega\hat{a}} = \|\hat{x}\|^2 I = \|\mathbf{x}\|^2 I \; .$$

As a consequence, one concludes that the relations between x and x' represent an isometry leaving invariant the Δ axis (fixed axis) with a unit vector $\boldsymbol{\xi} \in E_3$. We want then to prove that

$$\hat{\xi} = e^{-\frac{1}{2}\omega\hat{a}}\hat{\xi}e^{\frac{1}{2}\omega\hat{a}} \; .$$

Indeed one has

$$\frac{1}{2}[e^{i\omega\hat{a}}, \hat{\xi}] = e^{i\omega\hat{a}} \wedge \boldsymbol{\xi} = 0 \; ,$$

which means that $\hat{\xi} = \lambda\hat{a}, \lambda \in \mathbb{R}$, corresponding to $\boldsymbol{\xi} = \lambda\mathbf{a}$. The rotation is completely defined in $\mathcal{E}_3$, once a point $0 \in \mathcal{E}_3$ on the axis of rotation has been chosen. It is also possible to obtain $R(\boldsymbol{\omega})\mathbf{x}$ in the form given in (a), but the calculations are lengthy.

As a final application of the quaternions, let us calculate the rotation angle ω'' and the axis $\mathbf{a}''$ of a rotation product of two rotations of angles ω, ω' of axes $\mathbf{a}$ and $\mathbf{a}'$. To the rotation $\omega'', \mathbf{a}''$ corresponds the factor $e^{\frac{1}{2}\omega''\hat{a}''}$, and to the product of rotations the factor $e^{\frac{1}{2}\omega\hat{a}}e^{\frac{1}{2}\omega'\hat{a}'}$. Then since the three exponential are unit quaternions, one has

$$I\cos\frac{\omega''}{2} + i\hat{a}''\sin\frac{\omega''}{2}$$

$$= \left(I\cos\frac{\omega}{2} + i\hat{a}\sin\frac{\omega}{2}\right)\left(I\cos\frac{\omega'}{2} + i\hat{a}''\sin\frac{\omega'}{2}\right) \; .$$

Expanding the product, one gets

$$I\cos\frac{\omega''}{2} + ia^k\sigma_k\sin\frac{\omega''}{2} = I\cos\frac{\omega}{2}\cos\frac{\omega'}{2} + ia'^i\sigma_i\cos\frac{\omega}{2}\sin\frac{\omega'}{2}$$

$$+ ia^j\sigma_j\sin\frac{\omega}{2}\cos\frac{\omega'}{2} - a^i a'^j\sigma_i\sigma_j\sin\frac{\omega}{2}\sin\frac{\omega'}{2}$$

and has to identify the real and imaginary parts of the two sides. An elegant way of doing so is to proceed as follows. Consider the last term,

$$a^i a'^j \sigma_i\sigma_j = a^i a'^j\left(\frac{1}{2}[\sigma_i,\sigma_j]_+ + \frac{1}{2}[\sigma_i,\sigma_j]_-\right)$$

$$= \mathbf{a}\cdot\mathbf{a}' + i\varepsilon_{ij}{}^k\sigma_k a^i a'^j = \mathbf{a}\cdot\mathbf{a}' + i(\mathbf{a}\wedge\mathbf{a}')^k\sigma_k \; .$$

Taking this last result into account, we carry out the Tr operation on the last but one result and get

$$\cos\frac{\omega''}{2} = \cos\frac{\omega}{2}\cos\frac{\omega'}{2} - \mathbf{a}\cdot\mathbf{a}\sin\frac{\omega}{2}\sin\frac{\omega'}{2}\ .$$

A second relation can be obtained by multiplying with σ'_m the two sides of the equation used and taking the Tr of both sides. Since

$$\sigma_m^2 = I\ ,$$

$$Tr\{\sigma_i\sigma_j\} = 2\delta_{ij}\ ,$$

$$\sigma_j\sigma_m = i\sigma_n\ ,$$

where for the last relation j, m, n represent a circular permutation of 1 2 3, one readily obtains

$$a''^m\sin\frac{\omega''}{2} = a'^m\cos\frac{\omega}{2}\sin\frac{\omega'}{2} + a^m\sin\frac{\omega}{2}\cos\frac{\omega'}{2}$$

$$= (\mathbf{a}\wedge\mathbf{a}')^m\sin\frac{\omega}{2}\sin\frac{\omega'}{2}\ ,$$

or, in vector form,

$$\mathbf{a}''\sin\frac{\omega''}{2} = \mathbf{a}\cos\frac{\omega}{2}\sin\frac{\omega'}{2} + \mathbf{a}\sin\frac{\omega}{2}\cos\frac{\omega'}{2}$$

$$- \mathbf{a}\wedge\mathbf{a}'\sin\frac{\omega}{2}\sin\frac{\omega'}{2}\ .$$

| Problem 2.10. | *Some usual groups of matrices* |

Let us first review some notations about the matrix calculus. We consider real or complex matrices with N rows and N columns (called $N \times N$ matrices) $A, B, \ldots$ with elements $A_j^i, B_j^i \ldots$ where i refers to the rows and j to the columns. A^* is the complex conjugate of A; A^T is the transposed matrix of A (rows exchanged with columns); $\tilde{A}$ is the transposed complex conjugate of $A : \tilde{A} = (A^*)^T = (A^T)^*$, i.e. $\tilde{A}_j^i = (A_i^j)^*$; $\det A$ is the determinant of A. The adjoint A of a real matrix is A^T. If $|\det A| = 1, A$ is said to be unimodular.

(a) Representation of the translation group in homogeneous coordinates in $\mathcal{E}_3$.

(b) Consider the group O(2) of all real 2×2 orthogonal matrices ($A^T = A^{-1}$); they define a linear isometry in a two-dimensional vector space $\mathcal{E}_2$ with $(+,+)$ as signature. Properties of the O(2) group. Determine the number of its parameters. Same questions for SO(2).

(c) Same questions for O(3) and SO(3).

(d) Same question for U(1) and SU(1). Consider then the C^N vector space endowed with the scalar product

$$x \cdot y = \langle x, y \rangle = \sum_{k=1}^{N} x^{k*} y^{k}$$

and the corresponding length of vector x

$$\|x\|^2 = \langle x, x \rangle = \sum_{k=1}^{N} |x^k|^2 \;.$$

Specialize N to 2: the 2×2 unitary matrices which map $C^2 \to C^2$, build up the group U(2) and define a linear isometry in C^2 with the above metric. Structures of the matrices of U(2) and SU(2). Group homeomorphism between SU(2) and SO(3). Relation between the Lie algebras of SO(3) and SO(2).

Solution:

(a) For any $\mathbf{x} \in E_2$, $\mathbf{X}$ *the translated* $\mathbf{x}$ *vector is*

$$\mathbf{X} = \mathbf{x} + \mathbf{a} \begin{cases} X = x + a_x \;, \\ Y = y + a_y \;, \end{cases}$$

in homogeneous coordinates $(x', y', t), (X', Y', t)$,

$$x' = tx \;, \quad y = ty, t \;; \quad X' = tX \;, Y' = tY, t \;;$$

one has

$$\begin{aligned} X' &= tX = t(x + a_x) = x' + ta_x \\ Y' &= tY = y' + ta_y \;, \end{aligned}$$

or

$$\begin{pmatrix} X' \\ Y' \\ t \end{pmatrix} = \begin{pmatrix} 1 & 0 & a_x \\ 0 & 1 & a_y \\ 0 & 0 & 1 \end{pmatrix} \begin{pmatrix} x' \\ y' \\ t \end{pmatrix} \;.$$

The generalization to a $\mathcal{E}_N$ space is straightforward.

(b) We consider first the group O(2) *(the proofs that the orthogonal matrices and special orthogonal matrices form the groups* O(2) *and* SO(2) *are elementary and supposed to be known by the reader). Since the matrices of this group are orthogonal then each of these matrices conserves the length of any vector of $\mathcal{E}_2$:*

$$\|A\mathbf{x}\| = \langle A\mathbf{x}, A\mathbf{x} \rangle = \langle A^T A, \mathbf{x} \rangle = \|\mathbf{x}\|^2 \ ,$$

a relation which in its turn implies $A^T A = I$, i.e., $A^{-1} = A^T$. As a first consequence, one has

$$\det\{A^T A\} = |\det A|^2 = 1 \ .$$

As a second consequence, the orthogonality condition gives rise to the following three equations,

$$(a_1^1)^2 + (a_1^2)^2 = 1 \ , \quad a_1^1 a_2^1 + a_1^2 a_2^2 = 0 \ ,$$
$$(a_2^1)^2 + (a_2^2)^2 = 1 \ .$$

We choose any one of the unknowns, say $a_1^1 = \pm\cos\varphi$, then one finds $a_1^2 = -a_1^1 = \pm\sin\varphi, a_2^2 = \cos\varphi$, the signs being chosen in such a way that $\det A = \pm 1$. One can then distinguish three cases of importance:

$$\det A = 1 : \ A(\varphi) = \begin{pmatrix} \cos\varphi & -\sin\varphi \\ \sin\varphi & \cos\varphi \end{pmatrix} = R(\varphi) \ ,$$

$$\det A = -1 \begin{cases} A(\varphi) = \begin{pmatrix} -\cos\varphi & \sin\varphi \\ \sin\varphi & \cos\varphi \end{pmatrix} = \begin{pmatrix} -1 & 0 \\ 0 & 1 \end{pmatrix} R(\varphi) \ , \\[12pt] A(\varphi) = \begin{pmatrix} \cos\varphi & -\sin\varphi \\ -\sin\varphi & -\cos\varphi \end{pmatrix} = \begin{pmatrix} 1 & 0 \\ 0 & -1 \end{pmatrix} R(\varphi) \ , \end{cases}$$

where $R(\varphi)$ is a rotation of angle φ,

$$\mathbf{x} = \begin{pmatrix} x \\ y \end{pmatrix} = \begin{pmatrix} \|\mathbf{x}\|\cos\alpha \\ \|\mathbf{x}\|\sin\alpha \end{pmatrix} \ ; \ \mathbf{x}' = R(\varphi)\mathbf{x} = \begin{pmatrix} \|\mathbf{x}\|\cos(\varphi+\alpha) \\ \|\mathbf{x}\|\sin(\varphi+\alpha) \end{pmatrix} \ .$$

The second and the third ones are respectively the product of a symmetry with respect to Oy *and* Ox *and a rotation angle φ. One may note that there is a fourth one,*

$$A(\varphi) = \begin{pmatrix} -1 & 0 \\ 0 & -1 \end{pmatrix} R(\varphi) = R(\varphi + \pi) \ ,$$

the product of a central symmetry and a rotation of angle φ: it is another rotation of angle $\varphi + \pi$.

If one considers an infinitesimal rotation $\varphi = \delta\varphi$, $R(\varphi)$ can be written as

$$R(\varphi) = I + \begin{pmatrix} 0 & -\delta\varphi \\ \delta\varphi & 0 \end{pmatrix} \ .$$

$R(\varphi)$ belongs to a group which includes rotations infinitesimally close to the 1 matrix: $R(\varphi)$ is an element of a Lie group called SO(2), but transformations with symmetries are not elements of any Lie group.

To any value of φ, there is a corresponding rotation; but since $R(\varphi + 2\pi) = R(\varphi)$, it is enough to limit the variation of φ between 0 and 2π (or $-\pi$ to $+\pi$). However if one wants to have a one-to-one correspondence (bijection) between $(0, 2\pi)$ and the rotation group, one of the two end-points must be excluded (semi-open interval). If, furthermore, one uses the language of manifolds and charts, we have already pointed out in Problem 2.1 that one chart is not enough to cover the group SO(2), one needs at least two charts.

(c) Consider first the group O(3), its elements being orthogonal matrices, we have as in question (b),

$$A^T A = I : \ A^{-1} = A^T$$
$$|\det A|^2 = 1 \ .$$

The formal structure of the matrices of O(3) being definitely more involved than that of the matrices of O(2), we shall look for their diagonalized form and investigate their eigenvalues λ^i. We notice first that the λ^i's are given by a 3×3 determinant with $A^i_j - \lambda\delta^i_j$ as elements and the condition that it vanishes. This process then gives rise to a third degree equation: such an equation has at least one real root, since the corresponding third degree polynomial takes for $\lambda = \pm\infty$ the values $\pm\infty$ and, on the other hand, complex roots are not at all excluded.

Let now $\mathbf{x}$ be an eigenvector of A:

$$A^i_j x^j = \lambda^j = \lambda^i x^i \ ,$$

then the relation between norms leads to

$$\lambda^{i*} \lambda^i \|\mathbf{x}\|^2 = \|\mathbf{x}\|^2 : |\lambda^i|^2 = 1 \ .$$

As a consequence of $|\det A|^2 = 1$ and $|\lambda^i| = 1$, we may point out the following three cases: either $\lambda^1 = \lambda^2 = \lambda^3 = 1$, corresponding to the unit element of the group, or $\lambda^1 = \lambda^2 = -1, \lambda^3 = 1$, such transformations involving symmetries, or finally

$$\lambda^1 = e^{i\alpha}, \quad \lambda_2 = e^{-i\alpha}, \quad \lambda^3 = 1 : \alpha \in \mathbb{R}.$$

It should also be remarked that each eigenvector defines a fixed vectorial axis of E_3.

We have still to solve the question of the number of independent elements of the matrices of $O(3)$: writing down the expression of $A^T A = 1$, we find 6 relations for the 9 elements of the given matrix. We are then left with 3 independent parameters. We now consider the group $SO(3)$ built up by 3×3 orthogonal matrices with $\det A = 1$: any of its elements is parameterized by the triplet $\boldsymbol{\omega} = \{\omega^1, \omega^2, \omega^3\}$, the angle $\omega = \|\boldsymbol{\omega}\|$ belonging to the semi-interval $0, 2\pi (0 \le \omega < 2\pi)$ as noticed before. The matrix corresponding to $\boldsymbol{\omega} = 0$ should be the unit matrix and elements of $SO(3)$ should exist in the neighborhood of $\boldsymbol{\omega} = 0$. Such a group of matrices has been obtained in Problem 2.9: its elements are matrices $e^{\Omega} = e^{\omega^k J_k}$ as defined there. This set builds up a group. Indeed, any of its matrices has a determinant equal to 1. Using a result proved in question 6 of (a) in Problem 2.11:

$$\det e^{\omega^k J_k} = e^{\omega^k \operatorname{Tr} J_k} = 1,$$

since $\operatorname{Tr} J_k = 0$. Furthermore,

$$(e^{\omega^k J_k})^T = e^{\omega^k (J_k)^T} = e^{-\omega^k J_k},$$

because of the antisymmetric character of the matrices $J_k : J_k{}^T = -J_k$. This is a property which shows that e^{Ω} is an orthogonal matrix.

We have then sketched a preliminary intuitive approach to the theory of the Lie group $SO(3)$.

$SO(2)$ and $SO(3)$ are also often called rotation groups in 2 or 3 dimensions. It should be remarked that the geometric rotations they represent are around the origin or an axis going through the origin.

(d) The groups C and $U(1)$ are respectively the groups of all complex numbers z and all complex numbers such that $|z| = 1$. The group $U(1)$ is the set of all complex numbers $z = e^{i\omega^0}, \omega^0 \in R$ (one real parameter Lie

group). We consider the next group $U(2)$: *the unitarity condition* $\tilde{U}U = 1$, *i.e.,*

$$\tilde{U}U = \begin{pmatrix} a^* & c^* \\ b^* & d^* \end{pmatrix} \begin{pmatrix} a & b \\ c & d \end{pmatrix} = \begin{pmatrix} 1 & 0 \\ 0 & 1 \end{pmatrix}$$

gives rise to the following 3 complex relations,

$$|a|^2 + |c|^2 = 1 , \quad a^*b + c^*d = 0 ,$$
$$|b|^2 + |d|^2 = 1 .$$

The first and last equations are real, the middle one complex and gives rise to 2 real equations. But U *depends on 8 real quantities: therefore the group* $U(2)$ *is a 4 real parameters group.*

An additional simplification will occur for $SU(2)$ *defined as the set of matrices* $\hat{U}$ *such that* $\det \hat{U} = 1$, *the unitarity condition which can also be written as* $\hat{U}^{-1} = \tilde{\hat{U}}$ *and brought into the following form:*

$$\frac{1}{\Delta} \begin{pmatrix} d & -b \\ -b & a \end{pmatrix} = \begin{pmatrix} a^* & c^* \\ b^* & d^* \end{pmatrix} : \quad \Delta = \det \hat{U} = 1 .$$

The matrix $\hat{U}$ *can be written as*

$$\hat{U} = \begin{pmatrix} a & b \\ -b^* & a^* \end{pmatrix} : \quad \det \hat{U} = |a|^2 + |b|^2 = 1 .$$

We can now make two remarks. Put first $a = x + iy, b = z + it$ *with* x, y, z, t *real numbers, then the condition* $\det \hat{U} = 1$ *amounts to*

$$x^2 + y^2 + z^2 + t^2 = 1 ,$$

which is the equation of a sphere S^3. *The second remark, also important, is more far-reaching. Define 3 real numbers* $\{\omega^1, \omega^2, \omega^3\}$ *and* $\omega^2 = (\omega^1)^2 + (\omega^2)^2 + (\omega^3)^2$. *Since* $\hat{U}$ *depends on 3 real parameters, we may write*

$$a = \cos\omega + i\frac{\sin\omega}{\omega}\omega^3 , \quad b = (\omega^2 + i\omega^1)\frac{\sin\omega}{\omega} ,$$

then $\hat{U}$ *takes the form*

$$\hat{U} = \begin{pmatrix} \cos\omega + i\frac{\sin\omega}{\omega}\omega^3 & (\omega^2 + i\omega^1)\frac{\sin\omega}{\omega} \\ -(\omega^2 - i\omega^1)\frac{\sin\omega}{\omega} & \cos\omega - i\frac{\sin\omega}{\omega}\omega^3 \end{pmatrix} ,$$

and the condition $\det U = 1$ *is clearly satisfied.*

On the other hand consider the three Pauli Hermitian matrices

$$\sigma_1 = \begin{pmatrix} 0 & 1 \\ 1 & 0 \end{pmatrix} \;, \quad \sigma_2 = \begin{pmatrix} 0 & -i \\ i & 0 \end{pmatrix} \;, \quad \sigma_3 = \begin{pmatrix} 1 & 0 \\ 0 & -1 \end{pmatrix} \;.$$

These matrices have well-known properties:

$$\text{Tr } \sigma_i = 0 \;, \qquad\qquad [\sigma_i, \sigma_j]_+ = \sigma_i\sigma_j + \sigma_j\sigma_i = 2\delta_{ij} \;,$$
$$\text{Tr}\{\sigma_i\sigma_j\} = 2\delta_{ij} \;, \qquad\qquad [\sigma_i, \sigma_j]_- = 2i\varepsilon_{ij}k_{\sigma_k} \;,$$
$$\sigma_i\sigma_j = i\sigma_k \;.$$

In the last formula the indices i, j, k build up a circular permutation of 1,2,3. We now study the matrix $e^{i\omega^k \sigma_k}$. It is a well-defined matrix for any value of the real triplet $\{\omega^1, \omega^2, \omega^3\}$.

We notice that

$$(\omega^i \sigma_i)^2 = \omega^2 I : \; \omega = \|\boldsymbol{\omega}\| = \sqrt{(\omega')^2 + (\omega^2)^2 + (\omega^3)^2} \;,$$

and that $e^{i\omega^k \sigma_k}$ is defined by the convergent series

$$e^{i\omega^k \sigma k} = I + i\omega^k \sigma_k - \frac{(\omega^k \sigma_k)^2}{2!} - \frac{i(\omega^k \sigma_k)^3}{3!} + \cdots \;,$$

which is the sum of two series: an even and an odd one. The even one reads

$$I - \frac{\omega^2}{2!}I + \frac{\omega^4}{4!}I - \ldots (\cos\omega)I \;,$$

while the odd one is

$$i\left(\omega^k \sigma_k - \frac{(\omega^k \sigma_k)^3}{3!} + \frac{(\omega^k \sigma_k)^5}{5!} - \cdots\right)$$
$$i\omega^k \sigma_k \left(1 - \frac{\omega^2}{3!} + \frac{\omega^4}{6!} - \cdots\right) = i\omega^k \sigma_k \frac{\sin\omega}{\omega} \;,$$

and finally

$$e^{i\omega^k \sigma_k} = I\cos\omega + i\omega^k \sigma_k \frac{\sin\omega}{\omega} \;.$$

We may now use the above given expressions of the Pauli matrices and readily show that

$$\hat{U} = e^{i\omega^k \sigma_k} \;.$$

One has to admit that the proof of this last equality appears as definitely ad hoc but we will introduce this same relation in the next problem using Lie group theory.

We may conclude with some considerations about the U(2) group. Since its elements depend on 4 parameters, any matrix of this group can be obtained from a matrix of SU(2) by multiplication with the matrix $e^{i\omega^0 I}, \omega^0 \in \mathbb{R}$. Then any matrix $U \in U(2)$ has the form

$$U = e^{i\omega^0 I} e^{i\omega^k \sigma_k} = e^{i(\omega^0 I + \omega^k \sigma_k)} \; .$$

since I commutes with all matrices. It then appears that

$$SU(2) = U(2)/U(1) \; ,$$

i.e. SU(2) is the quotient group of U(2) by U(1) (which is also called the phase group by physicists).

We now examine the relation between SU(2) and SO(3). We first notice that any Hermitian 2×2 matrix is of the form

$$H = \begin{pmatrix} a & c+id \\ c-id & b \end{pmatrix} \; ,$$

where a, b, c, d are real numbers. Then a traceless Hermitian 2×2 matrix corresponds to $b = -a$.

Define now a map h between $\mathbb{R}^3$ and the vector space of traceless Hermitian, 2×2 matrices such that to any vector $\mathbf{x} \in \mathbb{R}^3, \mathbf{x} = \{x, y, z\}$, there corresponds the matrix

$$\mathbf{x} \rightarrow h(\mathbf{x}) = \begin{pmatrix} z & x+iy \\ x-iy & -z \end{pmatrix} \; ,$$
$$\|\mathbf{x}\|^2 = -\det h(\mathbf{x}) \; .$$

Consider a second map $r : \mathbb{R}^3 \overset{r}{\rightarrow} \mathbb{R}^3$ such that to any $\mathbf{x} \in \mathbb{R}^3$ and $\hat{U} \in$ SU(2) there corresponds an $\mathbf{x} \in \mathbb{R}^3$:

$$\mathbf{x}' = r\mathbf{x} : \; h(\mathbf{x}') = \hat{U} h(\mathbf{x}) \hat{U}^{-1} \; ,$$
$$\|\mathbf{x}'\|^2 = -\det h(\mathbf{x}') = -\det \hat{U} \det h(\mathbf{x}) (\det \hat{U})^{-1}$$
$$= -\det h(\mathbf{x}) = \|\mathbf{x}\|^2 \; .$$

As a consequence r is a rotation in $\mathbb{R}^3$, an element of $SO(3)$; and also an element of a group of homomorphisms between $SU(2)$ and $SO(3)$.

The map so defined is clearly a group homomorphism, but it is not a one-to-one correspondence; it is, as we shall see, a two-to-one correspondence. Choose indeed an $\mathbf{x} \in \mathbb{R}^3$ and a rotation r to which may correspond one or more matrices $\hat{U} \in SU(2)$. Let us specialize r to be the identical rotation and $\hat{U}_I$ one of the corresponding elements of $SU(2)$, then

$$\mathbf{x}' = \mathbf{x} : \quad h(\mathbf{x}) = \hat{U}_I h(\mathbf{x}) U_I^{-1} ,$$

and U_I commutes with all the Hermitian traceless matrices. It then commutes with the three Pauli matrices defined in the present question (d). Then, it may be inferred from Schurr's lemma (well-known to all physicists) that U_I is a multiple λ of the unit matrix. But since $U_I U_I = I$: $\lambda \lambda^* = 1, \lambda = \lambda^*, U_I = \pm 1$. For a physicist, this is a well-known property of the $1/2$ representation of the rotation group: a 2π rotation changes the sign of a Pauli spinor.

Finally, we turn our attention to the Lie relations for $SO(3)$ and $SU(2)$. In Problem 2.9(a), we wrote the commutation relations of the J_k matrices:

$$[J_i, J_j] = -\varepsilon_{ij}{}^k J_k ,$$

where all J_k are real matrices. In order to compare these relations with the ones relative to $SU(2)$, we introduce complex matrices

$$J_k^{(e)} = -iJ_k ,$$

and the commutation relations become

$$[J_i^{(c)}, J_j^{(c)}] = i\varepsilon_{ij}{}^k J_k .$$

The J_k matrices, since they are linearly independent and form the basis of a certain 3-dimensional vector space which, endowed with the internal operation [,] becomes an algebra: such an algebra is the Lie algebra of $SO(3)$ with structure coefficients $\varepsilon_{ij}{}^k$ (or $i\varepsilon_{ij}{}^k$). But, as noticed in Sec. 26, Level 2, under certain conditions, the Lie algebra of a group determines uniquely the group (see, for instance, remark 4 after formula (26.29)). Consequently, starting only from commutation relations we then may speak of the Lie group $SO(3)$.

We now come to the group SU(2), *the corresponding rotation is* $e^{i\omega^k \sigma_k}$. *In the commutator of the* σ_i *matrices,*

$$[\sigma_i, \sigma_j] = 2i\varepsilon_{ij}{}^k \sigma_k \ ,$$

we get rid of the factor 2 by defining

$$\tau_k = \frac{\sigma_k}{2} \ ,$$

then

$$[\tau_i, \tau_j] = i\varepsilon_{ij}{}^k \tau_k \ .$$

The τ_i *again build an algebra, the Lie algebra of* SU(2), *and have exactly the same commutator as the* $J_i^{(c)}$: SO(3) *and* SU(2) *are representations of the rotation group in a 3-dimensional vector space with the metric* $(+, +, +)$. *But there is an important difference between these two representations. Consider any matrix* $U \in$ SU(2). *To a rotation angle* ω, *there corresponds the rotation*

$$\boldsymbol{\omega} = \mathbf{n}\|\boldsymbol{\omega}\| = \mathbf{n}\omega \ : \ \mathbf{n}^2 = 1$$

and $\hat{U}$ *is now*

$$\hat{U}(\omega) = e^{i\omega^k \sigma_k} = e^{i\omega \mathbf{n} \cdot \boldsymbol{\sigma}} e^{i2\omega \mathbf{n} \cdot \boldsymbol{\tau}} \ .$$

We define: $\theta = 2\omega, \boldsymbol{\theta} = \theta\mathbf{n}$, *then* $\hat{U}$ *can also be written as a function of* θ:

$$\hat{U}(\theta) = e^{i\theta \mathbf{n} \cdot \boldsymbol{\tau}} \ .$$

We see that if $\theta = 0 : U(\omega = 0) = U(\theta = 0) = I$, *if* $\theta = 2\pi: U(\omega = \pi) = -I$ *while* $U(\theta = 2\pi) = I$, *and we find again the two-to-one correspondence as previously described.*

| Problem 2.11. | *Lie group of matrices*

All the groups studied in the preceding problem are Lie groups. We want to investigate their main properties, using now the vocabulary and concepts of paragraph 26, Level 2. We will first collect in question (a), some elementary and known properties of matrices.

(a)(i) Consider the set of $N \times N$ matrices, real or complex; they can be considered as elements of a certain vector space: endow such a space with a norm and a distance, and it becomes a metric vector space and a topological vector space.

(ii) Let A be any $N \times N$ matrix; show that the series corresponding to the matrix e^{tA} with $t \in \mathbb{R}$ is convergent and differentiable.

(iii) If A and B commute and α, β are real or complex numbers then

$$e^{\alpha A + \beta B} = e^{\alpha A} e^{\beta B} \ .$$

(iv) The matrix $A(t) = e^{tA}$ is invertible and

$$A^{-1}(t) = e^{-tA} \ .$$

(v) A^T being the transposed of the matrix A,

$$\left(e^{tA}\right)^T = e^{tA^T} \ .$$

(vi) If A is diagonalizable, then

$$\det e^{tA} = e^{t \mathrm{Tr} A} \ .$$

(b) Following the concepts in Sec. 26, Level 2, develop the general theory of Lie groups of matrices and their Lie algebras.

(c) Define the Lie groups:

$$
\begin{array}{llll}
\mathrm{GL}(N, \mathbb{R}) \ , & \mathrm{SL}(N, \mathbb{R}) \ , & \mathrm{GL}(N, \mathbb{C}), & \mathrm{SL}(N, \mathbb{C}) \\
\mathrm{O}(N) \ , & \mathrm{SO}(N) \ , & \mathrm{U}(N) \ , & \mathrm{SU}(N) \\
\mathrm{O}(p, q) & \mathrm{SO}(p, q) \ , & \mathrm{U}(p, q) \ , & \mathrm{SU}(p, q)
\end{array}
$$

and find their Lie algebras using the considerations developed in (b).

(d) Use the method developed at the beginning of (b) to discuss the following groups: SO(3), U(3), SU(3), SL(2, $\mathbb{R}$), SL(2, $\mathbb{C}$) and their generators and structure coefficients. Comment on the groups SU(3) and its physical applications.

(e) Using again the method of (b), seek a representation of the Lie algebra of SO(3) by means of differential operators. Physical meaning.

Solution:

(a) Let $\hat{E}$ be the vector space of all $N \times N$ matrices. We consider a bijective map $\varphi : \hat{E} \to \mathbb{R}^{N^2}$ such that to each matrix A there corresponds the set of all its elements $\{A_k^j \; j = 1 \ldots N, k = 1 \ldots N\}$. We define the norm $\|A\|$ by

$$\|A\|^2 = \sum_{j,k} |A^j{}_k|^2 \; .$$

It satisfies the characteristic conditions on the norm

$$\|A\| > 0 \;, \|A + B\| \leq \|A\| + \|B\| \;,$$
$$\|\alpha A\| = |\alpha| \|A\| \;,$$

α being a real or complex number. It furthermore satisfies also

$$\|AB\| \leq \|A\| \|B\| \;,$$

AB being the product of A by B following the multiplication law of matrices. The proofs of the first three inequalities are straightforward, the proof of the fourth inequality results from the Lagrange inequality for the two vectors x and y:

$$|\langle x, y \rangle|^2 < \|x\|^2 \|y\|^2 \; .$$

From the definition of the norm, one deduces the definition of the distance d,

$$d(A, B) = \|A - B\| \;,$$

and also the definition of open balls B in $\hat{E}$ of radius r,

$$B(A, r) = \{X \in \hat{E}; \; d(A, X) < r\} \;,$$

which represent neighborhoods of A.

(ii) We want to show that the series

$$e^{tA} = \sum \frac{t^n}{n!} A^n$$

is convergent. Consider any of its partial sums (up to p, for instance) and use the inequalities of (i) to show that

$$\| \sum_{}^{p} \frac{t^n}{n!} A^n \| \leq \sum_{}^{p} \frac{|t|^n}{n!} \|A\|^n \; .$$

The right-hand side tends to $e^{|t| \|A\|}$ for $p \to \infty$, then the left-hand side is convergent and absolutely convergent. The proofs of (iv) and (v) are straightforward.

We now come to (vi) where we supposed A to be diagonalizable (an Hermitian matrix for instance), then there exists an invertible matrix S such that SAS^{-1} is a diagonal matrix A_d. As a consequence, one has

$$Se^{tA}S^{-1} = \sum \frac{t^n}{n!} \underbrace{SSA^{-1}SAS^{-1}\ldots}_{n} = e^{tA_d}$$

and

$$\det\{Se^{tA}S^{-1}\} = \det S \det e^{tA} \det S^{-1} = \det e^{tA} \;,$$

since $\det S^{-1} = (\det S)^{-1}$, and $\mathrm{Tr}\{SAS^{-1}\} = \mathrm{Tr}\{SS^{-1}A\} = Tr\,A$, a relation which means that

$$\det e^{tA_d} = \det e^{tA} \;.$$

The matrix e^{tA_d} is diagonal with diagonal terms $e^{it\lambda_k}$, where λ_k is one of the terms of A_d. However, the determinant of a diagonal matrix is the product of its diagonal terms:

$$\det e^{tA} = \det e^{tA_d} = \prod_k e^{t\lambda_k}$$

$$= e^{t\Sigma\lambda k} = e^{t\mathrm{Tr}\,A} \;.$$

(b) We adopt the methods and vocabulary of Sec. 26, Level 2 and begin with the physicist's point of view as developed in Part A of this paragraph. Consider the set of $N \times N$ matrices $\{A_{(a)} = A_{(a^1\ldots a^r)}\}$, the r-uplet $\{a^1 \ldots a^r\}$ is real and the number r cannot exceed N^2 (which is the number of elements of $A_{(a)}$) since the parameter $\{a^1 \ldots a^r\}$ should be essential as mentioned in the same paragraph 26. The matrices $\{A_{(a)}\}$ are supposed to be elements of a group with respect to the multiplication law of matrices and we choose the parameters such that $A_{(0)} \neq I$, the unit element of the group. It is then clear that we have to require $\det A_{(a)} \neq 0$ for $A_{(a)}^{-1}$ to exist.

We now consider the corresponding Lie algebra. As we noticed in the comments following formula (26.28), since the structure coefficients are independent of the a^k we may choose $a^1 = a^2 = \ldots a^r = 0$, i.e., we study the Lie algebra in the neighborhood of $A_{(0)} = I$. Any element of the group in this neighborhood corresponds to the infinitesimal value. $\delta a = \{\delta a^1 \ldots \delta a^r\}$ of a. Then one has

$$A(\delta a) = I + \delta a^\lambda \frac{\partial A(a)}{\partial a^\lambda}\bigg|_{a=0} = I + \delta a^\lambda X_\lambda \;.$$

The generators X_λ are thus the $N \times N$ matrices

$$X_\lambda = \left. \frac{\partial A(a)}{\partial a^\lambda} \right|_{a=0} .$$

We now come to the mathematician's point of view as developed in Part (B), Sec. 26. The making of a Lie group G from a set of matrices needs the introduction of two structures on their set: first a group structure which is easy and obvious to obtain, and also the structure of a differentiable manifold. The group structure has been defined above; the definition of a differentiable manifold $\mathcal{M}$ whose points are the matrices $\{A, A', A'' \ldots\}$ of the group requires some consideration.

We parametrize the point $A \in \mathcal{M}$ by its matrix elements A^i_j, then to each point of $\mathcal{M}$ there corresponds a set of N^2 coordinates $\{A^i_j : i = 1 \ldots N, j = 1 \ldots N\}$. The number N is reduced when there are auxiliary conditions on the matrices $A, A' \ldots$ (such as $\det A = 1$ in the case of SO(3)). We may otherwise express this correspondence by endowing $\mathcal{M}$ with a homeomorphism φ such that

$$\mathcal{M} \xrightarrow{\varphi} \mathrm{I\!R}^{N^2} : \; \varphi(A) = \{A^1_1 \ldots A^N_N\} ,$$

and the manifold acquires then N^2 as a dimension.

We already used that kind of parametrization in the same Sec. 26, Level 2: see the GL(N, R) example after formula (26.69) where we defined $\mathcal{M}_{N^2}$ as a metric vector space as it was done in question (a)(i).

Our next task will be the definition of the tangent vector space $T_\mathcal{M}(A)$ at any point $A \in \mathcal{M}_{N^2}$ and, more particularly, at the point $A_{(0)} = I \in \mathcal{M}_{N^2}$. We remark that if we suppose the r-uplet a to be a function of a real parameter t, then to any value of t there corresponds a point $A_{(a(t))}$ of $\mathcal{M}^2_N$:

$$t \in \mathrm{I\!R} \to A_{(a(t))} \in \mathcal{M}_{N^2} .$$

Following the definition we gave before (see formula (2.5)), we describe in this way a curve $\Gamma \subset \mathcal{M}_{N^2}$. Any point of Γ has for coordinates the N^2 functions

$$A^i_j(a(t)) = A^i_j(a^1(t) \ldots a^r(t)) : \; i, j = 1, \ldots N ,$$

and the tangent to Γ at the point $A_{a(t)}$ has locally the components

$$\frac{dA^i_j(a(t))}{dt} = \frac{\partial A^i_j(a)}{\partial a^\lambda} \frac{da^\lambda(t)}{dt} : \; \begin{matrix} i, j = 1, \ldots N \\ \lambda = 1 \ldots r \end{matrix} .$$

Any element of the tangent vector space at $A_{(a(t))}$ has for local components linear combinations of the r vectors

$$X^i_{j\lambda}(t) = \frac{\partial A^i_j(a)}{\partial a^\lambda} \ .$$

In other words, the r vectors x_λ with components $X^i_{j\lambda}(t)$ generate the tangent space at $A_{(a(t))} \in \mathcal{M}_{N^2}$, i.e., with $t = 0$ the vectors x_λ with $X^i_{j\lambda}(0)$ as components generate that vector space and the Lie algebra of the group. We may also notice that for t belonging to a certain neighborhood of $t = 0$, one has

$$A^i_j(a(t)) = \delta^i_j + t \frac{dA^i_j(a(t))}{dt}\Big|_{t=0} \ ,$$

which shows that any matrix of the group can be written in an exponential form

$$A_{(a(t))} = I + \mathcal{A} = e^{\mathcal{A}}$$

in a neighborhood of the matrix I.

(c) *For the study of the Lie algebras of the groups under consideration in the present question, we shall use the preceding remark.*

(i) GL*(N, $\mathbb{R}$) and SL(N, $\mathbb{R}$): groups of $N \times N$ matrices such that*

$$\det A \neq 0 \text{ for GL}(N, \mathbb{R})$$
$$\det A = 1 \text{ for S}(N, \mathbb{R}) \ ,$$

i.e., regular real matrices. Such groups are respectively N^2 and $N^2 - 1$ parameters groups. In the neighborhood of the element I, any matrix $\mathcal{A}$ is an element of the Lie algebra of GL(N, $\mathbb{R}$). For SL(N,$\mathbb{R}$) we remark that

$$\det A = e^{\text{Tr}\,\mathcal{A}} \ ,$$

which means that only traceless $N \times N$ matrices are elements of the Lie algebra of SL(N, $\mathbb{R}$).

(ii) *Groups GL(N, C) and SL(N, C): groups of all $N \times N$ complex matrices with respectively $\det A \neq 0$ and $\det A = 1$. They depend repectively on N^2 and $N^2 - 1$ complex parameters. For the characterization of their Lie groups, same conclusions as before.*

(iii) *Groups O(N) and SO(N): they are generalization of the O(2) and SO(2) groups previously studied. The O(N) group is defined by*

$$A^T A = I : \ |\det A|^2 = 1 \ ,$$

and for the SO*(N) group the last condition is replaced by*

$$\det A = 1 .$$

The number of its independent (essential) parameters can be determined as follows: the condition $A^T A = I$ *gives rise to* N^2 *equations:*

$$(A^T)^i_j A^j_{\ k} = \delta^i_k .$$

Among them, N *equations concern the columns (or rows) of* A:

$$(A^T)^i_j A^j_i = 1 \quad \textit{no summation on i! },$$

while the others,

$$(A^T)^i_j A^j_k = 0 : \ j = k ,$$

are not all independent: the equations in (i, k) *or* (k, i) *are the same. This leaves us with*

$$\binom{N}{2} = \frac{N(N-1)}{2}$$

distinct equations. We then have

$$N + \frac{N(N-1)}{2} = \frac{N(N+1)}{2} ,$$

conditions for N^2 *unknown quantities; finally there remain*

$$N^2 - \frac{N(N+1)}{2} = \frac{N(N-1)}{2} ,$$

independent parameters. The SO*(N) group has the same number of parameters. The* SO*(N) group has the same number of parameters.*

The Lie algebra of O*(N) should satisfy*

$$I = A^T A = A A^T = (e^{\mathcal{A}})^T e^{\mathcal{A}} = e^{\mathcal{A}^T} e^{\mathcal{A}} ,$$

which implies that $\mathcal{A}^T$ *should commute with* $\mathcal{A}$, *then*

$$e^{(\mathcal{A} - \mathcal{A}^T)} = I .$$

Thus, the elements of the Lie algebra of O*(N) are skew-symmetric matrices* $(\mathcal{A} = -\mathcal{A}^T)$ *and the same conclusion applies to* SO*(N).*

(iv) Groups U*(N) and* SU*(N): generalization of* U*(2) and* SU*(2). Both are characterized by*

$$\tilde{U}U = I \ , \quad |\det U|^2 = 1$$

and SU*(N) is subject to the supplementary condition* $\det U = 1$. *If we count as in (iii) the number of independent parameters which result after the condition* $\tilde{U}U = I$ *being satisfied, we find*

$$N + 2\frac{N(N-1)}{2} = N^2$$

complex parameters or $2N^2$ *real ones. For* SU*(N) we have one more condition to take into account, which reduces the number of parameters to* $N^2 - 1$ *complex independent ones. The conditions* $\tilde{U}U = I$ *or* $\tilde{\hat{U}}\hat{U} = I$ *for* SU*(N) have as consequences*

$$\sum_{i,j} |U_j^i|^2 = 1 \ ,$$

$$\sum_{i,j} |\hat{U}_j^i|^2 = 1 \ ,$$

which require that the matrix elements $|U_j^i|$ *and* $|\hat{U}_j^i|$ *are* ≤ 1 *for any i and j. Thus, the parameters of* U*(N) and* SU*(N) are restricted to vary over a finite range such that both groups are compact, i.e., none of their elements* $|U_j^i|$ *or* $|\hat{U}_j^i|$ *can grow indefinitely. Consequently, all their subgroup* O*(N) and* SO*(N) in particular are also compact.*

The elements of the Lie algebra of U*(N) denoted by* $\mathcal{A}$ *are such that*

$$I = U\tilde{U} = \tilde{U}U = e^{\mathcal{A}}e^{\tilde{\mathcal{A}}} = e^{\tilde{\mathcal{A}}}e^{\mathcal{A}} = e^{(\tilde{\mathcal{A}}+\mathcal{A})} \ ,$$

where we have used the property that $\mathcal{A}$ *and* $\tilde{\mathcal{A}}$ *should anticommute. Then* $\tilde{\mathcal{A}} = -\mathcal{A}$ *and the elements of the Lie algebra of* U*(N) are anti-hermitian matrices. Physicists prefer to deal with hermitian matrices, so they define* $a = iA$, *then* $\tilde{a} = a$ *and a becomes hermitian. It should also be noticed that all the above-mentioned groups are also Lie algebras since the commutator of two matrices is an internal operation of the algebra.*

The groups U*(p, q) and* O*(p, q) represent special applications of a more general case which can be stated as follows: consider the column matrices*

$$x = \begin{pmatrix} x^1 \\ \vdots \\ x^N \end{pmatrix} \ , \quad y = \begin{pmatrix} y^1 \\ \vdots \\ y^N \end{pmatrix} \ ,$$

where the x^k and y^k are complex or real numbers. Their scalar product is defined in matrix notation as

$$\langle x, y \rangle = \tilde{x} G y \ ,$$

for a given $N \times N$ real symmetric matrix G. The group $U(G)$ is the group of matrices $U \in U(G)$ such that if

$$x' = U x \ , \quad y' = U y \ ,$$

then the scalar product is preserved, i.e.,

$$\langle x', y' \rangle = \langle x, y \rangle \ .$$

It can be easily checked that such a condition implies that U should satisfy the relation*

$$\tilde{U} G U = G \ .$$

Let u be any element of the Lie algebra of such a group; using the exponential representation of U, one has

$$U = e^{\mathcal{U}} = I + \mathcal{U} \ .$$

The next to last one relation shows that

$$\tilde{\mathcal{U}} G + G \mathcal{U} = 0 \ .$$

But since G is real, symmetric one may also write

$$(G U)^{\sim} + G U = 0 \ .$$

The meaning of the last formula is simple: it means that there exists a map $\mathcal{U} \to \mathcal{U}(g)$ between the Lie algebra of $U(G)$ and the set of anti-hermitian matrices. If we restrict ourselves to real matrices $A \in O(G)$, the relation corresponding to the preceding one is

$$(G A)^T + G A = 0 \ ,$$

and the preceding map is between the Lie algebra of $O(G)$ and the set of skew-symmetric matrices.

---

*For U real, the conditions $U^T G U = G$, $G^T = -G$ define a real symplectic matrix U.

The groups $O(p,q)$ and $U(p,q)$ are special cases corresponding to G being a diagonal matrix with p positive elements and $q = N - p$ negative ones.

For the groups $SU(p,q)$ and $SO(p,q)$ we have to add the respective conditions

$$\mathrm{Tr}\,\mathcal{U} = 0 \;, \quad \mathrm{Tr}\,\mathcal{A} = 0 \;,$$

expressing their unimodularity.

We note also that a Lorentz matrix is an element of $O(1,3)$ corresponding to diagonal G with terms $(1,-1,-1,-1)$ (see Problem 1.18, question (e)).

(d) The present question concerning $SO(3)$ and $SU(2)$ has been practically solved in the two preceding problems and we may here be rather concise. Matrices of $SO(3)$ can be brought into the form $e^{\omega^i J_i}$ and, following the method elaborated at the beginning of question (b), the generators of the Lie algebra are

$$X_k = \frac{\partial}{\partial \omega^k}\{e^{\omega^i J_i}\}|_{\omega=0} = J_k$$

and the structure coefficients are

$$c_{ij}{}^k = -\varepsilon_{ij}{}^k \;.$$

The same considerations apply to $SU(2)$ defined on a complex manifold.

The group $SL(2,\mathbb{R})$ has 3 independent real parameters $\lambda^1, \lambda^2, \lambda^3$. Any matrix $\mathcal{A}$ of its Lie algebra should be traceless. From the Pauli matrices we derive the real traceless matrices

$$s_1 = i\sigma_1 = \begin{pmatrix} 0 & 1 \\ 1 & 0 \end{pmatrix} \;, \quad s_2 = i\sigma_2 = \begin{pmatrix} 0 & 1 \\ -1 & 0 \end{pmatrix} \;, \quad s_3 = \sigma_3 = \begin{pmatrix} 1 & 0 \\ 0 & -1 \end{pmatrix} \;.$$

Any matrix of the Lie algebra of $SL(2,\mathbb{R})$ has the form (neighborhood of 1),

$$e^{i\lambda^k s_k} \;.$$

Its generators are s_k and the structure coefficients can easily be calculated. The $SL(2,\mathbb{C})$ group has 3 independent parameters; any matrix of its Lie algebra should be traceless and a system of complex generators can be obtained from

$$s_1^{(c)} = \sigma_1 + i\sigma_1 \;, \quad s_2^{(c)} = \sigma_2 + i\sigma_2 \;, \quad s_3^{(c)} = \sigma_3 + i\sigma_3 \;.$$

The three complex parameters are equivalent to 6 real ones, to which corresponds the system of generators

$$\sigma_1, \ \sigma_2, \ \sigma_3; \ \ i\sigma_1, \ i\sigma_2, \ i\sigma_3 \ .$$

We now consider the SU*(3) group: it depends on* $9 - 1 = 8$ *independent complex parameters. For more details about the exponential form of the matrices of this group, the reader should look up any textbook on elementary particles. At least, we may try to explain why the* SU*(3) group plays such an important part in physics. Consider first the* SO*(3) matrix group. None of its generators* J_1, J_2, J_3 *commute with the others: such a group is called a group of rank 1. This is not the case of* SU*(3): one can diagonalize simultaneously two of its generators since one can find two generators which commute and* SU*(3) is said to be of rank 2. This property has an important physical consequence, it means that we can measure precisely two of its generators. Now, in quantum mechanics, a generator (defined as Hermitian) is an observable and also in elementary particles theory one assigns to each matrix group a family of particles. The one assigned to* SU*(3) is the family of quarks (the ultimate brick stones of matter). They are defined by 2 observables (among others!): one is the isospin of each member of the family, the other is the hypercharge (or the strangeness) of the particle, this is the property which explains the part played by* SU*(3) in elementary particles theory.*

(e) In Problem 2.9, question (e), we showed that for a rotation SO*(3) of vector* **δω***, the introduction of the skew-symmetric tensor* $\delta\omega_{ij}$ *such that*

$$\delta\omega_{ij} = \frac{1}{2}\varepsilon_{ijk}\delta\omega^k$$

permits us to write the components of the vector **x**′*, image of the vector* **x***, as*

$$x'^k = x^k + \delta\omega_j^k x^j \ .$$

This relation can also be brought in the form

$$
\begin{aligned}
x'^k &= x^k + \delta\omega^i{}_j x^j \partial_i\{x^k\} \\
&= x^k + \frac{1}{2}\delta\omega_j^i\{x^j\partial_i - x^i\partial_j\}x^k \\
&= x^k \sum_{i<j}\delta\omega^i{}_j\{x^j\partial_i - x^i\partial_j\}x^k
\end{aligned}
$$

and is to be compared with Eq. (26.13), Level 2: one sees that $x^j\partial_i - x^i\partial_j$ is a generator of the Lie algebra of SO(3). One generally defines this generator by the triplet of differential operators

$$L_k(\mathbf{x}, \partial) = \frac{1}{2}\varepsilon^k{}_j{}^i\{x^j\partial_i - x^i\partial_j\}\ ,$$

where $\varepsilon^k{}_j{}^i$ is one of the numerical Levi-Civita tensors. The commutator of two such generators can be evaluated as follows. We notice that

$$L_k(\mathbf{x}, \partial) = x^j\partial_i - x^i\partial_j\ ,$$

where k, j, i represent a circular permutation of 1,2,3. Simple, but somewhat tedious calculations show that

$$[L_i, L_j] = -\varepsilon_{ij}{}^l L_l\ .$$

In quantum mechanics the L_k represent the angular momenta of a particle and it is well-known that their representation through differential operators allows only integer angular momenta, not half-integer ones. (See also Problem. 1.19.)

| Problem 2.12. | Gauge theories and covariant derivations
(See also [Bibl. 12].)

This problem represents the physicist's approach to fibre bundles theories (see next chapter) and justifies its motivation and interests in the study of that chapter of pure mathematics.

(a) Let us come back to formula (5.12), Level 1: it defines the behaviour of the Christoffel symbols under a change of chart $x \to x' = x'(x)$. Show that (5.12) can be written as

$$\Gamma'^\beta{}_{\gamma\alpha}(x') = \frac{\partial x'^\beta}{\partial x^\nu}\frac{\partial x^\rho}{\partial x'^\gamma}\frac{\partial x^\sigma}{\partial x'^\alpha}\Gamma^\nu{}_{\rho\sigma}(x) + \frac{\partial x'^\beta}{\partial x^\mu}\frac{\partial x^\sigma}{\partial x'^\alpha}\partial_\sigma\frac{\partial x^\mu}{\partial x'^\gamma}\ .$$

We now define two $N \times N$ matrix fields $A_\alpha(x)$ and $A'_\alpha(x')$ such that

$$\Gamma^\beta{}_{\gamma\alpha}(x) = (A_\alpha(x))^\beta_\gamma\ ,\quad \Gamma'^\beta{}_{\gamma\alpha}(x') = (A'_\alpha(x'))^\beta_\gamma\ .$$

Show then that the first formula can be written in matrix form:

$$A'_\alpha(x') = \frac{\partial x'^\beta}{\partial x^\nu}\frac{\partial x^\rho}{\partial x'^\mu}\frac{\partial x^\sigma}{\partial x'^\alpha}(A_\sigma(x))^\nu_\rho + \frac{\partial x'^\beta}{\partial x^\mu}\frac{\partial x^\sigma}{\partial x'^\alpha}\partial_\sigma\frac{\partial x^\mu}{\partial x'^\gamma}\ .$$

The properties of the Jacobian matrices have been studied in Sec. 3 of Level 1 where it was shown that

$$\frac{\partial x'^\alpha}{\partial x^\beta} = \left(\frac{\mathbf{D}(x')}{\mathbf{D}(x)}\right)^\alpha_\beta \; , \quad \frac{\partial x^\beta}{\partial x'^\alpha} = \left(\frac{\mathbf{D}(x)}{\mathbf{D}(x')}\right)^\beta_\alpha \; .$$

Show that one can write

$$A'_\alpha(x') = \left[\frac{\mathbf{D}(x')}{\mathbf{D}(x)} A_\sigma(x) \frac{\mathbf{D}(x)}{\mathbf{D}(x')} + \frac{\mathbf{D}(x')}{\mathbf{D}(x)} \partial_\sigma \frac{\mathbf{D}(x)}{\mathbf{D}x')}\right] \left(\frac{\mathbf{D}(x)}{\mathbf{D}(x')}\right)^\sigma_\alpha \; .$$

The field $A_\alpha(x)$ is called a **gauge field**. Let us call $\hat{A}_\alpha(x)$ the field defined by the bracket in the above formula:

$$\hat{A}_\sigma(x) = \frac{\mathbf{D}(x')}{\mathbf{D}(x)} A_\sigma(x) \frac{\mathbf{D}(x)}{\mathbf{D}(x')} + \frac{\mathbf{D}(x')}{\mathbf{D}(x)} \partial_\sigma \frac{\mathbf{D}(x)}{\mathbf{D}(x')} \; .$$

One says that the gauge field $A_\alpha(x)$ is transformed into the gauge field $\hat{A}_\alpha(x)$ and the bracket defines a **gauge transformation**.

One also has

$$A'_\alpha(x') = \hat{A}_\sigma(x) \left(\frac{\mathbf{D}(x)}{\mathbf{D}(x')}\right)^\sigma_\alpha = \frac{\partial x^\sigma}{\partial x'^\alpha} \hat{A}_\sigma(x) \; ,$$

a formula which shows that under a change of chart, a gauge field behaves as a covariant vector field. One can then formulate the following statement: the variance of a vector field under a change of chart is the product of a gauge transformation by a change of chart of this field.

Show finally that the covariant derivative of a contravariant vector field $\psi(x)$ as given by formula (5.16), Level 1 can be expressed by

$$\nabla_\alpha \psi(x) = \partial_\alpha \psi(x) + A_\alpha(x)\psi(x) \; .$$

The results presented thus far in the present problem are considered as a shorthand version of complicated formulae, with a considerable number of indices. Our next task will be to fit the previous remarks into an abstract and organized frame, the frame of **gauge theories**. Nevertheless, it should be emphasized that all the considerations in the present problem are strictly local, valid in the neighborhood of a certain point $x \in \mathbb{R}^N$.

Let $\psi(x)$ and $\Psi(x)$ be elements of a certain vector space E, g a smooth map $E \to E$ such that

$$\psi(x) \to \hat{\psi}(x) = g \circ \psi(x) , \quad \Psi(x) \to \hat{\Psi}(x) = g \circ \Psi(x) .$$

Let $\mathcal{L}$ be a function (a polynomial in most cases) of ψ and Ψ such that it remains invariant when $\psi \to \hat{\psi}, \Psi \to \hat{\Psi}$:

$$\mathcal{L}(\psi, \Psi) = \mathcal{L}(\hat{\psi}, \hat{\Psi}) .$$

(b) We suppose that for an $x \in \mathbb{R}^N$, ψ represents a set of C^∞ real or complex functions $\psi(x) = \{\psi^i(x), i = 1 \ldots n\}$, and $\Psi(x)$ is defined as the set

$$\Psi_\alpha(x) = \{\partial_\alpha \psi^1(x) \ldots \partial_\alpha \psi^n(x) : \ \alpha = 1 \ldots N\} .$$

Then at a point x, $\mathcal{L}(\psi, \partial\psi)$ can be considered as a Lagrangian density from which one can deduce Euler-Lagrange equations (also called field equations) following the classical method of Problem 0.22.

Suppose that $g(x)$ belongs to a Lie group G of $n \times n$ matrices (for instance, GL(n, $\mathbb{R}$) or SL(n, C)) with elements C^∞ functions of $x \in \mathbb{R}^N$; consider $\hat{\psi}(x) = g(x)\psi(x)$, with the derivative $\partial_\alpha \psi(x)$ which may or may not be transformed into $\partial_\alpha \hat{\psi}(x)$. Suppose the second assumption to be true, we may then ask the question: how do we define a linear operation symbolized by D obeying the Leibniz rule and such that

$$\partial_\alpha \hat{\psi}(x) = g(x) D_\alpha \psi(x) \ ?$$

Then the new Lagrangian is, following our assumption, identical to the original one and the field equations remain unchanged (are covariant as one often says), if everywhere in the Euler-Lagrange equations the derivative ∂_α is replaced by the differential operator D_α.

The definition of $D\psi(x)$ is clearly an open choice. Among all possible choices, physics suggests the following arbitrary one which will be illustrated by several examples.

We now specialize E to be E_n with elements

$$\psi(x) = \{\psi^1(x) \ldots \psi^n(x)\} ,$$

being C^∞ or smooth functions. The elements of the Lie group G are the $n \times n$ matrices $g(x)$. Consider its Lie algebra $\mathbf{G}$ and N of its elements:

$$\{A_1(x) \ldots A_N(x)\} ,$$

being C^∞ or smooth **G** valued functions: they are $n \times n$ matrices, called **connections, gauge** or **Yang-Mills fields**. The **covariant derivative** $D_\alpha^{(A)}\psi(x)$ is defined by

$$D_\alpha^{(A)}\psi(x) = \partial_\alpha\psi(x) + A_\alpha(x)\psi(x) : \alpha = 1\ldots N .$$

We assume the following arbitrary transformation laws, called **gauge transformations**, which build up the **gauge group**.

$$\psi(x) \to \hat{\psi}(x) = g(x)\psi(x) ,$$
$$A_\alpha(x) \to \hat{A}_\alpha(x) = g(x)A_\alpha(x)g^{-1}(x) - (\partial_\alpha g(x))g^{-1}(x) .$$

One also assumes that under a change of charts ($x' = x'(x)$) the set $\{A_\alpha(x)\}$ (and also $\{\hat{A}_\alpha(x)\}$) behaves as a covariant vector field, for instance,

$$\hat{A}_\alpha(x) \to \hat{A}'_\alpha(x') = \frac{\partial x^\rho}{\partial x'^\alpha}\hat{A}_\rho(x) .$$

Show that

$$D_\alpha^{(\hat{A})}\hat{\psi}(x) = g(x)D_\alpha^{(A)}\psi(x)$$

and conclude that if one defines $\hat{\mathcal{L}}(\psi, \partial\psi) = \mathcal{L}(\hat{\psi}, D\hat{\psi})$ then the Lagrangian is invariant and the field equations covariant (they keep their form). Show that there exists a $h(x) \in G$ such that the transformation law of A_α can be brought into the form

$$\hat{A}_\alpha(x) = h^{-1}(x)A_\alpha(x)h(x) + h^{-1}(x)\partial_\alpha h(x) .$$

Conversely, if the law of transformation of ψ and A are the ones given above and if we suppose that between $D_\alpha^{(A)}$ and $D_\alpha^{(A)}$ exists the relation

$$D_\alpha^{(\hat{A})}\hat{\psi}(x) = g(x)D_\alpha^{(A)}\psi(x) ,$$

we also have

$$D_\alpha^{(\hat{A})} = \partial_\alpha + \hat{A}_\alpha(x) .$$

This last property is referred to as the **gauge invariance** (one should rather say covariance) of the covariant derivative.

(e) Calculate the commutator $[D_\alpha, D_\beta]\psi(x)$ and show that there exists a set of $n \times n$ matrices $r_{\alpha\beta}(x)$ with α and β running from 1 to N[*]:

$$[D_\alpha, D_\beta]\psi(x) = r_{\alpha\beta}(x)\psi(x) ,$$

---

[*]We omitted the index (A). This formula is sometimes called the Ricci identity.

where

$$r_{\alpha\beta}(x) = \partial_\alpha A_\beta(x) - \partial_\beta A_\alpha(x) + [A_\alpha(x), A_\beta(x)] \ ,$$

and $r_{\alpha\beta} = -r_{\beta\alpha}$ should not be confused with the Ricci tensor or the curvature form. Show furthermore that under a gauge transformation,

$$\hat{r}_{\alpha\beta} = g(x) r_{\alpha\beta} g^{-1}(x) \ ,$$

and that under a change of charts, $r_{\alpha\beta}$ behaves like a twice-covariant tensor:

$$r_{\alpha\beta}(x) \to r'_{\alpha\beta}(x') = \frac{\partial x^\mu}{\partial x'^\alpha} \frac{\partial x^\nu}{\partial x'^\beta} r_{\mu\nu}(x) \ .$$

For further study of these points, the reader should refer to Part (B) of the next chapter.

We want now to give a few examples which will make clear the versatility of the gauge transformation method.

(d) Our first example will deal with concepts of differential geometry and we will study again the problems of question (a). We consider a manifold $\mathcal{M}_N$ and let $\{x^1 \dots x^N\}$ be the coordinates of any of its points M, and consider also a $N \times N$ matrix

$$g(x) = \frac{\mathbf{D}(x'(x))}{\mathbf{D}(x)}$$

with elements

$$g_\beta^\alpha(x) = \partial_\beta x'^\alpha(x) \ .$$

We perform a gauge transformation (see toward the end of question (b)) with

$$h(x) = g^{-1}(x) = \frac{\mathbf{D}(x)}{\mathbf{D}(x')} \ .$$

The g and h matrices are elements of the GL$(N, \mathbb{R})$ Lie group.

Show that the formula giving $A'_\sigma(x')$ in question (a) can easily be brought into the simple form

$$\hat{A}_\sigma(x) = h^{-1}(x) A_\sigma(x) h(x) + h^{-1}(x) \partial_\sigma h(x) \ ,$$

and show finally that the components of the Riemann tensor field are

$$R^\beta{}_{\alpha\mu\nu}(x) = (r_{\mu\nu}(x))_\alpha^\beta \ .$$

Let us add two remarks. We notice first that a covariant field with

$$A_\alpha(x) = 0$$

identically vanishing components is a gauge field and that to the covariant vector field with components $A_\alpha(x)$ which are smooth functions there corresponds a matrix representing a gauge field

$$A_\alpha(x)\delta^\beta{}_\gamma = \Gamma^\beta{}_{\gamma\alpha}(x) \ .$$

The conclusion which can be drawn from the preceding remark is that gauge theories form a powerful unifying and versatile tool in differential geometry.

The next questions are devoted to physical situations: the non-physicist reader should refer to any textbook on quantum field theory for the few elementary concepts which will be needed.

(e) Let $\psi(x)$ be a complex smooth field and consider the following density Lagrangian,

$$\mathcal{L}(\psi, \psi^*, \partial\psi, \partial\psi^*) = -\frac{1}{2}(\eta^{\mu\nu}\partial_\mu\psi\partial_\nu\psi^* - m^2\psi^*\psi) \ ,$$

where $x \in \mathbb{R}^4$ represents the coordinates of the particle; m, its mass, is a real number and the $\eta_{\mu\nu}$'s refer to the Lorentz metric. The field ψ and ψ^* should be treated as independent functions: show then that the Euler-Lagrange equations read

$$\{\eta^{\mu\nu}\partial_\mu\partial_\nu + m^2\}\psi(x) = 0 \ , \quad \{\eta^{\mu\nu}\partial_\mu\partial_\nu + m^2\}\psi^*(x) = 0 \ .$$

Take G as being the group $U(1)$, and show that it is a 1-parameter Lie group and that the elements of its Lie algebra are pure complex functions $i\,F(x), F(x)$ being a real smooth function. Follow then the general procedure of question (b) and let $\{ieA_\mu(x), \mu = 0, 1, 2, 3\}, A_\mu(x)$ being real smooth functions, represent the gauge fields. We define the gauge transformation by

$$g\varphi(x) = e^{ie\varphi(x)} : \ e \in \mathbb{R} \ ,$$

where $\varphi(x)$ is a real smooth function, the charge e is a real number and the covariant derivatives of ψ and ψ^* are defined as follows:

$$D^{(A)}_\alpha\psi(x) = \partial_\alpha\psi(x) + ieA_\alpha(x)\psi(x) \ ,$$
$$D^{(A)}_\alpha\psi^*(x) = \partial_\alpha\psi^*(x) - ieA_\alpha(x)\psi^*(x) \ .$$

Starting from the Lagrangian density $\mathcal{L}(\psi'\psi^*, \partial\psi, \partial\psi^*)$ show that the field equations deduced from $\mathcal{L}$ are indeed gauge invariant. These equations represent the interaction of charged massive particles with photons.

Often, physicists distinguish between **global gauge** transformations, where the function $\varphi(x)$ is replaced by a constant, and the **local** ones, which have been previously described. Global gauge invariance leads, via Noether's theorem, to the definition and conservation of the current due to the massive particles.

(f) In an elementary approach, the Dirac equation for free electrons may be considered as a matrix equation:

$$\{\gamma^\mu \partial_\mu + mI\}\psi(x) = 0 \ ,$$

where the γ^μ's ($\mu = 0, 1, 2, 3$) are 4×4 hermitian matrices such that

$$[\gamma^\mu, \gamma^\nu]_+ = \gamma^\mu\gamma^\nu + \gamma^\nu\gamma^\mu = 2\eta^{\mu\nu} \ ,$$

$m \in \mathbb{R}_+$, and $\psi(x)$ is a 4-column matrix whose elements are complex functions of x. $\psi(x)$ is called **a Dirac spinor**. It has a specific transformation law (its variance) under a Lorentz transformation which is studied in all textbooks of quantum mechanics. Furthermore, if $\tilde{\psi}(x)$ represents the adjoint matrix (a row matrix whose elements are the complex conjugates of ψ), one defines $\overline{\psi}(x) = \gamma^0\tilde{\psi}(x)$ and one can easily see that $\overline{\psi}(x)$ satisfies the following equation:

$$\overline{\psi}(x)\{\gamma^\mu\underset{\leftarrow}{\partial_\mu} + mI\} = 0 \ .$$

One chooses as equation for electrons described by $\psi(x)$ interacting with photons described by the electromagnetic potential $eA^\mu(x)$ the equation[*]

$$\{\gamma^\mu D_\mu^{(A)} + mI\}\psi(x) = 0 \ ,$$

where $D_\mu^{(A)}$ is defined as in question (e).

Using the same procedure as for the previous question: one chooses for the group G the Lie group U(1), then $ieA_\alpha(x)$, with $\{A_\alpha(x)\} = A(x)$, are

---

[*]In the case of interacting fields, the principle which postulates the replacement of ∂_α by D_α is called the "principle of minimal interaction".

elements of its Lie algebra (as in question (e)), and the gauge transformation is defined by

$$\psi(x) \to \hat{\psi}(x) = g_\varphi(x)\psi(x) = e^{ie\varphi(x)}\psi(x) ,$$

$\varphi(x)$ being a real function and $e \in \mathbb{R}$.

Study and comment on the gauge invariance of the interacting Dirac equation. An important physical problem concerns the interaction of bosons of isospin 1 (pions $\pi^\pm, \pi^0$ for instance) with particles of spin 1/2 (nucleons for instance). But such a problem is too technical to be developed here. The reader may consult any textbook on quantum field theory.

Note finally that similar considerations apply to $G = SO(2)$ which is also a 1-parameter Lie group. Complements on Yang-Mill's fields can be found in Chap. 4, Sec. 12.

Remark 1: In the next chapter (Part (B)) we shall use the transformation law of the Yang-Mills field in another form. Let us start from the last but two equations of question (b) and introduce the 1-form

$$\hat{A}_\mu(x)\underline{dx}^\mu = g^{-1}(x)A_\mu(x)\underline{dx}^\mu + g^{-1}(x)\partial_\mu g(x)\underline{dx}^\mu$$
$$= g^{-1}\underline{A}(x)g(x) + g^{-1}(x)\underline{dg}(x) = \hat{A}(M) .$$

Its value for $\varepsilon_\alpha^M = \partial_\alpha$ is then

$$\hat{A}(\varepsilon_\alpha^M) = g^{-1}(x)A_\alpha(x)g(x) + g^{-1}(x)\partial_\alpha g(x) = \hat{A}_\alpha(x) .$$

The transformed Yang-Mills field $A'_\alpha(x)$ should also belong to the Lie algebra **G**: this is clearly a physical requirement which can be easily proved. The first term is an element of **G**: indeed, denoting by $e + A_\alpha(x) \in G$ (since $A_\alpha(x)$ is an element of G and e represents the identity element of G), then

$$g^{-1}(x)(e + A_\alpha(x))g(x) = e + g^{-1}(x)A_\alpha(x)g(x) \in G$$

is a relation which shows that $g^{-1}A_\alpha g$ necessarily belongs to **G**. The second term is another element of **G**: indeed, $g(x) + \partial_\alpha g(x) \in G$, then

$$g^{-1}(x)(g(x) + \partial_\alpha g(x)) = e + g^{-1}(x)\partial_\alpha g(x) \in G$$

and again this relation proves our assertion for the second term. As a conclusion, one may say that ω is a differential 1-form valued in the Lie algebra $\mathbf{G}$ of the Lie group G. The 1-form ω which is called the **connection form** will play a central part in Sec. 8 of the next chapter.

Remark 2: Although from a physicist's point of view we considered only the Lie group depending on a finite number of essential parameters (Sec. 26, Part (A)), the mathematician uses a less restrictive hypothesis. As a matter of fact, the set of $g(x)$ matrices of question (d) depends on a number of arbitrary analytic functions $f(x)$: the ones defining the point transformation

$$x^k \to f^k(x^1 \ldots x^N) = x'^k(x) , \quad k = 1 \ldots N .$$

This is also true for the Lie group U(1).

As a further remark, we notice that the set of matrices $g(x)$ (d) could also be considered as depending on the two sets of parameters x and x': then the set $\{g(x', x) = \mathbf{D}(x')/\mathbf{D}(x)\}$ is not a group but a pseudo-group with respect to those parameters, since following the chain rule for Jacobian matrices only products of matrices having one argument in common make sense. One has, indeed,

$$g(x', x)g(x, x'') = g(x', x'') , \quad g^{-1}(x', x) = g(x, x') .$$

Several properties of Lie group can be extended to pseudogroups: one may for instance define generators. Since $g(x, x) = I$ and denoting $\delta x = x' - x$, one may write up to the first order in δx

$$g(x', x) = g(x, x) + \delta x^j \left(\frac{\partial g(x', x)}{\partial x'^j} \right)_{x'=x}$$
$$= I + \delta x^j X_j(x) .$$

The $X_j(x)$ matrices are then the generators of the pseudo-group.

Solution:

(a) *The last term of formula (5.12), Level 1 can be written as*

$$\frac{\partial x'^\beta}{\partial x^\mu} \frac{\partial^2 x^\mu}{\partial x'^\gamma \partial x'^\alpha} = \frac{\partial x'^\beta}{\partial x^\mu} \frac{\partial x^\sigma}{\partial x'^\alpha} \partial_\sigma \frac{\partial x^\mu}{\partial x'^\gamma} ,$$

which is the required expression. The derivation of all the other formulae of this question is straightforward enough.

(b) The proof is as follows:

$$D_\alpha^{(\hat{A})}\hat{\psi}(x) = \partial_\alpha\hat{\psi}(x) + \hat{A}_\alpha(x)\psi(x)$$
$$= \partial_\alpha\{g(x)\psi(x)\} + [g(x)A_\alpha(x)g^{-1}(x) - (\partial_\alpha g(x))g^{-1}(x)]g\psi(x)$$
$$= g(x)D_\alpha^{(A)}\psi(x) .$$

We also notice that $\partial_k\{g(x)g^{-1}(x)\} = 0$ leads to

$$(\partial_k g(x))g^{-1}(x) = -g(x)\partial_k g^{-1}(x) .$$

Denoting then $g^{-1}(x)$ by $h(x)$, we find the last formula of (a).
(c) Indeed
$$[D_\alpha, D_\beta] = [\partial_\alpha + A_\alpha, \partial_\beta + A_\beta] ,$$

a simple expansion that leads to the required result. In order to prove the third formula of (c), note that $r_{\alpha\beta}\psi$ is an element of E; then under a gauge transformation,

$$r_{\alpha\beta}\psi(x) \to g(x)r_{\alpha\beta}(x)\psi(x)$$
$$= g(x)r_{\alpha\beta}(x)g^{-1}(x)g(x)\psi(x)$$
$$= g(x)r_{\alpha\beta}(x)g^{-1}(x)\hat{\psi}(x) .$$

For the last formula one uses the property of A_α to behave like a covariant vector under a change of chart.
(d) The first part of question (d) is very simple indeed.
One obtains the answer to its last part by remarking that

$$(r_{\mu\nu}(x))_\alpha^\beta = \partial_\mu\{A_\nu(x)\}_\alpha^\beta - \partial_\nu\{A_\mu(x)\}_\alpha^\beta + [A_\mu(x), A_\nu(x)]_\alpha^\beta .$$

We then introduce $\Gamma_{\gamma\alpha}^\beta$ as defined before to get formula (7.4), Level 1 which gives the components of the Riemann tensor.
(e) The required field equations are deduced from $\mathcal{L}$ following a straightforward application of the method of Problem 0.22. The elements $g_\varphi(x)$ of U(1) can be written indeed as $g_\varphi(x) = e^{i\varphi(x)}$: U(1) is a 1-parameter Lie group. Its Lie algebra can be obtained by considering the expansion of the element $g_{\varphi+\delta\varphi}(x)$ (see formula (26.15), Level 2):

$$g_\varphi + \delta_\varphi(x) = e^{i(\varphi(x)+\delta\varphi(x))}$$
$$= (1 + i\delta\varphi)e^{i\varphi(x)} .$$

There is a single generator $X = i$ and the elements of the corresponding Lie algebra are pure complex functions $iF(x)$($F(x)$ real smooth function). Following the method of (a), we will consider 4 smooth functions $\{ieA_0(x), ieA_1(x), ieA_2(x), ieA_3(x); e \in \mathbb{R}\}$, being covariant components of a covector under a change of chart.

We define the covariant derivative as in question (d) and form the Lagrangian density

$$\hat{L}(\hat{\psi}, \hat{\psi}^*, \partial\hat{\psi}, \partial\hat{\psi}^*) = \mathcal{L}(\hat{\psi}, \hat{\psi}^*, D^{(\hat{A})}\hat{\psi}, D^{(\hat{A})}\hat{\psi}^*)$$
$$= -\frac{1}{2}\eta^{\mu\nu}(\partial_\mu\hat{\psi}^* - ie\hat{A}_\mu\psi^*)(\partial_\nu\hat{\psi} + ie\hat{A}_\nu\hat{\psi}) + \frac{1}{2}m^2\hat{\psi}^*\hat{\psi} .$$

This is a formula which leads to the Klein-Gordon field equation

$$\eta^{\mu\nu}\{\partial_\mu + ie\hat{A}_\mu\}\{\partial_\nu + ie\hat{A}_\nu\}\hat{\psi}(x) - m^2\hat{\psi}(x) = 0$$

and its complex conjugate.

Under a gauge transformation $\psi = e^{ie\varphi}\psi$, A should be transformed as follows:

$$\hat{A}_\mu(x) = A_\mu(x) + ie\psi(x) ,$$

and, as is well-known, Maxwell's equations are precisely covariant (they keep their form) under such a gauge transformation, i.e., the original electric and magnetic fields are identical to the gauge transformed ones.

(f) All the methods used in the preceding question are also valid for the present question. One has

$$D_\alpha^{(\hat{A})}\hat{\psi}(x) = \partial_\alpha\hat{\psi}(x) + ie\hat{A}_\alpha(x)\hat{\psi}(x) ,$$

and the interacting Dirac equation reads

$$\{\gamma^\mu\partial_\mu + ie\gamma^\mu\hat{A}_\mu(x) + mI\}\hat{\psi}(x) = 0 .$$

Under a gauge transformation on ψ:

$$\psi(x) \rightarrow \hat{\psi}(\alpha) = e^{ie\varphi(x)}\psi(x) ,$$

the Dirac equation becomes

$$e^{-ie\varphi}\{\gamma^\mu\partial_\mu + ie\gamma^\mu(\hat{A}_\mu(x) - \partial_\mu\varphi(x)) + mI\}\hat{\psi}(x) = 0 .$$

The potential A(x) is transformed as

$$A_\alpha(x) \rightarrow A_\alpha(x) = A_\alpha(x) + \partial_\alpha \varphi(x) \ ,$$

and the gauge covariance of Maxwell's equations leads to the property of $\hat{A}$ and A to define identical electric and magnetic fields. See complements Chap. 4, Sec. 12.

Remark: In the next chapter, we will define geometrically a curvature form on $\mathcal{M}_N$: the 2-form $\underline{F}$, given by formula (9.4) may be connected with the tensor $r_{\alpha\beta}$ of question (b) as follows. Consider the 2-form

$$
\begin{aligned}
&r_{\alpha\beta}(x)\underline{dx}^\alpha \wedge \underline{dx}^\beta \\
&= \partial_\alpha A_\beta(x)\underline{dx}^\alpha \wedge \underline{dx}^\beta - \partial_\beta A_\alpha(x)\underline{dx}^\alpha \wedge \underline{dx}^\beta + [A_\alpha(x), A_\beta(x)]\underline{dx}^\alpha \wedge \underline{dx}^\beta \ .
\end{aligned}
$$

The first two terms combine to give $2\underline{d}\,\underline{A}(M)$ and the third term is usually written as $[\underline{A}(M) \wedge \underline{A}(M)]$.

Considering the 2-form

$$\underline{F}(M) = \frac{1}{2} r_{\alpha\beta}(x)\underline{dx}^\alpha \wedge dx^\beta \ ,$$

one obtains formula (9.4) of Chap. 4,

$$\underline{F}(M) = \underline{dA}(M) + \frac{1}{2}[\underline{A}(M) \wedge \underline{A}(M)] \ .$$

With a different interpretation and other notations: $\underline{F}$ being replaced by Ω_z and $\underline{A}$ by ω_z, the preceding formula will appear as formula (9.2), Chap. 4.

CHAPTER IV
Fibre Spaces

Most of the concepts we have developed so far can be generalized to other situations. Rather than introducing these new ideas via axiomatics, we will try to be guided by what we already know and are used to. In particular, we shall define several of these new concepts using the process of abstraction from the knowledge of moving frames (see Sec. 4, Level 2), vectors and linear connections (Sec. 3 and Sec. 18 to 20, Level 2).

It should be made clear at the very beginning that the present chapter is not supposed to present an exhaustive study of fibre bundles but a simple and preliminary introduction. We shall skim most of the subjects, urging the reader to consult the bibliography ([Bibl. 16] and following).

A. Fibre Bundles
1. *The Bundle of Linear Frames*[*]

Let $\mathcal{M}_N$ be a differentiable manifold of dimension N and let $\mathcal{P}$ be the space of all linear frames[**] of $\mathcal{M}_N$, i.e., a point $z \in \mathcal{P}$ represents the following couple: {a point $M \in \mathcal{M}_N$ and a basis of $T_{\mathcal{M}}(M)$}.

---

[*] As it was pointed out in the acknowledgements paragraph of the introduction, the help of Prof. R. Coquereaux has been invaluable for the completion of the present chapter.

[**] At a given point M, a frame is indeed a basis: we will denote a basis by $\{z_i\}$ instead of the previous notation $\{e_i\}$ used elsewhere in this book. Notice also that the $\{z_i\}$ should not be confused with any kind of covariant coordinates of $z \in \mathcal{P}$.

There is a natural map $\mathcal{P} \to \mathcal{M}_N$ which associates to any frame $z \in \mathcal{P}$ the point $M \in \mathcal{M}_N$ where the frame is defined: $M = \pi(z)$. The counter image $\pi^{-1}(M)$ is the set of all linear frames at M; it is called the "**fibre of $\mathcal{P}$ above M$\in \mathcal{M}_N$**" and will be denoted by $F_M = \pi^{-1}(M)$.

$\mathcal{P}$ is thus the collection built up by the point M itself and all the possible basis at M. Our next considerations will be local, valid for M and its neighborhood. Suppose that we choose a particular frame z_0 at M, then it is clear that we can get any frame z at M by the action of the linear group $\mathrm{GL}(N, \mathbb{R})$: there is indeed a group element $L \in \mathrm{GL}(N, \mathbb{R})$ such that $z = z_0 \cdot L$ ($z_i = (z_0)_k L^k{}_i$, where z_i is the i-th vector of the basis z); conversely any group element L determines a frame z by the above transformation. It is therefore obvious that provided we choose a particular frame z_0 at M, there exists a one-to-one correspondence between the group $\mathrm{GL}(N, \mathbb{R})$ and the fibre F_M. However, such a correspondence is not canonical since it depends upon the choice of z_0. We may now repeat the same observation for any point of a neighborhood of M. Let z_0^M, i.e., $\{(z_0^M)_i, i = 1 \ldots N\}$ be a moving frame in a neighborhood of M. We know that it is usually impossible to choose a globally defined moving frame, the same for all points of $\mathcal{M}_N$, and this is clearly why we restrict ourselves to a neighborhood of M. In other words, since the couple $\{M, z\}$ can be replaced by the couple $\{M, L\}$, we can identify the fibre F_M with the linear group $\mathrm{GL}(N, \mathbb{R})$ and may write:

$$\pi^{-1}(U_M) = (\text{union of fibres above } U_M) = \text{set of all linear}$$
$$\text{frames at the points } P \in U_M \times \mathrm{GL}(N\mathbb{R}) \, .$$

When the choice of a global moving frame is possible, we may choose U_M as $\mathcal{M}_N$ itself and write

$$\mathcal{P} = \mathcal{M} \times \mathrm{GL}(N, \mathbb{R}) \, .$$

In such a case $\mathcal{P}$ is called **trivial**, whereas in the previous situation it was only **locally trivial**, i.e., $\mathcal{P}$ itself could not be written as a direct product but it could be covered with neighborhoods $\pi^{-1}(U)$, each of which is a product. Although the concept of **principal bundle** has not yet been defined let us mention that $\mathcal{P}$ is a principle bundle (the bundle of linear frames of $\mathcal{M}_N$) and pause a while just to see what pieces of the structure are involved. We have the following.

(a) Two spaces $\mathcal{P}$ and $\mathcal{M}_N$ and $\mathcal{P} \xrightarrow{\pi} \mathcal{M}_N$. The triplet $\{\mathcal{P}, \mathcal{M}_N, \pi\}$ will be called a fibre bundle, $\mathcal{P}$ the **total space**, $\mathcal{M}_N$ the **base space** and π the **projection map**.

(b) The counter images $\pi^{-1}(M)$, called the fibres, are everywhere diffeomorphic to a **typical fibre F**. In the present case the typical fibre is $GL(N, \mathbb{R})$ which acts from the right (by convention) on F. In this situation we say that $\mathcal{P}$ is the **principal bundle**.

(c) $\mathcal{P}$ cannot usually be written as the direct product (base times a typical fibre). This is only locally possible.

From the terminological viewpoint, we also made use of a local moving frame which is to be viewed as a map $\mathcal{M} \to \mathcal{P}$ such that $\pi(z_M) = M$. In the general case a locally defined moving frame will be called a **local section** and any change of moving frame will be called a **gauge transformation**.

2. *Examples of Bundles*

(a) The bundle of oriented linear frames.

Let us suppose that one could assign the number $+1$ smoothly and continuously all over the manifold $\mathcal{M}$ (this would be impossible for a Möbius strip, see footnote*), $\mathcal{M}_N$ is then said to be orientable (and oriented by the choice of $+1$) and we may consider the space $\mathcal{P}$ of all frames of $\mathcal{M}_N$ which have a positive orientation. The reader could go through Sec. 1 once more and replace $GL(N, \mathbb{R})$ by $GL(N, \mathbb{R})_+$ (regular matrices with positive determinant) each time that $GL(N, \mathbb{R})$ enters in the line of reasoning. One may also remark that the hypothesis of $\mathcal{M}_N$ being orientable amounts to finding a global smooth map from $\mathcal{M}_N$ to $GL(N, \mathbb{R})/GL(N, \mathbb{R})_+ = \mathbb{Z}_2$ which denotes the set $\{-1, +1\}$; this is a statement to be elaborated on later.

(b) The bundle of linear frames with fixed volume.

Let us suppose $\mathcal{M}_N$ to be oriented and pick one particular frame at M; such a frame determines a volume element (or rather its dual basis at M). We can proceed globally and consider the space of frames corresponding to a given volume element: the group acting on such frames will be the group

---

*As pointed out in Sec. 12, Level 2, such an assignment can be made as follows: let $M^{(\alpha)}$ be a set of points of $\mathcal{M}_N$ and φ^α a set of homeomorphisms between open sets $U_\mathcal{M}(M^{(\alpha)})$ and $\mathbb{R}^N$, such that $\{U_\mathcal{M}(M^{(\alpha)}), \varphi^{(\alpha)}\}$ is an atlas. Suppose that in every non-empty intersection $U_\mathcal{M}(M^{(\alpha)}) \cap U_\mathcal{M}(M^{[\beta]})$ the Jacobian $\partial(x')/\partial(x)$ is positive, $\{x^k\}$ and $\{x'^k\}$ being coordinates in $U_\mathcal{M}(M^{(\alpha)})$ and $U_\mathcal{M}(M^{[\beta]})$ respectively, then $\mathcal{M}_N$ will be said to be oriented and characterized by $+1$. Notice that the Möbius strip obtained from $\mathbb{R}^2$ with identification of the points $\{x, y\}$ and $\{x + 2\pi, y\}$ is not orientable.

SL(N, $\mathbb{R}$)* and the reader is urged to read Sec. 1 again, replacing GL(N, $\mathbb{R}$) by SL(N, $\mathbb{R}$). We remark that the choice of a volume element means that we have been able to choose a global smooth map from $\mathcal{M}_N$ to GL(N, $\mathbb{R}$)/SL(N, $\mathbb{R}$) = R, a statement which can also be refined.

(c) The bundle of orthonormal frames.

We now choose a metric G on $\mathcal{M}_N$, then an orthonormal basis is clearly defined and we restrict our attention to the space of orthonormal frames for the specified metric. The reader can go back again to Sec. 1, where he will replace the group GL($N/\mathbb{R}$) by SO(N). Notice that there exists as many bundles of orthonormal frames as there are metrics. We also remark that the choice of a metric amounts to the choice of $N(N+1)/2$ (symmetry of the metric!) real numbers at each point $M \in \mathcal{M}_N$, but since

$$\dim\{\text{GL}(N, \mathbb{R}/\text{SO}(N))\} = N^2 - \frac{N(N-1)}{2} = \frac{N(N+1)}{2} \; ,$$

the choice of a metric implies the existence of a smooth continuous map from $\mathcal{M}_n$ to the homogeneous space* GL($N/\mathbb{R}$)/SO(N).

(d) Kähler manifold.

Suppose N to be even: $N = 2\nu$ and let $\mathcal{M}_N$ be endowed with complex structures. In some situations the structure group can be reduced from SO(N) to $U(\nu\mu)$. The reader may show in particular that $N(N-1)/2 - \nu = \nu(\nu-1)$ real numbers should be given at each $M \in \mathcal{M}_N$.

Remarks: Some comments on group theory may be of use to non-high energy physicists. **Group homomorphism** is an important concept. Let us consider two groups: $G = \{e, g_1, g_2 \dots\}$ and $G' = \{e', g'_1, g'_2, \dots\}$. Any map of G onto G' which preserves the group law in G', in other words, such that

$$(g_i g_j)' = g'_i g'_j \; , \quad (g'_i)^{-1} = g'^{-1}{}_i \; ,$$

is called an **homomorphism**. It should be pointed out that, in contradistinction to **group isomorphism**, several elements of G may have the same image in G'. We notice by the way that the image of e is e' and consider the subset H of G : $H = \{e, h_1, h_2, \dots\}$ of elements of G having e' as image, i.e.,

$$h'_1 = h'_2 = \dots e' \; ,$$

---

*Any space $\mathcal{M}$ is said to be homogeneous if there is a group G acting transitively on $\mathcal{M}$ (any 2 points of $\mathcal{M}$ can be connected by a group transformation). For instance, the sphere S^1 is homogeneous since by an appropriate rotation of $\mathbb{R}^2$ we may transform any of its points to another.

H is called the **kernel** of G. We remark that any kernel of a group is itself a group, indeed,

$$(h_i, h_j)' = h_i' h_j' = e' .$$

H is then a subgroup of G, but it possesses a further property: H is an **invariant subgroup** of G.

Let us define that concept in general: let $G^{(1)}$ be a subgroup of G with elements $\{e, g_1^{(1)}, g_2^{(1)}, \ldots\}$, then for any $g \in G$ one defines the coset $gG^{(1)}$ by

$$gG^1 = \{gg_1^{(1)}, gg_2^{(1)} \ldots\} .$$

If it happens that the coset $gG^{(1)}g^{-1}$ is such that

$$gG^1 g^{-1} = G^1 ,$$

then $G^{(1)}$ is called an invariant subgroup of G : H is indeed an invariant subgroup of G, since for any $h \in H$,

$$(ghg^{-1})' = g'g'^{-1} = e' .$$

ghg^{-1} is consequently an element of H and

$$gHg^{-1} = H \leftrightarrow gH = Hg .$$

That last property will lead to the definition of the **quotient** or **factor group**. Consider all the cosets having H as a common factor:

$$\{g_1 H, g_2 H, \ldots\} .$$

Since H is an invariant subgroup, we may endow such a set with a product law, indeed,

$$(g_i H)(g_j H) = g_i(H g_j)H = g_i g_j(HH) = g_i g_j H ,$$

and H plays the part of the identity of the former cosets group; we thus take as definition of the quotient group

$$G|H = \{H, g_1 H, g_2 H, \ldots\} .$$

But we notice that

$$\begin{aligned}
(G|H)' &= \{H', g_1' H', g_2' H', \ldots\} \\
&= \{e', g_1', g_2', \ldots\} = G' ,
\end{aligned}$$

and also that the map $G/H \Rightarrow G'$ is an isomorphism. As a conclusion, we have the important theorem

$$G/H = G' \ .$$

High energy physicists make constant use of that result, applying it to matrix groups. For such a group with g as an element, the map $g \Rightarrow \det g$ is indeed a homomorphism. Consider the case (a) of the present paragraph: we may map each $g \in \mathrm{GL}(N, \mathbb{R})$ into ± 1 following the sign of $\det g$; this is clearly a homomorphism. The set $\{+1, -1\}$ is the group Z_2 and to its unit element $+1$ there corresponds the subgroup $\mathrm{GL}(N, \mathbb{R})_+$. Then a straightforward application of the above theorem leads to

$$\mathrm{GL}(N, \mathbb{R})/\mathrm{GL}(N, \mathbb{R})_+ = Z_2 \ .$$

For the other examples of that paragraph and the following ones, the type of reasoning remains the same. See also Sec. 4 for a generalization.

We finally add two or more definitions to the preceding considerations: a **simple group** has no invariant subgroup and a **semi-simple group** has no Abelian (commutative invariant) subgroups.

3. *Axiomatic Definition of a Principal Bundle*

The set $\{\mathcal{P}, \mathcal{M}, \pi, G\}$ is a **right principal bundle** with a base* $\mathcal{M}$, a **total space $\mathcal{P}$,** a **projection** π and a **structural group,** G, iff the following conditions are satisfied.

(1) We consider first $\mathcal{P}$ and $\mathcal{M}$ which are smooth manifolds** with $\mathcal{P} \xrightarrow{\pi} \mathcal{M}$, π being a smooth map onto (a smooth surjective map). For any $z \in \mathcal{P}$, one has $M = \pi(z) \in \mathcal{M}$.

Since π is surjective, we may define $F_M \subset \mathcal{P}$, the **fibre above** M as a set of all $z \in \mathcal{P}$ having M as a counter image and we shall write

$$\pi^{-1}(M) = F_M \tag{3.1}$$

and also

$$\mathcal{P} = \bigcup_M F_M \ . \tag{3.2}$$

---

*The base is often denoted by B and $\{\mathcal{P}, \mathcal{M}, \pi, G\}$ is also referred as a $\mathcal{P}$-right principal bundle.

**The term "smooth" means of class C^p with p large enough for the statement concerning the manifold to be meaningful. At Level 0, we used the term "honest function" as an equivalent term.

(2) Let $P \in \mathcal{M}, U_{\mathcal{M}}(P)$ — denoted by U — be a neighborhood of P and $M \in U_{\mathcal{M}}(P)$. The set of fibres above all points of U will be represented by $\pi^{-1}(U)$ and we suppose that there exists a diffeomorphism ψ:

$$\pi^{-1}(U) \to \psi U \times G , \qquad (3.3)$$

such that for any point z of any of the elements F_M of $\pi^{-1}(U)$ one has

$$\psi(z) = \{M, g(z)\} = \{\pi(z), g(z)\} . \qquad (3.4)$$

This is the condition of **local triviality**. In other words, there exists a differentiable homeomorphism (i.e. a diffeomorphism) between each of the fibres F_M and G which is called **typical fibre** (see also comments about Fig. 4.1).[*]

(3) As defined in (2), g is the map $\pi^{-1}(U) \to G$. We shall further assume that G is a Lie group acting smoothly on $\mathcal{P}$ and transforming each of the fibres F_M into itself. Then, for any point z of any of the fibres $\pi^{-1}(U)$, the couple $\{z, g\} \in \mathcal{P} \times G$ is transformed by the action of g into $zg \in \mathcal{P}$ in a way compatible with (3.4), i.e.,

$$\psi(zg) = \{\pi(z), g(z)g\} . \qquad (3.5)$$

A local **section** σ is a map $M \to \mathcal{P}$ (i.e. $\sigma(M) = z$) such that

$$\pi_{0\sigma} = I_{\mathcal{M}} , \qquad (3.6)$$

where $I_{\mathcal{M}}$ is the identical map on $\mathcal{M}$.

The following intuitive point of view may be of help to the reader: any $\mathcal{P}$-principal fibre bundle can be looked upon as a collection of fibres (all of the same type) glued together and parametrized by a space $\mathcal{M}$ called the base. All the fibres should be diffeomorphic to a given Lie group G which acts on $\mathcal{P}$ from the right and locally each fibre F should be equivalent to the product $\mathcal{M} \times G$.

One often refers to the base $\mathcal{M}$ as an **horizontal** set and each fibre F_{M_0} is pictured as being **vertical**; the group action is also said to be vertical. One may also define **fibre bundles** which are not necessarily principal ones by relaxing one of the previously stated conditions. We no longer assume

---

[*]For a general fibre bundle, the typical fibre may not be identical to the structural group G.

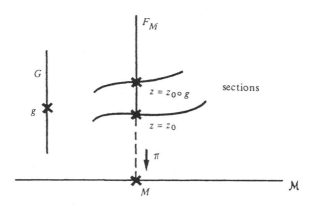

Fig. 4.1.

that the typical fibre F is a group, or if it is one we do not require that the group in question acts transitively on the fibres.

We then have to distinguish between a typical fibre F homeomorphic to each fibre F_M and the group G acting on the fibres F_M : $(F_M \xrightarrow{G} F_M)$, and hence on F. In that case, $\mathcal{P}$ is replaced by a topological space called E the (new) total space, π is a surjective map $E \xrightarrow{\pi} \mathcal{M}$ and one calls the set $\{E, \mathcal{M}, \pi, G\}$ a **fibre bundle**.

As a conclusion let us add some comments which will be extremely useful later on. Suppose that we choose a definite section σ: then any point $M_0 \in \mathcal{M}$ is mapped into the corresponding point $z_0 \in F_{M_0} \subset \mathcal{P}$: $z_0 = \sigma(M_0)$. But since G maps F_{M_0} into F_{M_0}, any other point $z \in F_{M_0}$ can be deduced from z_0 by $z = z_0 g$. In other words, the choice of a section σ allows us to parametrize any point $z \in F_{M_0}$ by M_0 and g, and for any point $z \in F_M$ we may write

$$z = \{M, g\} \in \mathcal{M} \times G \ . \tag{3.7}$$

This is the generalization of a fundamental remark made at the beginning of Sec. 1.

4. *More Examples of Principal Bundles*

We may add to the examples given in Sec. 2 the following ones. A particularly important class of examples can be formulated using the following construction: let K be a Lie group and H a closed not invariant Lie subgroup of K (with respect to the induced topology), then the quotient K/H is a **homogeneous space** not a group, in contradistinction with the definition of the factor group given in Sec. 2, Remarks. One has indeed

$$K/H = \{kH : k \in K\} ,\qquad (4.1)$$

and we may write the following coset decomposition of K as

$$K = \{H \cup k_1 H \cup k_2 H \cup \ldots\} ,\qquad (4.2)$$

with

$$k_1 \notin H , \quad k_2 \notin \begin{cases} H \\ k_1 H \end{cases} , \quad k_3 \notin \begin{cases} H \\ k_1 H, \ldots \\ k_2 H \end{cases} .\qquad (4.3)$$

In other words, K can be written as a union of spaces $\{k_i H\}$, all of them diffeomorphic with H and parametrized by the coset space K/H : K being a principal fibre bundle with structure group H and base K/H. Furthermore, the group K acts transitively on the elements of K/H: any two of its elements can be connected by a transformation, by element of K and this is precisely the characteristic property of a homogeneous space. See also footnote of Sec. 2.

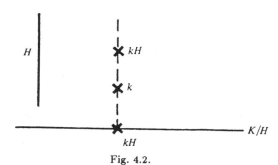

Fig. 4.2.

For instance SU(2) is a U(1)-principal bundle above S^2 (2-dimensional sphere). Indeed

$$S^2 = \mathrm{SU}(2)/U(1) .\qquad (4.4)$$

and S^2 is invariant under the action of SU(2). SU(3) is an SU(2)-principal bundle above the sphere S^5:

$$S^5 = \mathrm{SU}(3)/\mathrm{SU}(2) .\qquad (4.5)$$

SO(6) is an SO(5)-principal bundle above S^5:

$$S^5 = \mathrm{SO}(6)/\mathrm{SO}(5) \ . \tag{4.6}$$

It is important to draw the reader's attention to the following point: two principal fibre bundles with the same base and the same typical fibre may be very different (hence the usefulness of the concept of fibre bundle!). For instance, $P_1 = S^2 \times \mathrm{U}(1)$ is a trivial U(1)-principal bundle over S^2, but $P_2 = \mathrm{SU}(2)$ is also a principal bundle over S^2 (as noticed above) and obviously $P_2 \neq P_1$. It is therefore natural to classify those principal bundles with a given base and a given structure group. We shall see, later on (see Secs. 11 and 12), that the classification of U(1)-bundles over S^2 relies on the value of an integer n which is, from a physicist's point of view, a monopole charge. Furthermore, principal fibre bundles are also at the root of all the constructions involving Yang-Mills fields which play a central part in the theory of elementary particles.

As a final remark, we defined right principal bundles by assuming a right action of G on $\mathcal{P}$, but we could also define similarly left principal bundles.

5. *Vectors and Tensors Associated Vector Bundles*

Using the new set of concepts which we have been investigating we want to define, once more, vectors and tensors.

The notion of vectors has been twice presented previously: at Level 1, a vector was defined as a collection of numbers (or functions) with a given law of transformation (its variance); at Level 2 as a linear map $C^\infty(U_\mathcal{M}) \to C^\infty$. (See (3.3), Level 2.) We want now to examine an alternative definition abstract enough to unify the concepts of vectors and tensors.

We consider an abstract object V called a vector whose components are a set of real numbers $\{v^1 \ldots v^N\}$ denoted by v in the frame $z = \{z_1 \ldots z_n\}$. Let us take two frames z and z' and use the following notation,

$$V = z \cdot v = z_i v^i = z' \cdot v' = z_i' v_i' \ . \tag{5.1}$$

But since z' is also a frame, there exists a $L \in \mathrm{GL}(N, \mathbb{R})$ such that

$$z_i' = z_j L_i^j \ . \tag{5.2}$$

By bringing this relation into the definition of V, we may find (as we did before) the transformation law of the components[*] v^i:

$$L^j{}_i v'^i = v^j \Leftrightarrow v'^i = (L^{-1})^i{}_j v^j \ , \tag{5.3}$$

---

[*]The present variance of v will agree with the one common to Levels 1 and 2, if $L = D(x)/D(x')$.

and, finally, replacing L by g, the definition of V can be alternatively written as

$$V = z \cdot v = z' \cdot v' = zg \cdot g^{-1}v \ , \tag{5.4}$$

a relation which is to be read as follows: the vector V has components v in the frame z and components $g^{-1}v$ in the frame $z' = zg$.

For an immediate generalization we may proceed as follows: in the case considered above we had $v \in \mathbb{R}^N$, but $\mathbb{R}^N$ is a representation space for $GL(N, \mathbb{R})$ and ν acts on $\mathbb{R}^N$ via a representation ρ (for instance ρ could be any tensor product of matrices L). Then we would define a tensor v of type ρ by

$$V = z \cdot v = zg \cdot \rho(g^{-1})v \ , \tag{5.5}$$

meaning that the object V has components v in the frame z, and $\rho(g^{-1})v$ in the frame $z' = zg$. We also remark that g acts on the frame z while $\rho(g^{-1})$ acts on the set of numbers symbolized by v.

The second part of the present paragraph deals with associated vector bundles, a notion closely connected with the generalization of the preceding intuitive definition of vectors and tensors. We start by considering a principal vector bundle $\{\mathcal{P}, \mathcal{M}, \pi, G\}$ as defined in Sec. 3, also a certain vector space F with elements denoted $f, f' \ldots$, and let ρ be a representative of G on F. We proceed by endowing the space product $\mathcal{P} \times F$ with the following equivalence relation: any two couples $\{z, f\}$ and $\{z', f'\}$, elements of $\mathcal{P} \times F$, are equivalent if

$$\{z, f\} \sim \{z', f'\} \Leftrightarrow z' = zg, f' = \rho(g^{-1})f \ . \tag{5.6}$$

We may now build up the space $E = \mathcal{P} \overset{\rho}{\times} F$, a coset space, with the above equivalence classes as elements. Such a space will be the **associated vector bundle** $\{E, \mathcal{M}, \pi, G\}$ with a typical fibre F and structural group G. Then any of its elements $u \in E$:

$$u = \text{class}\{z, f\} \ , \tag{5.7}$$

will be denoted by

$$u = z \cdot f \ , \tag{5.8}$$

and because of the equivalence relation we also have $u = zg \cdot \rho(g^{-1})f$, an object which will play the same part as $z \cdot f$ in our introductory considerations. The object $u = z \cdot f$ is now a geometrical entity which has components f in the frame z and $\rho(g^{-1})f$ in the frame $z' = zg$: it is a **generalized vector (tensor)**. The situation can be summarized by the following figure:

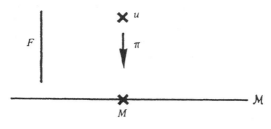

Fig. 4.3.

We also notice that dim $E = $ dim $\mathcal{M}+$ dim F and that E is indeed a fibre bundle $\{E, \mathcal{M}, \pi, G\}$ with the projection map $\pi : E \xrightarrow{\pi} \mathcal{M} : \pi(u) = M$.

We also remark that E is a fibre bundle, but not a principal one, since F is no more a Lie group acting on E, but a linear space: E can be thought of as a collection of linear spaces parametrized by the base $\mathcal{M}$. We will say that the **vector bundle** E is associated with the principal bundle $\mathcal{P}$.

It is also worthwhile noticing that the method of fibre bundles allows the unification of the taxonomic approach of Level 1 and the intrinsic one of Level 2, since the new definition of vectors and tensor fields takes into account their specific transformation laws (their variances) without fixing *a priori* any given frame.

Let us now look for some examples. As a first example, we consider the **tangent vector bundle** at a manifold $\mathcal{M}_N$: let $\mathcal{P}$ be the bundle of linear frames studied in Sec. 1: $\{\mathcal{P}, \mathcal{M}_N, \pi, \mathrm{GL}(N, \mathbb{R})\}$, where π maps any frame $z \in \mathcal{P}$ onto $\mathcal{M}_N$ and $\mathrm{GL}(N, \mathbb{R})$ is the group acting on linear frames. We choose the fundamental representation ρ of $\mathrm{GL}(N, \mathbb{R})$ on $F = \mathbb{R}^N$, then the associated vector bundle $\{T_\mathcal{M}, \mathcal{M}_N, \pi, \mathrm{GL}(N, \mathbb{R})\}$, where $T_\mathcal{M} = \mathcal{P} \overset{\rho}{\times} \mathbb{R}^N$, is the tangent bundle to $\mathcal{M}_N$ and may be called the right $T_\mathcal{M}$-bundle (not a principal one!), ρ is simply the identity if g, as in (5.4) belongs to the fundamental representation of $\mathrm{GL}(N, \mathbb{R})$, and π represents the projection $T_\mathcal{M} \rightarrow \mathcal{M}$ (no more $\mathcal{P} \rightarrow \mathcal{M}_N$!). Its elements are tangent vectors (or simply vectors) to $\mathcal{M}_N$ and any section $\sigma(\mathcal{M}_N \rightarrow T_\mathcal{M})$ defines a vector field. Starting with the same principal bundle, we could choose a tensorial representation ρ of $\mathrm{GL}(N, \mathbb{R})$ and therefore construct the **bundle of tensors of type $\boldsymbol{\rho}$**.

Suppose now that we choose a metric on $\mathcal{M}_N$. We may then look at the bundle of orthonormal frames (see Sec. 1); the group G is thus $SO(N)$ and we may consider its spinorial representations.[*] In the same way as before, one builds up the bundle of spinors whose sections are spinor fields. We could also take $\mathcal{P}$ as an SU(3)-principal bundle on $\mathcal{M}$ and consider the representation ρ of SU(3) on $\mathbb{C}^3$ (the fundamental representation). Then elements of the vector bundle $\mathcal{P} \overset{\rho}{\times} \mathbb{C}^3$ may be thought of as scalar quarks (blue, white, red).

We finally look at the problem of local parametrization in a vector bundle $E = \mathcal{P} \overset{\rho}{\times} F$. Suppose that we choose a local section $\mathcal{M} \overset{\sigma}{\to} \mathcal{P} : M \to z_0^M$ on the principal bundle $\mathcal{P}$ then we can express any vector $u \in E$ as

$$u = z_0^M \cdot f . \tag{5.9}$$

We denote u by its only components $\{f\}$ since z_0 has been arbitrarily chosen once and for all.

6. Non-Vectorial Associated Bundles

In our preceding considerations, we never made use of the vectorial character of F. Only one point was of importance: the hypothesis that there is an action (described by ρ) of G on F. Such an action has been supposed to be in the previous paragraph, a linear one: actually a representation ρ of G on the vector space F. But we can repeat all the constructions of Sec. 5 without assuming the linearity of ρ. Let ρ denote then a smooth action of G on F (for instance the action of SU(2) on the sphere S^2): if $g \in G$ and $f \in F$ then $\rho(g)f \in F$. We now repeat the construction developed in the previous paragraph: we define $E = \mathcal{P} \overset{\rho}{\times} F$ and write

$$u = z(g) = zg \cdot \rho(g^{-1})f . \tag{6.1}$$

There are no other changes: E is a fibre bundle with F as a typical fibre (not necessarily a vector space), E is said to be associated with $\mathcal{P}$ and **locally** E can be considered as the product $\mathcal{M} \times F$.

Let us look for examples: let us suppose $\mathcal{P}$ to be a principal bundle with G as structure group and let H be a subgroup of G. We may choose

---

[*]The vocabulary we are using is sloppy indeed: spinors are not representations of $SO(N)$ but of its double covering group spin (ν), and there may be some topological difficulties to the present construction in some circumstances.

$F = G/H$ and denote by ρ the obvious action of G on G/H,

$$\{g, g_1 H\} = g g_1 H \ , \tag{6.2}$$

with g and g_1 being elements of G, and not of H. Then $E = \mathcal{P} \overset{\rho}{\times} F$ is a bundle with the fibre G/H above $M \in \mathcal{M}$. We may, for instance, choose $\mathcal{P}$ be the group SU(3) expressed as SU(2) principal bundle over S^5 (see end of Sec. 4) and ρ to be the action SU(2) over $S^2 = $ SU(2)/U(1); we then build $E = \mathcal{P} \overset{\rho}{\times} S^2$, and S^2 bundle over S^5.

As another example, we may consider a principal bundle $\mathcal{P}$ with a structure group G and choose for F the group G itself. We then define as ρ the adjoint action of G on itself, i.e.,

$$\rho(g) h = h g h^{-1} \ , \tag{6.3}$$

g and h being elements of G. Consider $E = \mathcal{P} \overset{\text{ad}}{\times} G$: it is an associated bundle, called the **adjoint bundle**, although its typical fibre is G, it is not a principal bundle. In physics of elementary particles, a section I of E is called gauge transformation.*

The following theorem, which we shall state without proof, is fundamental for the explanation of the examples of Sec. 1:

"The structure group G of a principal fibre bundle may be restricted to a subgroup H if and only if the associated fibre bundle $E = \mathcal{P} \overset{G}{\times} G/H$ admits a global section".

One can then verify that this is indeed the case for all the examples given in Sec. 1.

---

*However it is not a trivial task to show that this notion coincides with the one introduced in Sec. 1.

B. Connections

7. *Some Introductory Comments*

What is the need and the use of the general notion of **connection**? As we saw in Part A, the definition of a moving frame as a local section of a principle bundle $\mathcal{P}$ is general, and geometric too. If the bundle is somehow related to the tangent vector space of a manifold $\mathcal{M}$, we obtain the concept of the "usual" moving frame. However, if the structure group G has no relation whatsoever with the tangent vector space (as for instance $G = \mathrm{SU}(3)$ in chromodynamics), one obtains an "internal" moving frame following the physicist's vocabulary, called also a moving frame in the "internal space": so is the case of the complex space C^3 spanned by the "blue, white, red" quarks. In any case, there remains the question to be answered: given a point $M_0 \in \mathcal{M}$ in the base manifold and given a frame z_0 at the same point, how do we associate an infinitesimal displacement of z_0 to a given infinitesimal displacement of M_0? There is an infinite number of ways and a given choice corresponds to a given connection for such an association.

At Level 1, we already examined one of these possibilities (see formulae (7.5) to the end of the paragraph 7) and also, in the case where $\mathcal{M}$ is Riemannian, we were able to express the Riemann tensor and the Christoffel symbols by means of the metric tensor (see Sec. 8,9,10). However, such methods were not general and geometric enough: indeed the use of a metric is unnecessary (since the definition of a connection does not involve higher derivatives of the $g_{\lambda\nu}$) and also too restrictive. Indeed, how do we describe the connection if the space of frames has nothing to do with the tangent vector space, but builds up a principal bundle as in the case of an internal space (for an elementary particle)?

The purpose of Part B is to investigate how to define a connection in general (along with the formalism of covariant derivative, curvature,...) and to recover several results obtained elsewhere using other formulations.

8. *Connection on Principal Bundles*

In order to take advantage of notions which were previously acquired, we shall use a method which starts from the results and conclusions of Problem 2.12. We then proceed along two steps: step 1 will be practically a rewriting of several formulae, proved in that problem, while in step 2, we shall look for an appropriate reformulation using the concepts and language of fibre bundle theory.

We consider a smooth manifold $\mathcal{M}_N$ whose points M have $\{x^1 \ldots x^N\}$ as coordinates; $\psi(x) = \{\psi^i(x), i = 1 \ldots n\}$ will be a vector field on $\mathcal{M}_N$, G a Lie group of elements $\{g(x), h(x), \ldots\}$ with its Lie algebra $\mathbf{G}$ and $\{A_\alpha(x), \alpha = 1 \ldots N\}$ the set of Yang-Mills fields valued in $\mathbf{G}$ (each $A_\alpha(x)$ is an element of $\mathbf{G}$). It is then clear that $\psi(x)$ is a column matrix of n elements, $g(x)$ an $n \times n$ matrix, and so are the N matrices $A_\alpha(x)$. We recall the three fundamental results of Problem 2.12 (questions (a) and (b)): under a transformation

$$\psi(x) \to \psi'(x) = g^{-1}(x)\psi(x) , \tag{8.1}$$

the corresponding Yang-Mills field of components $A_\alpha(x)$, which will be called **connections** from now on, undergoes the following transformation:

$$A_\alpha(x) \to A'_\alpha(x) = g^{-1}(x)A_\alpha(x)g(x) + g^{-1}(x)\partial_\alpha g(x) , \tag{8.2}$$

and the corresponding covariant derivative is defined by

$$D_\alpha^{(A)}\psi(x) = \partial_\alpha\psi(x) + A_\alpha(x)\psi(x) . \tag{8.3}$$

We may also notice that since $A_\alpha(x)$ is valued in $\mathbf{G}$, the Lie algebra of G, we may also write

$$A_\alpha(x) = A_\alpha^q(x)X_q , \tag{8.4}$$

where X_q are the generators of $\mathbf{G}$.

A familiar and simple example was also given in question (d) of Problem 2.12 where we took $G = \mathrm{GL}(N, \mathbb{R})$ and $g(x) = \mathbf{D}(x')/\mathbf{D}(x)$. We noticed, by the way, that all the matrices one comes across in that case are $N \times N$. The matrix elements of the connection $A_\alpha(x)$ were denoted by

$$(A_\alpha(x)_\nu^\mu = \Gamma_{\nu\alpha}^\mu(x) . \tag{8.5}$$

One may also remark that since the dimension of the vector space built up by $N \times N$ matrices is clearly N^2, any matrix A_α can be written as

$$A_\alpha(x) = (A_\alpha(x))_\rho^\nu B_\nu^\rho = \Gamma^\nu{}_{\rho\alpha}(x)B_\nu^\rho , \qquad \begin{array}{l} \rho = 1 \ldots N \\ \nu = 1 \ldots N , \end{array} \tag{8.6}$$

and the N^2 matrices $B^\rho{}_\nu$ have all their elements vanishing except one which is equal to 1. But since $A_\alpha(x)$ is also an element of the Lie algebra of

$GL(N, \mathbb{R})$ which admits N^2 generators, we may label differently the $B^p{}_\nu$ matrices and denote them by $X_q, q = 1 \ldots N^2$; we are then back to formula (8.4).

Another significant example for a physicist corresponds to $G = SU(2)$ (see Problems 2.11 and 12); the generators of its Lie algebra are the Pauli matrices τ_q and the theory corresponds to the well-known **Yang-Mills theory**.

We may now turn our attention to step 2 and generalize formulae (8.2) and (8.3) following the remark at the end of Problem 2.12. We thus introduce Lie algebra-valued differential 1-forms. Instead of the formula (8.2) we write

$$A'_\alpha(x)\underline{d}x^\alpha = g^{-1}(x)A_\alpha(x)\underline{d}x^\alpha g(x) + g^{-1}(x)\partial_\alpha g(x)\underline{d}x^\alpha . \qquad (8.7)$$

We then define the Lie algebra-valued 1-forms:

$$\underline{A}'(x) = A'_\alpha(x)\underline{d}x^\alpha , \quad \underline{A}(x) = A_\alpha(x)\underline{d}x^\alpha , \qquad (8.8)$$

and we are able to give the transformation law of $\underline{A}(x)$ under a gauge transformation as follows:

$$\underline{A}'(x) = g^{-1}(x)\underline{A}(x)g(x) + g^{-1}\underline{d}g(x) . \qquad (8.9)$$

The second term in the right-hand side which is equal to $g^{-1}(x)\partial_\mu g(x)\underline{d}x^\mu$ is to be interpreted *cum grano salis* but this is a notation well adopted by physicists.[*]

We are now dealing with a connection $\underline{A}(x)$ which is a Lie algebra-valued 1-form whose components are precisely the Yang-Mills field $A_\alpha(x)$, elements of **G**. Our connection theory will consider (8.9) as a generalization of (8.2). In the same spirit, instead of considering D_α (given by (8.3)) we shall consider $\mathbf{D} = \underline{d}x^\alpha D_\alpha$ and replace (8.3) by

$$\mathbf{D}\psi(x) = \{\underline{d}x^\alpha \partial_\alpha + \underline{d}x^\alpha A_\alpha(x)\}\psi(x) = \{\underline{d} + \underline{A}(x)\}\psi(x) . \qquad (8.10)$$

We want now to set our former considerations in the frame of fibre bundles. We consider a right principal bundle $\{\mathcal{P}, \mathcal{M}, \pi, G\}$ and, as observed in the

---

[*]One may also present another justification of this notation: since $g(x)$ is a regular matrix, one can define $dg(x) = \partial_\alpha\ g(x)\ dx^\alpha$, then $g(x) + dg(x)$ is an element of the tangent vector space to **G** at the point $g\ (x)$. As a consequence, the group element g^{-1} $(g + dg)$ can be written as $e + g^{-1}\ dg$ and this shows that $g^{-1}\ dg$ is clearly an element of the Lie algebra **G**.

last comment of Sec. 3, we parametrize points $z \in \mathcal{P}$ by $z = \{M, g\} \in \mathcal{M} \times G$, which means in its turn that we choose an arbitrary section. Up to now all our considerations dealt with the 1-form $\underline{A}$ on $\mathcal{M}$ valued in $\mathbf{G}$; we want now to substitute to that notion a 1-form ω which is globally defined on $\mathcal{P}$ and is gauge covariant: ω_z will be called the **connection form**. We, hence, define a differential 1-form ω_z on $\mathcal{P}$, valued in $\mathbf{G}$ and depending on $z \in \mathcal{P}$: i.e., a map of the tangent vector space $T_{\mathcal{P}}(z)$ into the tangent vector space $T_G(e)$ which is the Lie algebra $\mathbf{G}$ of G. One may connect ω_z to $\underline{A}$ by the following definition:

$$\omega_z = g^{-1}\underline{A}g + g^{-1}\underline{d}g \; , \tag{8.11}$$

similar to (8.9) but with different meaning. The form ω_z possesses two important properties:

(a) It is globally defined and is gauge covariant: let z be a point of F_M; in the section σ, its coordinates are $\{M, g\}$ while in the section σ' its coordinates are $\{M, g'\}$.

Consider ω'_z in the section σ':

$$\omega'_z = g'^{-1}\underline{A}'g' + g'^{-1}\underline{d}g' \; . \tag{8.12}$$

Our assertion is that there exists an $h \in G$ such that if $g = hg'$ and $\underline{A}$ obeys (8.9) then

$$\omega_z = \omega'_z \; . \tag{8.13}$$

Indeed, formula (8.9), where the group element g is now called h, takes the form

$$\underline{A}' = h^{-1}\underline{A}h + h^{-1}\underline{d}h \; . \tag{8.14}$$

But since $g = hg'$,

$$\begin{aligned}
\omega_z &= g'^{-1}h^{-1}\underline{A}hg' + g'^{-1}h^{-1}\underline{d}\{hg'\} \\
&= g'^{-1}(h^{-1}\underline{A}h + h^{-1}\underline{d}h)g' + g'^{-1}\underline{d}g' \\
&= g'^{-1}\underline{A}'g' + g'^{-1}\underline{d}g' = \omega'_z \; .
\end{aligned} \tag{8.15}$$

A second property completes the preceding one.

(b) Let $z = \{M, g\}, z' = \{M, gg'\}$ be two points of F_M. One has

$$\begin{aligned}
\omega_{z'} &= (gg')^{-1}\underline{A}gg' + (gg')^{-1}\underline{d}\{gg'\} \\
&= g'^{-1}(g^{-1}\underline{A}g + g^{-1}\underline{d}g)g' + g'^{-1}\underline{d}g' \\
&= g'^{-1}\omega_z g' + g'^{-1}\underline{d}g' \; ,
\end{aligned} \tag{8.16}$$

a formula which defines the transformation law for ω_z when $z \to z'$.

All our attention was focused on the 1-form $\underline{A}$ (defined by (8.9) at $M \in \mathcal{M}$) and ω_z defined by (8.11), but there is an alternative geometric method where the basic concepts involve two kinds of vectors defined in $T_{\mathcal{P}}(z)$: the **horizontal** ones and the **vertical** ones. We first introduce a linear map $\lambda_z : T_{\mathcal{M}}(M) \to T_{\mathcal{P}}(z)$ called a **connection lift**, and let π denote, as before, the projection $\pi(z) = M$. We assume that

(i) λ_z is linear and depends differentiably on z;
(ii) $D\pi \circ \lambda_z = 1_{\mathcal{M}}$ (identity map on $T_{\mathcal{M}}(M)$);   (8.17)
(iii) if $z' = R_g z$, then $\lambda_{z'} = DR_{(g)} \circ \lambda_z$.

We have previously defined the differential map D in Sec. 6, Level 2 and also the right translation $R_{(g)}$ by formula (26.49) and following, Level 2.

We may now consider horizontal and vertical vectors: the space

$$H_z = \lambda_z(T_{\mathcal{M}}(M)) \tag{8.18}$$

is a subspace of $T_{\mathcal{P}}(z)$ (λ_z is indeed linear); its elements u_h are called **horizontal vectors** while the elements u_v of the space $\mathbf{V}_z$, tangent vector space to F_M (or G) at z,

$$\mathbf{V}_z = T_{F_M}(z) = T_G(z) \tag{8.19}$$

are called **vertical vectors**. One has

$$T_{\mathcal{P}}(z) = H_z \oplus \mathbf{V}_z , \tag{8.20a}$$

and a vector $u \in T_{\mathcal{P}}(z)$ can be decomposed as follows:

$$u = u_h + u_v . \tag{8.20b}$$

Vertical vector elements of $\mathbf{V}_z$ can be easily constructed as follows. For any $h \in G$ the point $R_{(h)}z$ can be parametrized by

$$R_{(h)}z = zh = \{M, g\}h = \{M, gh\} . \tag{8.21}$$

Such a field is globally defined (see Sec. 1 where one assumes that any point $zh \in \mathcal{P}$ is supposed to be globally defined). Furthermore, the differential map $DR_{(h)}$ is, as pointed out in Sec. 6, Level 2 (formula (6.3)), a diffeomorphism between the tangent vector space to G at h and the tangent vector space to $\mathcal{P}$ at hz. When h tends to the identity e of $G, DR_{(e)}$ will map the Lie algebra $\mathbf{G}$ of G (which is simply the tangent vector space $\mathbf{G}$ at e) to the tangent vector space to $\mathcal{P}$ at hz. If X_q is any generator of $\mathbf{G}, DR_{(e)}(X_q) = e^{(v)}{}_q(z)$ is a vertical vector. After all these preliminaries we are now able to define the **connection form** ω_z as follows: ω_z is the differentiable linear map $T_{\mathcal{P}}(z) \to T_G(e) = \mathbf{G}$ with the following properties:

(i) $\omega_z(u) = 0 \rightleftharpoons u \in H_z$,

(ii) $\omega_z(e_q{}^{(v)}(z)) = X_q$. $\hspace{4cm}$ (8.22)

We are now in possession of two definitions of ω_z: one linking ω_z to $\underline{A}$ and the one presented here: the interested reader will find in ([Bibl. 16], p. 81 and p. 204) the proof that both definitions are fully equivalent.

In the special and important case where the principal bundle is the bundle of frames on $\mathcal{M}_N$ with GL(N, $\mathbb{R}$) as structural group (as seen in Sec. 1), the connection defined for such a bundle is said to be **linear**.

In conclusion we add a few words about **Riemannian connections**: if $\mathcal{M}_N$ is endowed with a Riemannian metric, in which case one can define orthonormal frames, then the above structural group, GL(N, $\mathbb{R}$), is replaced by $G = $ SO(N) as seen in Sec. 1. One can check that there is one connection (the Levi-Civita one) expressed by the Christoffel symbols $\gamma^\mu{}_{\nu\rho}$ as introduced in Part F, Level 2. For a complete study of these points, see [Bibl. 9], p. 300 and [Bibl. 17].

9. *The Curvature 2-Form and Torsion 2-Form*

We saw in Sec. 8 that a connection is specified either by the choice of a Yang-Mills field $A_\alpha(x)$ defined on the base $\mathcal{M}_N$ or by a globally defined connection 1-form ω_z defined on the principal bundle $\mathcal{P}$. Here we shall define globally either a 2-form $\underline{\Omega}$ on $\mathcal{P}$ or a 2-form $\underline{F}$ (to compare with $r_{\alpha\beta}$ in Problem 2.12) on $\mathcal{M}_N$, where $\underline{\Omega}$ as well as $\underline{F}$ will be valued on the Lie algebra $\mathbf{G}$ of the structural group G (as were ω_z, A_α). We intend to be brief and summarize (even without proof) the results, referring the reader to textbooks quoted in the bibliography. We start with the definition of the curvature 2-form Ω_z for which one usually uses the concept of the **exterior covariant derivative** $\hat{\mathbf{D}}$ **on bundles**, and we shall take a shortcut, skip the general definition of $\hat{\mathbf{D}}$ and write directly

$$\Omega_z = \hat{\mathbf{D}}\omega_z = \underline{d}\omega_z(u \otimes v) + \frac{1}{2}[\omega_z \wedge \omega_z](u \otimes v) , \hspace{1.5cm} (9.1)$$

where u and v are elements of the vector space $T_{\mathcal{P}}(z)$. One may also write[*]

$$\Omega_z = \underline{d}\omega_z + \frac{1}{2}[\omega_z \wedge \omega_z] \ . \tag{9.2}$$

It can be shown that at a point $z' = \{M, gg'\} = zg'$, one has

$$\Omega_{z'} = g'^{-1}\Omega_z g' \tag{9.3}$$

and that $\underline{\Omega}$ has a tensorial character (which is not the case of ω_z). By using methods similar to those used in the preceding paragraph, we can go over from Ω_z to the 2-form $\underline{F}$ expressed[*] by the 1-form $\underline{A}$:

$$\underline{F} = \underline{d}\underline{A} + \frac{1}{2}[\underline{A} \wedge \underline{A}] \ . \tag{9.4}$$

At the end of Sec. 8, we considered the bundle of frames at $M\{z^M\}$ of $\mathcal{M}_N$, where each frame $z^M{}_{(k)}$ of $T_{\mathcal{M}}(M)$ may be considered as a regular linear map (an isomorphism) $T_{\mathcal{M}}(M) \to \mathbb{R}^N$. Such a bundle has $G = \mathrm{GL}(N, \mathbb{R})$ as the structural group.

One may then recover all the well-known results about the Riemann tensor and curvature form obtained in Part F, Level 2, and also the ones obtained in question (c), Problem 2.12.

One can also introduce the concept of **torsion 2-form** (see Secs. 20, 21 of Level 2). For a study of these problems, the interested reader is referred to the textbooks quoted in the bibliography of this chapter.

Remark: Formula (9.2) indeed requires a comment. Consider two Lie algebra-valued forms: a p-form α and a q-form β. The set $\{X_n, n = 1 \ldots r\}$ being a basis of $\mathbf{G}$, α and β can be expressed by the summations:

$$\alpha = X_k\alpha^k \ , \quad \beta = X_{k'}\beta^{k'} \ ,$$

where the sets $\{\alpha^k\}$ and $\{\beta^{k'}\}$ are respectively p and q-forms, and X_k and $X_{k'}$ are generators of $\mathbf{G}$.

By definition,

$$[\alpha \wedge \beta] = [X_k, X_{k'}]\alpha^k \wedge \beta^{k'} \ , \tag{9.5}$$

where the bracket is the Lie-bracket. It follows from the anti-symmetries of the factors of (9.5) and formula (8.17), Level 2 that

$$[\beta \wedge \alpha] = [X_{k'}, X_k]\beta^{k'} \wedge \alpha^k = -[X_k, X_{k'}](-1)^{pq}\alpha^k \wedge \beta^{k'} \ .$$

[*]See the remarks at the end of Problem 2.12 and at the end of the present paragraph.

Finally we get the law

$$[\beta \wedge \alpha] = (-1)^{pq+1}[\alpha \wedge \beta] \ . \tag{9.6}$$

If $\beta = \alpha$ and p^2 is odd, there is no contradiction in supposing $[\alpha \wedge \alpha] \neq 0$, which is the case of the 1-form ω_z.

One may also notice that a straightforward application (26.69), Level 2 leads to

$$[\alpha \wedge \beta] = c^j{}_{kk'} X_j \alpha^k \wedge \beta^{k'} \ . \tag{9.7}$$

10. *Connection Form and Curvature Form in Associated Vector Bundles*

The introduction of associated vector bundles as defined in Sec. 5 is important: physicists call their sections matter fields. It will also be interesting to define on such bundles the connection form ω (also $\underline{A}$) and the curvature 2-form $\underline{F}$ on the base $\mathcal{M}$.

The total space is $E = \mathcal{P} \times V$, where ρ is a representation of the structural group G (hence of its Lie algebra $\mathbf{G}$) on the vector space* V with elements v. We may then write

$$\underline{A}(x) = A^q_\mu(x)\underline{d}x^\mu \rho(X_q) \ ,$$
$$\underline{F}(x) = \frac{1}{2}F^q{}_{\mu\lambda}(x)\underline{d}x^\mu \wedge \underline{d}x^\lambda \rho(X_q) \ ,$$

where we have used the same symbols A and F with diferent indices in order to avoid the introduction of too many new notations. The reader should be careful and avoid any possible confusion with the notations used before.

On the other hand, let $\{\tau_\alpha\}$ be a basis of V. Since $\rho(X_q)$ is a linear operator on V, one may write

$$\rho(X_q)\tau_\alpha = (\rho(X_q))^\nu_\alpha \tau_\nu$$

and

$$\underline{A}(x)\tau_\alpha = A^q_\mu(x)\underline{d}x^\mu(\rho(X_q))^\nu_\alpha \tau_\nu = \underline{A}^\nu_\alpha(x)\tau_\nu \ ,$$
$$\underline{F}(x)\tau_\alpha = \frac{1}{2}F^q{}_{\mu\nu}(x)\underline{d}x^\mu \wedge \underline{d}x^\lambda(\rho(X_q))^\nu_\alpha \tau_\nu = \underline{F}^\nu_\alpha(x)\tau_\nu \ .$$

---

*In Sec. 5, the vector space V was noted by F (elements f) but we denote here as in the previous paragraph by $\underline{F}$ the curvature 2-form on $\mathcal{M}_N$, hence the choice V instead of F.

The matrix with 1-form elements

$$\underline{A}^{\nu}_{\alpha}(x) = A^{q}{}_{\mu}(x)\underline{d}x^{\mu}(\rho(X_q))^{\nu}_{\alpha}$$

is called the **connection matrix**. A similar interpretation is valid for the matrix with elements $\underline{F}^{\nu}_{\alpha}(x)$.

The following comment is also of importance: let $\mathcal{P}$ be the bundle of linear frames (or one of its sub-bundles) and E its tangent bundle (or any tensor product of the tangent bundle by itself), then V coincides with $\mathbb{R}^N$ and can be identified with the tangent space to $\mathcal{M}_N$ itself. It is then natural to choose the same basis on both vector spaces (i.e., to identify the μ index with α); one gets a connection matrix ω^{ρ}_{ν} with the 2-forms $\omega^{\rho}_{\mu\nu}$ as elements and curvature matrix $\underline{F}^{\rho}_{\sigma}$ with the 2-forms $F^{\rho}_{\mu\nu\sigma}$ as elements. This represents an abstract and elaborate way of introducing the Riemann tensor $R^{\rho}_{\mu\nu\sigma}$; we chose in Problem 2.12 a more naive introduction.

11. *Gauge Group — Characteristic Classes*

The set of all connections that one can consider on a given principal bundle space is itself an (infinite dimensional) space which plays a central part in the quantization of Yang-Mills fields, and also of the gravitational field. Although this is clearly a subject out of our scope, let us first mention that this space is acted upon by the **gauge group**, which one should distinguish carefully from the structural group (denoted by G) even if the the two notions get mixed in the physics literature. The structural group previously defined is a finite dimensional Lie group (SU(3) for instance). The gauge group can be defined in the case of a trivial bundle ($\mathcal{M} \times G$) as follows: it is the set of all maps from the base $\mathcal{M}$ to the group G, an infinite dimensional group, provided that $\mathcal{M}$ is not a finite set.

Let us now present a few comments about characteristic classes. The problem, roughly speaking, is the following: we gave at the end of Sec. 4 different examples which showed that two principal bundles with the same base and the same typical fibre can indeed be very different (for instance the trivial bundles $S^2 \times S^1$ and the so-called Hopf fibration of S^3 as a S^1 bundle over S^2). As pointed out in Sec. 4, one would like to have a criterion for characterizing such situations.

It turns out that there exist some polynomials adapted to such a purpose. Let us consider as an illustration the case of SU(N)-bundles. We denote by $p(B)$ an invariant polynomial on the algebra of all $N \times N$ complex matrices: in other words $p(B)$ satisfies the relation

$$p(B) = p(TBT^{-1})$$

for every non-singular matrix T. This is the case for the trace or the determinant function of a matrix. We may choose a bundle endowed with a curvature form $\underline{F}$, and it makes sense to replace in $p(B)$ the matrix element B_β^α (numbers) by the matrix elements $\underline{F}_\beta^\alpha$ which are 2-forms on the base (see the previous paragraph). We then get an object $p(\underline{F})$ which is a sum of exterior forms of various degrees. For instance, taking $p(A) =$ Tr A^k (k number) then $p(\underline{F}) =$ Tr $\underline{F}^k$ is another $2k$-form on the base. The first non-trivial result is that the differential form $p(\underline{F})$ is closed, i.e.,

$$\underline{d}\, p(\underline{F}) = 0 \ .$$

A second result is the following: let us choose another connection with $\underline{F}'$ as a curvature form, then $\underline{d}p(F') = 0$ but one also has

$$p(\underline{F}') = p(\underline{F}) + \underline{d}\hat{f} \ ,$$

where $\hat{f}$ is a certain function. In other words $p(\underline{F})$ and $p(\underline{F}')$ are cohomologous (see Sec. 17, Level 2).

As an application, we may choose $p(A) = (\text{coef})\, A^N$. If we suppose dim $\mathcal{M}_N = 2n$, we also have

$$\int_{\mathcal{M}_N} \pi(\underline{F}) = \int_{\mathcal{M}_N} \pi(\underline{F}') \ ,$$

since using Stokes' theorem,

$$\int_{\mathcal{M}_N} \underline{d}\hat{f} = \int_{\partial \mathcal{M}_N} \hat{f} = 0 \ .$$

A final non-trivial result is that the common value of the two integrals above is a real integer k. One may then assert that $p(A)$ determines a **characteristic class**. In the case of SU(N) bundles, the corresponding class is called a **Chern class** but there exist other characteristic classes which play an important part in differential geometry (Stiefel-Whitney classes, Pontryagin classes, Euler classes,...). See also Sec. 17, Level 2.

12. *Physical Outlook*

We begin by adding some complements to questions (e) and (f) of Problem 2.12.

Consider first the gauge group $G = U(1)$; it concerns the scalar charged field, the electromagnetic and Dirac fields. Any elements $g(x)$ of $U(1)$ is of the form $g(x) = e^{ie\varphi(x)}$, where the charge e and the gauge field $\varphi(x)$ are real. Let us take the electromagnetic field with vector potential $A_\alpha(x), \alpha = 0, 1, 2, 3$, then we define two 1-forms

$$\underline{A}(x) = A_\mu(x)\underline{d}x^\mu \; , \quad g^{-1}(x)\underline{d}g(x) = ie\underline{d}\varphi(x) \; .$$

The gauge transformation laws for $\psi(x)$, the electron field, and $A(x)$, the electromagnetic field, are

$$\psi'(x) = g^{-1}(x)\psi(x) \; , \quad \underline{A}'(x) = \underline{A}(x) + ie\underline{d}\varphi(x) \; ,$$

and the covariant differential of ψ is

$$D\psi(x) = \underline{d}\psi(x) + ie\underline{A}(x)\varphi(x) \; .$$

The field strength is given by (9.4) and we remark that its second term vanishes since the corresponding Lie algebra is commutative (9.7),

$$\mathbf{D}\underline{A}(x) = \underline{d}\,\underline{A}(x) + \frac{1}{2}[\underline{A}(x) \wedge \underline{A}(x)]$$
$$= \partial_\mu A_\nu(x)\underline{d}x^\mu \wedge \underline{d}x^\nu + \frac{1}{2}A_\mu(x)A_\nu(x)\underline{d}x^\mu \wedge \underline{d}x^\nu \; .$$

We are then left with

$$\mathbf{D}\underline{A}(x) = \frac{1}{2}(\partial_\mu A_\nu - \partial_\nu A_\mu)\underline{d}x^\mu \wedge \underline{d}x^\nu \; ,$$
$$= \sum_{\mu<\nu}(\partial_\mu A_\nu - \partial_\nu A_\mu)\underline{d}x^\mu \wedge \underline{d}x^\nu$$

and the electromagnetic field strength appears in the parentheses of the summation. One furthermore verifies the so-called Bianchi identity

$$\mathbf{D}^2\underline{A}(x) = 0 \; .$$

We now consider the Yang-Mills fields: the structural group G is now SU(2), the generators of its Lie algebra are the Pauli matrices $\tau_k, k = 1, 2, 3$, the isospin matrices. The Yang-Mills field are denoted by $A^k{}_\alpha(x), k = 1, 2, 3, \alpha = 0, 1, 2, 3$; for the physicists k is the isospin index and α the spin index, acted upon by any Lorentz transformation.* To the $A^k{}_\alpha(x)$, there corresponds the 1-form $\underline{A}^k(x) = A^k_\alpha(x)\underline{d}x^\alpha$ and, also, the Lie algebra-valued 1-form $\underline{A}(x) = \underline{A}^k(x)\tau_k$.

Any element of **G** is of the form

$$g(x) = e^{iW^k(x)\tau_k} = e^{i\mathbf{W}(x)\cdot\boldsymbol{\tau}} ,$$

where $W^k(x)$ are real vector fields.** We first notice that

$$g^{-1}(x)\underline{d}g(x) = ie^{-i\mathbf{W}\cdot\boldsymbol{\tau}}\tau_k e^{i\mathbf{W}\cdot\boldsymbol{\tau}}\underline{d}W^t ,$$

and then that, using (8.9),

$$\underline{A}'(x) = g^{-1}(x)\underline{A}(x)g(x) + g^{-1}(x)\underline{d}g(x) .$$

With the help of (8.10) one also has

$$D\psi(x) = \underline{d}\psi(x) + \underline{A}(x)\psi(x) .$$

The other meaningful quantities are the field strengths from (9.4),

$$\underline{F}(x) = \hat{\mathbf{D}}\underline{A}(x) = \underline{d}\,\underline{A}(x) + \frac{1}{2}[A(x) \wedge A(x)] .$$

We shall find results in a form more familiar to the physicist if we decompose both sides in their isospin and spin indices. We first write

$$\underline{F}(x) = \sum_{\mu<\nu} F^k_{\mu\nu}(x)\tau_k\underline{d}x^\mu \wedge \underline{d}x^\nu ,$$

then, using (9.7), we have

$$[\underline{A}(x) \wedge \underline{A}(x)] = \sum_{\mu<\nu} 2c^j_{kk'}A^k_\mu(x)A^{k'}_\nu(x)\tau_j\underline{d}x^\mu \wedge \underline{d}x^\nu$$

---

*The $A^k{}_\alpha$ fields are coupled to a matter field $\psi(x)$, but we shall focus our attention on the $A^k{}_\alpha$ fields.

**Physicists consider $G = \text{SU}(2)\times \text{U}(1)$, then there is one more field $B(x)$ to be added. See for instance, CERN lecture notes [Ref. 82-04] of 1981, notes by C. Jarlskog, p. 63 and following.

and also

$$dA(x) = d\{A_\mu^k(x)\tau_k dx^\mu\} = \sum_{\mu<\nu}(\partial_\mu A_\nu^k(x) - \partial_\nu A_\mu^k(x))\tau_k dx^\mu \wedge dx^\nu .$$

Collecting all these terms together, we have

$$F_{\mu\nu}^k(x) = \partial_\mu A_\nu^k(x) - \partial_\nu A_\mu^k(x) + c_{kk'}^j A_\mu^k(x)A_\nu^{k'}(x) .$$

We shall not pursue further our study of Yang-Mills fields, urging the interested reader to consult any specialized book or monograph.

There are at least two topics which have been notably developed by physicists of elementary particles and which represent important applications of differential geometry: these are the theories of monopoles and instantons.

A. Monopoles: It was shown by Dirac that the existence of isolated magnetic charges or monopoles lead to the quantization of the electric charge (through the quantization of the magnetic flux). The classification of such monopoles provides an interesting application of the formalism of fibre bundles.

Classically, the motion of a test particle in the field of a monopole is described as the motion on an S^2 sphere centered at the monopole: the problem has indeed spherical symmetry. The electromagnetic field is determined by the vector potential $A(x)$ and we are therefore led to considering U(1) bundles over the sphere S^2. One can then show that such bundles are classified by the first Chern class $C_1(p) = -[F/2\pi]$, where F is the curvature 2-form for an arbitrary connection. One also shows that

$$\int_{S^2} C_1(p) = -n ,$$

n being an integer giving the value of the monopole charge. If $n = 0$, there is no monopole charge and the principal $\mathcal{P}$ is trivial, $\mathcal{P} = S^2 \times$ U(1). The case $n = 1$ is more complex to consider: it corresponds to the Hopf fibration of SU(2) over S^2, the base of the bundle.

We limit ourselves to these brief considerations.

B. Instantons: Again as in A, we shall remain very general with our considerations.

In quantum electrodynamics (QED), a theory on the interactions of electrically charged particles and photons, as well as in chromodynamics

(QCD), a theory of the interaction of quarks and gluons, one has not been able to construct the exact solutions (except for trivial cases) of the field equations: the only recourse is, for the time being, to look for a perturbative expansion around the vacuum state, characterized in QED by $A_\mu(x) = 0, \mu = 0, 1, 2, 3$.

In general, gauge theories are described by connections on principal bundles which are given by ω_z or $\underline{A}(x)$. Consider the case of Euclideanized QED and QCD. In both cases, the base is the sphere S^4, but their respective structural groups are U(1) and SU(3). Now, it can be proved that any U(1) bundle over S^4 is trivial; therefore in the QED case $\mathcal{P} = S^4 \times$ U(1). The vacuum corresponds to $\omega_z = 0$, and also $\underline{A}(x) = 0$ or* $\underline{A}(x) = d\phi(x)$; then the curvature $\underline{F}(x) = 0$ and Maxwell's equations are trivially satisfied.

In contradistinction to QED, owing to the properties of the SU(3) group for QCD there exist infinite, topologically distinct SU(3) bundles over S^4. One can then show that any of these bundles can be characterized by the triplet $\{S^4, \mathrm{SU}(3), n\}$, where n is an integer. It is called the **instanton number** specific to the bundle under study.

Finally, the perturbative process goes as follows: after the choice of the bundle (i.e., of n), one looks for the classical solution of the equations of motion and follow the perturbative approach around it: such a classical solution is called the "vacuum of instanton with number n". This last step represents the technical and most difficult part of the approximation process. Its study is out of the scope of the present paragraph: we limit ourselves to these sketchy considerations and urge the interested reader to consult specialized monographs and papers.

---

*Since Poincaré's lemma is valid.

BIBLIOGRAPHY

The list of references should not by any means be considered exhaustive. The only textbooks quoted are those which have been consulted while completing the present book.

Chapter I — Level 0 — Theory of surfaces

[1] G. Darboux: Leçons sur la théorie génerale des surfaces (4 volumes). The first edition appeared in 1896 and was followed by several others – Gauthier-Villars.

[2] V.I. Smirnov: A course of Higher Mathematics (5 volumes). See volume 2, Chapter 5 – Pergamon Press 1964.

[3] J.J. Stoker: Differential geometry — Wiley Interscience 1969.
For further reading, see most of the textbooks on classical analysis which appeared before 1950 or some textbooks on applied mathematics. One can also mention some internationally known French classics in Analysis, for instance. E. Goursat (many editions from 1905 on), G. Valiron (1945).

[4] J. Dieudonné and collaborators: Abrégé d'histoire des mathématiques (2 volumes), chapter on Géometrie différentielle (P. Libermann) in volume 2, p. 177 – Hermann 1978.

Chapter II — Level 1 – Tensor analysis

[5] L.P. Eisenhart: Riemannian geometry, first German edition 1925, English version Princeton 1964.

E. Schrödinger: Space time structure. A short account on the introduction of Christoffel symbols on a non-Riemannian space and application to a unitary theory of gravitation. University Press 1950.

All textbooks on general relativity include an account of tensor analysis – For instance.

[6] L.D. Landau, E.M. Lifshitz: Course of Theoretical Physics (8 volumes). See volume 2: The classical theory of fields, chapter 10, Secs. 83-84-85-86; Secs. 91-92 — Pergamon Press.

[7] S. Weinberg: Gravitation and Cosmology. See chapter 4, tensor analysis Secs. 2-3-4-5-6.

Refs 6 and 7 can be considered as classics.

[8] M. Carmeli: Classical Fields, General Relativity and Gauge Fields. Chapter 2 — J. Wiley 1982.

Chapter III — Level 2 — Intrinsic formulation

[9] Y. Choquet-Bruhat, C. de Witt-Morette, M. Dillard-Bleick: Analysis, Manifolds and Physics – North-Holland 1977.

[10] Y. Choquet-Bruhat: Géométrie différentielle et systèmes extérieurs – Dunod 1968.

[11] M. Spivak: A comprehensive introduction to differential geometry (5 volumes) with several problems and examples – Publish or Perish 1979.

[12] B. Doubrovine, S. Novikov, A. Fomenko: Géométrie contemporaine. Translated from Russian by V. Kotliar, volumes 1, 2 and 3 – Editions Mir, Moscou 1979.

Some textbooks on physics use also intrinsic formulation.

[13] C.W. Misner, K.S. Thorne, J.A. Wheeler: Gravitation – W.H. Freeman and Company 1973 – This book uses a formulation which is a compromise between Levels 1 and 2.

[14] S.W. Hawking and Ellis: The large scale structure of space-time – Cambridge 1973.

[15] V. Arnold: Méthodes mathématiques de la Mécanique classique. Translated from Russian by D. Embarek. Editions Mir, Moscou 1974.

All works quoted in Chapter III: [Bibl. 9] to [Bibl 15].

Chapter IV — Fibre bundles

[16] W. Drechsel, M.E. Mayer: Fiber bundles techniques in gauge theories – Springer Verlag 1977.

[17] R. Coquereaux and A. Jadczyk: Riemannian geometry, fibre bundles, Kaluza-Klein theories and all that – World Scientific Publications 1987.

[18] D. Husemoler: Fibre bundles – Springer Verlag 1975.

[19] S. Kobayashi and F. Nomizu: Foundations of differential geometry (2 volumes) – Wiley 1969.

[20] N. Steerod: The topology of fibre bundles – Princeton 1951.

[21] A. Trautman: Differential geometry for physicists – Bibliopolis 1984.

INDEX

INTRODUCTORY DIFFERENTIAL GEOMETRY FOR PHYSICISTS

This book has its origin in a set of lectures given by the late Professor A. Visconti who could not see it through to its final form.

As it is, it ought to be useful to a wide variety of readers, since the exposition has been deliberately kept close to geometrical and physical intuition.

INTRODUCTION

"The axiomatisation and algebraisation of mathematics, which has been taking place in the last 50 years, has made many mathematical texts so illegible that the old danger of complete loss of contact with physics and with natural sciences has become a reality"

V. I. Arnold

At first sight, such a statement may seem exaggerated. I believe however that it is a faithful description of reality. Indeed a characteristic of mathematicians and physicists of the XIXth century was the existence of a common language. There was consequently no linguistic barrier between them, the only obstacles to comprehension being the ones that arose from an incomplete mastery of the subject matter or from the requirements of rigour.

The landscape changes at the beginning of the XXth century. The development of set theory and the resolution of paradoxes that arose there forced the mathematicians to work with ever increasing precision and led eventually to the necessity of a new language. The ideal solution would have been to coin new names for new objects; however the vocabulary is not infinitely extensible, and the new objects had to be described by old words. The language was changing and the physicists were confused by the use of terms which did not mean any more what they had meant for the last three centuries. In such a way a barrier started arising.

This barrier was even more important because of the fact that the points of departure of the mathematical reasoning were not any more the ones that had been found spontaneously starting from the immediate interpretation of experience. The new framework did allow the reaching of traditional conclusions which had been verified many times by experience, but this was often at the price of a certain lengthiness and of intellectual contorsions. The gap between physics and mathematics, which was apparent already at the end of the XIXth century became an abyss after 1950.

It is good to remark, in order to excuse a certain inertia, that this new language introduced an additional difficulty in the application of the "Correspondence Principle". This principle is a criterion of validity for a new discipline (for example, *Special Relativity*) which extends and generalizes a discipline that is older (for instance, *Newtonian Mechanics*); it states that there exists a common domain where the predictions of the first are in accordance with the statements of the second. For instance, the intrinsic formulation of *Differential Geometry* (which will be described in this work) will be shown at Level 2 to extend the taxonomic formulation described at Level 1. All the valid results of the first Level should be found as statements of the second. The verification of this assertion is however not always as straightforward as one might wish!

The present work has no other ambition than to be useful: its aim is to show to the Physicist that his discussions are sometimes unfinished in their logical structure, and to make the Mathematician aware of certain advantages to the scientific community if he changed his methods of exposition so as to become more easily understandable to his colleague the Physicist.

The plan of the present work is extremely simple; it is divided into three levels corresponding to three chapters: Level 0 – the nearest to intuition and to geometrical experience – is a short summary of the theory of curves and surfaces as it was developed in the XIXth century. In Level 0, metric concepts play a central part, indeed the most elementary experiment involving a curved surface consists of measuring lengths of arcs of curves of the surface with a simple string. Starting from there, one is led straight away to the fundamental notions of Christoffel symbols, curvature, Riemann tensor and geodesics. The landscape changes at Level 1 (Chapter II): the concept of the variance of a field (i.e. description of how a field changes under a point transformation) allows the classification of tensor fields, then the introduction – independent of any metrical considerations – of Christoffel symbols or affine connection coefficients lead to the notion of absolute or covariant derivative. Given an hypersurface, suppose that a vector field describes under certain conditions (parallel transport) a closed curve of that surface: it may happen that its initial value differs from the final one when the closed curve has been described. That difference can be used to define the curvature of the hypersurface. All these properties have nothing to do with the metrical properties of the surface, the study of which constitutes a separate chapter of that Level 1 (Riemannian geometry). At Level 2 (Chapter III) the point of view is again different from the previous one: the fundamental concept here is the one of manifold. On a manifold, one introduces the notion of tangent vector attached to a given point and one defines a tangent vector as a vector of the manifold under study. One considers also linear forms attached to any point of a manifold, and using the technique of tensorial products one is finally able to define tensors (at a point) of different orders and types. After all these preliminaries one defines an affine connection on a manifold and the affine connection coefficients appear in the local expression of the connection. Again, metrical properties and riemannian geometry are introduced in a final step. But the main characteristic of the notions introduced at that Level 2 lies in their intrinsic presentation which avoids, as far as possible (in contrast to Levels 0 and 1), all recourse to special frames leading to particular coordinates of a point of the considered manifold or to components of the mathematical objects defined there.

There are advantages in this procedure, first intellectual ones: concepts loosely defined acquire a distinct and clear meaning and secondly formal ones: the avalanche of indices common to the methods of Levels 0 and 1 is by now stopped, formulas can be written in a compact form which can be used to simplify certain involved calculations. Indeed such an approach often appears clear and simple but this is not always the case. Some intuitive and fundamental concepts, which were near to everyday experiments presented within the framework of Level 0, become, of course, more general but also more abstract and the average physicist may find some difficulty in understanding them correctly when they are presented in that novel perspective.

The final Chapter IV is dedicated to the study of fibre spaces, and that constitutes another and new presentation of differential geometry. It had been for some years familiar to mathematicians, but particle physicists (the theoretical ones!) began using the language in their study of gauge theories. This chapter of mathematics received there a renewed attention from their part and the application of fibre space techniques to physics (and to general relativity which can be also considered as a gauge theory) is by now very extensive.

In the present textbook we give a short and elementary survey of the domain of fibre spaces: the interested reader should consult some of the references given in the bibliography.

Exercises and problems have been added at the end of Levels 0, 1, 2 (Chapters I, II, III): each problem has its own title which summarizes the subject under study. Each problem is followed by "Hints" or by a "solution" more or less complete depending on its difficulty.

Problems of Level 0 deal with the theory of surfaces using traditional notations, the study of the curvature of surfaces has been emphasized. We also devoted some attention to variational problems which are important for physical applications.

Exercises and problems of Level 1 (Chapter II) deal with tensor analysis using the standard presentation and notations which are familiar to all students of general relativity. A set of the problems of that level has been devoted to physical applications ranging from the study of Lorentz transformation and Maxwell equations to the Robertson–Walker metric of general relativity.

The exercises and problems of Level 2 are devoted to the study of the intrinsic presentation of differential geometry. Problems dealing with applications (to physics in particular) are not separated from the ones on pure mathematics as in the previous level. The reader will find, among the titles, applications to systems of partial differential equations of first order, dynamical systems, Maxwell equation. The solution of Einstein's equations as given by Robertson and Walker is once more considered using specific methods of Level 2. It should be emphasized that the problems are supposed to bring also complements to the main text: it is therefore important, for a dedicated reader, to consider that they represent an important step in the mastery of differential geometry.

Even though, we have been particularly interested in the structural aspects of our discipline, we may add that there exist many excellent detailed works on the subject. We have mentioned a few of them in the Bibliography, in particular ones we have actually used.

How should the present work be used? The preceding discussions show the way: a Physicist acquainted with the techniques that are generally used in his domain will read with profit the Levels 0 and 1 simultaneously with a work on *General Relativity* which happens to be familiar to him, and where he will find the study of some technical points which we had to omit for lack of space. It should be noticed that the third paragraph of Level 0 gives the fundamental ingredients of Einstein equations, developing them from concrete and particular examples. Level 1 repeats, comments and often develops the traditional method of discussion of *Tensor Algebra and Analysis*, which is used by the majority on works on general relativity. Finally Level 2 and Chapter IV constitute an introduction to the language of modern Differential Geometry. As a companion work we can recommend [Bibl. 9] which we have often used, and also [Bibl. 11].

Let me add that I would greatly appreciate any comment or rectification including corrections of misprints or other errors.

A. Visconti

ACKNOWLEDGEMENTS

While completing the preliminary french version of the present text-book, I enjoyed many inspiring talks and discussions with Professor A. Emeric (Université de Provence). Without the encouragement and the help of Professor A. Grossmann (Centre de Physique théorique de Marseille, Section II), I would never have found the resource and the courage to undertake the present translation. He has been kind enough to provide a draft of an english translation of the Introduction, of Chapters I and II and of important parts of Chapter III. On the other hand, I am the only one responsible for Chapter IV and all the problems (texts and solutions).

I have been introduced to the ideas which lie at the root of Chapter IV by Professor R. Coquereaux (Centre de Physique théorique de Marseille, Section II). He has kindly drawn my attention to the meaning and interest of fibre bundles techniques. I had direct access to and made liberal and extensive use of his private notes. I am thus deeply in debt for his help which was notable, his patience and his pedagogical skill. Thanks are finally due to the Administration of the Centre de Physique théorique for its kind cooperation.